Nanoscale Silicon Devices

Nanoscale Silicon Devices

EDITED BY

Shunri Oda
Tokyo Institute of Technology, Japan

David K. Ferry
Arizona State University, Tempe, USA

CRC Press
Taylor & Francis Group
Boca Raton London New York

CRC Press is an imprint of the
Taylor & Francis Group, an **informa** business

CRC Press
Taylor & Francis Group
6000 Broken Sound Parkway NW, Suite 300
Boca Raton, FL 33487-2742

First issued in paperback 2017

© 2016 by Taylor & Francis Group, LLC
CRC Press is an imprint of Taylor & Francis Group, an Informa business

No claim to original U.S. Government works

ISBN-13: 978-1-4822-2867-0 (hbk)
ISBN-13: 978-1-138-74932-0 (pbk)

Visit the Taylor & Francis Web site at
http://www.taylorandfrancis.com

and the CRC Press Web site at
http://www.crcpress.com

Contents

Preface...vii
Editors ix
Contributors xi

Chapter 1 Physics of Silicon Nanodevices..1

David K. Ferry and Richard Akis

Chapter 2 Tri-Gate Transistors 37

Suman Datta

Chapter 3 Variability in Scaled MOSFETs...53

Toshiro Hiramoto

Chapter 4 Self-Heating Effects in Nanoscale 3D MOSFETs................83

Tsunaki Takahashi and Ken Uchida

Chapter 5 Spintronics-Based Nonvolatile Computing Systems.........105

Tetsuo Endoh

Chapter 6 NEMS Devices...123

Yoshishige Tsuchiya and Hiroshi Mizuta

Chapter 7 Tunnel FETs for More Energy-Efficient Computing..........155

Adrian M. Ionescu

Chapter 8 Dopant-Atom Silicon Tunneling Nanodevices.................181

Daniel Moraru and Michiharu Tabe

Chapter 9 Single-Electron Transfer in Si Nanowires.......................207

Akira Fujiwara, Gento Yamahata, and Katsuhiko Nishiguchi

Chapter 10 Coupled Si Quantum Dots for Spin-Based Qubits 231

          Tetsuo Kodera and Shunri Oda

Chapter 11 Potential of Nonvolatile Magnetoelectric Devices for
          Spintronic Applications 255

          Peter A. Dowben, Christian Binek, and Dmitri E. Nikonov

Index 279

Preface

Semiconductor integrated circuits are essential in the Information Age, since they are used in various applications, including in PCs and mobile phones. CMOS (complementary metal oxide semiconductor) transistors are a key component for these integrated circuits. A continuous scaling down of transistor size has enhanced the performance of PCs and mobile phones.

Ten years ago, we published the book *Silicon Nanoelectronics* (Oda and Ferry, CRC Press, 2005). Technology has changed rapidly since. CMOS feature sizes reduced from 45 to 14 nm today. Scaling of CMOS devices requires much more than merely improvements in the technology of lithography. There is a need for new concepts such as strain, new materials such as high-k dielectrics and germanium, and new device structures in three dimensions.

This book deals with the recent advancements in, and future prospects for, nanoscale Si devices, including the fundamentals of scaled Si devices, 3D structures, variability of scaled transistors, thermally aware transistors, nonvolatile memories, NEMS devices, tunnel transistors for low power devices, atomic level doping control, silicon nanowires, coupled quantum dots for spin-based qubits, and magneto-electric devices. This book is a compact reference source for students, scientists, engineers, and specialists in the fields of electron devices, solid-state physics, and nanotechnology.

Shunri Oda
David K. Ferry

Editors

Shunri Oda received his BSc in physics in 1974 and MEng and DEng from the Tokyo Institute of Technology in 1976 and 1979, respectively.

He is a professor in the Department of Physical Electronics and Quantum Nanoelectronics Research Center, Tokyo Institute of Technology. His current research interests include fabrication of silicon quantum dots by pulsed plasma processes, single-electron tunneling devices based on nanocrystalline silicon, ballistic transport in silicon nanodevices, silicon-based quantum information devices, and NEMS hybrid devices. He has authored more than 700 technical papers in international journals and conferences, including 200 invited papers. He has edited the book (with D. K. Ferry) *Silicon Nanoelectronics* (CRC Press, 2005).

Professor Oda is a fellow of the Institute of Electrical and Electronics Engineers (IEEE) and the Japan Society for Applied Physics and a member of the Electrochemical Society and the Materials Research Society. He is a distinguished lecturer at the IEEE Electron Devices Society.

David K. Ferry received his BSEE and MSEE from Texas Tech University, Lubbock, Texas, in 1962 and 1963, respectively, and PhD from the University of Texas (UT), Austin, Texas, in 1966. Following this, he was a National Science Foundation post-doctoral fellow in Vienna, Austria. From 1967 to 1973, he was a faculty member with Texas Tech University and then joined the Office of Naval Research, Washington, DC. From 1977 to 1983, he was with Colorado State University at Fort Collins, Colorado, where he served as professor and chair of the Electrical Engineering Department. He then joined Arizona State University (ASU), Tempe, Arizona, where he served as director of the Center for Solid State Electronics Research from 1983 to 1989, as chair of electrical engineering from 1989 to 1992, and as associate dean for research from 1993 to 1995. His research interests include transport physics and modeling of quantum effects in ultrasmall semiconductor devices. He has published more than 800 articles, books, book chapters, and conference papers. Dr. Ferry was selected as one of the first Regents' Professors at ASU in 1988 and received the IEEE Cledo Brunetti Award for advances in nanoelectronics in 1999. He is a fellow of the American Physical Society, the Institute of Electrical and Electronics Engineers, and the Institute of Physics (United Kingdom). He also serves as editor of the *Journal of Computational Electronics* and is an admiral in the Texas Navy and the Tennessee Squire Association.

Contributors

Richard Akis
School of Electrical, Computer, and
 Energy Engineering
Arizona State University
Tempe, Arizona

Christian Binek
Department of Physics and Astronomy
University of Nebraska–Lincoln
Lincoln, Nebraska

Suman Datta
Department of Electrical Engineering
University of Notre Dame
Notre Dame, Indiana

Peter A. Dowben
Department of Physics and Astronomy
University of Nebraska–Lincoln
Lincoln, Nebraska

Tetsuo Endoh
Department of Electrical Engineering
Center for Innovative Integrated
 Electronic Systems
and
Center for Spintronics Integrated
 Systems
Tohoku University
Sendai, Japan

David K. Ferry
School of Electrical, Computer, and
 Energy Engineering
Arizona State University
Tempe, Arizona

Akira Fujiwara
NTT Basic Research Laboratories
Atsugi, Japan

Toshiro Hiramoto
Institute of Industrial Science
University of Tokyo
Tokyo, Japan

Adrian M. Ionescu
Institute of Electrical Engineering
Ecole Polytechnique Fédérale de
 Lausanne
Lausanne, Switzerland

Tetsuo Kodera
Quantum Nanoelectronics Research
 Center
Tokyo Institute of Technology
Tokyo, Japan

Hiroshi Mizuta
School of Materials Science
Japan Advanced Institute of Science
 and Technology
Ishikawa, Japan

and

Electronics and Computer Science
University of Southampton
Southampton, United Kingdom

Daniel Moraru
Research Institute of Electronics
Shizuoka University
Hamamatsu, Japan

Dmitri E. Nikonov
Intel Corporation
Hillsboro, Oregon

Katsuhiko Nishiguchi
NTT Basic Research Laboratories
Atsugi, Japan

Shunri Oda
Quantum Nanoelectronics Research
    Center
Tokyo Institute of Technology
Tokyo, Japan

Michiharu Tabe
Research Institute of Electronics
Shizuoka University
Hamamatsu, Japan

Tsunaki Takahashi
Department of Electronics and
    Electrical Engineering
Keio University
Yokohama, Japan

Yoshishige Tsuchiya
Electronics and Computer Science
University of Southampton
Southampton, United Kingdom

Ken Uchida
Department of Electronics and
    Electrical Engineering
Keio University
Yokohama, Japan

Gento Yamahata
NTT Basic Research Laboratories
Atsugi, Japan

1 Physics of Silicon Nanodevices

David K. Ferry and Richard Akis

CONTENTS

Abstract 1
1.1 Introduction 2
1.2 Small MOSFETs 2
    1.2.1 The Simple One-Dimensional Theory 3
    1.2.2 Ballistic Transport in the MOSFET 5
1.3 Granularity 8
1.4 Quantum Behavior in the Device 11
    1.4.1 The Effective Potential 11
        1.4.1.1 Effective Carrier Wave Packet 11
        1.4.1.2 Statistical Considerations 13
        1.4.1.3 Using the Effective Potential in Device Simulation 15
    1.4.2 The Quantum Potential 16
    1.4.3 Quantum Simulations 17
        1.4.3.1 The Device Structure 18
        1.4.3.2 The Wave Function and Technique 19
        1.4.3.3 Scattering in the Site Representation 22
        1.4.3.4 Results 24
1.5 Quantum Dot Single-Electron Devices 27
1.6 Many-Body Interactions 27
Acknowledgments 30
References 30

ABSTRACT

With the continuing advance of VLSI technology, the size of individual transistors has been reduced significantly. Today, chips are already made with transistors that have nanometer scale and are a driving force for nanotechnology. This will continue to be the case for many years in the future. Yet, new and novel silicon nanodevices have appeared and others have been suggested. The behavior of such small devices, with characteristic lengths on a 5–20 nm scale, is described by quantum mechanics, and this brings new limitations into play.

1.1   INTRODUCTION

For the past several decades, miniaturization in silicon integrated circuits has progressed steadily at an exponential scale described by Moore's law [1]. This incredible progress has generally meant that critical dimensions are reduced by a factor of 2 once in every 3 years, while chip density increases by a factor of 4 over this entire period. However, modern chip manufacturers have been accelerating this pace recently, and currently, chips are being made at the 14 nm node, although gate lengths can vary significantly depending upon the application for the specific chip. This node is significant, as Intel has moved away from the planar transistor with this node and has adopted the trigate technology [2]. More scaling is expected, however, and 10 nm gate lengths are scheduled for production within a few years. While the creation of these very small transistors is remarkable enough, the fact that they seem to operate in a quite normal fashion is perhaps even more remarkable.

Almost 45 years ago, the prospects of making such small transistors were discussed, and a suggested technique for a 25 nm gate length, Schottky source–drain device, was proposed [3]. At that time, it was suggested that the central feature of transport in such small devices would be that the microdynamics could not be treated in isolation from the overall device environment (of a great many similar devices). Rather, it was thought that the transport would by necessity be described by quantum transport and that the array of such small devices on the chip would lead to considerable coherent many-device interactions. While this early suggestion does not seem to have been fulfilled, as witnessed by the quite normal behavior of the aforementioned devices, there have been many subsequent suggestions for treatment via quantum transport [4–8]. Moreover, there is ample suggestion that the transport will not be normal, but will have significant ballistic transport effects [3], and this, in turn, will lead to quantum transport effects.

In this first chapter, the concept of ballistic transport will be reviewed, starting in the next section. We then turn to the most important aspect of small devices, and that is the breakdown of ensemble averaging, so that the role of discrete, localized impurities and fluctuations in sizes becomes important. Following this, we begin to discuss the role of quantization. First, we will review how it is found in large metal-oxide-semiconductor field-effect transistors (MOSFETs) and then turn to the much more important role in small transistors. We follow this with a discussion of the ultimately small device—the quantum dot and single electron tunneling. Finally, a discussion of many-body effects in such small devices is given. Each of these topics will be discussed in far greater detail in subsequent chapters, but here we hope to give an overall consolidated view on these topics.

1.2   SMALL MOSFETs

The MOSFET is created when the electric field between the gate and the semiconductor is such that an inverted carrier population is created to form a conducting channel between the source and drain electrodes. The transport through this channel is modulated by the gate potential. This much has been known since the first descriptive patent on the topic [9]. Indeed, the operation of the MOSFET is almost exactly as described in a simple 1D semiclassical treatment, and this approach has been modified and adapted continuously over the past few decades. However, it is now understood that there is quantization in the basic MOSFET, even for quite large gate lengths. This is because the gate field pulls the inversion channel carriers quite close to the oxide–semiconductor interface, and these carriers are confined between this interface and the potential in the bulk. This confinement is sufficient to cause quantization to occur in the direction normal to the oxide–semiconductor interface [10]. This quantization leads to a quasi-2D carrier gas in the plane of the channel [11]. While this effect is quite important, it is equally important to understand that the transport is in the plane of this quantized layer, and so is only weakly affected by this quantization. We will discuss this in more detail in a subsequent section.

As the channel length has become smaller, there has been considerable effort to incorporate a variety of new effects into the simple (as well as the more complex) models. These include short channel effects, narrow width effects, degradation of the mobility due to surface scattering, hot carrier effects, and velocity overshoot [10]. However, as gate lengths have become less than ca. 30 nm, the issue boils down to *ballistic* transport rather than these other problems. By ballistic transport, we refer to the situation in which the channel length is less than the mean-free path of the carriers, so that very little scattering occurs within the channel itself. If we take the thermal velocity of a carrier in Si as 2.5×10^7 cm/s at room temperature, a channel mobility of 300 cm^2/V s leads to a relaxation time of 5×10^{-14} s and a mean-free path of the order of 12×10^{-7} cm, or 12 nm. Thus, we might expect only a few scattering events in a channel length of 20–30 nm. While this is a very rough approximation, it points out that the properties of the carriers in these very small devices will be much different from those in larger devices. In this case, the "theory" of the device is actually much closer to that of the simple approach discussed in the following text, at least in conceptual detail. For this reason, we will review some simple interpretations of the 1D current equation and then develop the ballistic device theory. This becomes important, because the same intuitive ideas are extrapolated from the Landauer formula [12], which is often invoked in pure quantum transport situations.

1.2.1 The Simple One-Dimensional Theory

In general, the current through a semiconductor device is calculated by writing an equation for the differential voltage drop along a point in the channel in terms of the current and local conductance (this may be found in most elementary textbooks; see, e.g., Ref. [13]). This expression is then integrated over the length of the channel, with the following result (for the MOSFET)

$$I_D = \frac{eWC\mu}{L}\left(V_G - V_T - \frac{V_D}{2}\right)V_D, \qquad (1.1)$$

where
  I_D is the drain current
  W is the width of the channel
  C is the gate capacitance per unit area
  μ is the mobility of the carriers
  L is the electrical channel length
  V_G is the gate–source voltage
  V_T is the threshold voltage (at which the channel begins to form)
  V_D is the drain–source voltage

From this expression, the current rises almost linearly for small drain voltage, and then saturates at a value of drain voltage given by

$$V_{D,sat} = V_G - V_T,$$ (1.2)

which may be found by taking the derivative of Equation 1.1 and setting it to zero.

A more intuitive view of the current may be obtained by rewriting Equation 1.1 to separate the source originating current and the drain originating current as

$$I_D = \frac{eWC\mu}{2L}[(V_G - V_T)^2 - (V_G - V_T - V_D)^2] = I_{SD} - I_{DS}.$$ (1.3)

Now, it is clear that saturation sets in when the second term in the square brackets, the drain originating current (or reverse current), vanishes for the condition of Equation 1.2. In this equation, we can connect parts of the formula with particular physical effects. Here, we may connect

$$C(V_G - \cdots)$$ (1.4)

with the local carrier density (in carriers per unit area) in the channel and

$$\mu \frac{V}{2L},$$ (1.5)

which is the (average) velocity in the channel. Hence, we may rewrite Equation 1.3 once again as

$$I_D = eW[n_S v_S - n_D v_D],$$ (1.6)

where
  n_S and n_D are the 2D densities at the source and drain, respectively
  v_S and v_D are the velocities at these two points
  W is the width of the channel

The form of Equations 1.3 and 1.6 is quite reminiscent of the Landauer formula, as it is applied to mesoscopic devices. In a quasi-1D system, we often write the Landauer formula as an integral of the energy range of the transmitting modes times the conductance of each mode. In a sense, Equations 1.4 and 1.5 give us the conductance of the channel, and the difference between the two potential terms gives us the potential drop between the two sides of the device. Hence, we relate to the Landauer formula through the conductance, which includes the number of transverse modes in the structure. The major change as we reduce the size of the device, particularly the length of the channel, is a transition from scattering controlled velocities to barrier-controlled velocities. We already see in Equation 1.3 that the two currents are controlled by the height of the barriers at the two ends of the channel. But, as we go smaller, the mobility becomes a much less important parameter, and relation (1.5) is replaced by the rate of carriers tunneling through the barrier. It is this change that leads to the normal current of Equation 1.1 getting replaced by a form of the Landauer formula. We explore this in the next section.

1.2.2 BALLISTIC TRANSPORT IN THE MOSFET

In general, the potential profile through a MOSFET looks somewhat like that shown in Figure 1.1. From the source end, there is a small potential barrier between the source and the channel, and then the potential falls to the level of the drain potential (the energy is shown; this has a negative sign from the voltage). There are a number of important points that occur in this figure. First, near the source end (left side of the image), there is a potential barrier. This barrier serves a number of purposes. The barrier prevents most of the electrons in the source contact from entering the channel. Depending upon the height of the barrier, the number of injected carriers can range from zero (high barrier) to a level near the electron density in the source (very low, or zero, barrier). In addition, the barrier is modulated by the gate potential. This is the key point of the transistor, as the first term in Equation 1.3 arises from the gate potential effect on this barrier. Right at the source end, where the energy begins to rise, we must have a depletion of electrons, which means that this point occurs just at the end of the n-doped source of the device. This depletion is required for the discontinuity in the electric field that is evident in the figure. Note that this depletion extends just to the inflection point of the rising energy, which is the metallurgical junction between the source and the channel.

Away from the source end of the channel, the energy drops down toward the drain end of the channel. There is bending of this curve in the negative direction, which reflects the accumulation of negative charge in the channel region. The negative charge arises from the depletion of holes in the p-doped region as well as the accumulation of the injected electrons in the inversion layer. Some authors draw the curve as a straight line, giving a uniform electric field in this region. But, this is not correct, as a uniform electric field reflects a space-charge neutral region, and the channel is

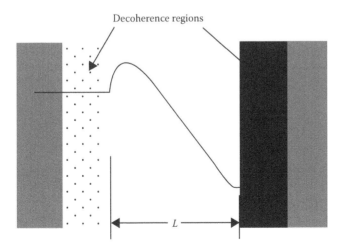

FIGURE 1.1 A conceptual device under bias. The source is at the left and the drain at the right, as indicated by the two gray areas, which may be considered to be the "contacts." The areas to the left and right of the traditional active length L, indicated here as the decoherence regions, must now be considered part of the active device.

anything but that. In the normal MOSFET, reflected in the aforementioned equations, the transport is governed by the mobility and an electric field that rises from the source to the drain. The drain end of the channel supports the highest electric field, especially when the device is in saturation. Because of the rising electric field, the velocity of the carriers increases as we move toward the drain. The current density through the transistor is constant; as the velocity rises, the sheet density must decrease. In fact, the nature of the total barrier shown in Figure 1.1 is that of a self-consistent potential subject to a constraint of the applied gate and drain voltages. The exact distribution of charge in the channel and in the drain will affect this potential barrier due to the nonlinear feedback from the solution of Poisson's equation. This has been shown already in some detail by many authors [14].

How might this change if we have ballistic transport, in which the carrier moves through the channel without scattering? If this is the case, the carriers move horizontally from the source to the drain in Figure 1.1. They enter near the top of the barrier shown by the source. The curve represents the potential energy throughout the device. Thus, a carrier moving horizontally in the figure converts its total energy from potential energy into kinetic energy. This means that the velocity of the electrons rises as it moves from the left to the right, exactly as in the mobility-dominated motion discussed in the last paragraph. As a result of the velocity increase, the sheet density must consequently decrease to maintain the constant current. As mentioned earlier, the nature of the total barrier shown in Figure 1.1 is that of a self-consistent potential subject to a constraint of the applied gate and drain voltages. The exact details of the potential variation will change with ballistic transport, because the exact charge density will vary slightly, but the general shape of the barrier remains about the same [15].

Natori [7] has given an another version of a ballistic transport treatment for the MOSFET and has used this to some success in fitting to experimental data [16]. Although Natori developed his expression with a full quantum mechanical basis, the approach is an outgrowth of the Duke tunneling formula [17] and the Landauer formula [12], and we can follow a variation of the semiclassical approach [13]. We will assume that the direction normal to the oxide–semiconductor interface (the y-direction) is quantized [11], and concern ourselves with integrations over the other two directions in reciprocal space. Then, the forward current may be written as

$$J_{SD} = 2e \sum_{valleys} \sum_{n_y} \iint \frac{dk_z dk_x}{4\pi^2} v_x(k_x) T(k_x) f(\varphi_{FS}, E)[1 - f(\varphi_{FD}, E)]. \qquad (1.7)$$

The integer n_y runs over the occupied subbands in the inversion layer, the first summation runs over the six equivalent valleys of the conduction band, and the total energy is

$$E = E_x + E_z = \frac{\hbar^2}{2}\left(\frac{k_x^2}{m_x} + \frac{k_z^2}{m_z}\right). \qquad (1.8)$$

The valley summation is necessary, since the mass that is appropriate for the two coordinate axes is different in each of the three pairs of valleys (this will be discussed further in a later section).

In a similar manner to Equation 1.7, we may also write the reverse current (that flowing from the drain to the source) as

$$J_{DS} = 2e \sum_{valleys} \sum_{n_y} \iint \frac{dk_z dk_x}{4\pi^2} v_x(k_x) T(k_x) f(\varphi_{FD}, E)[1 - f(\varphi_{FS}, E)]. \qquad (1.9)$$

We may then write the total current as

$$I_{SD} = 2eW \sum_{valleys} \sum_{n_y} \int \frac{dk_z dk_x}{4\pi^2} v_x(k_x) T(k_x)[f(\varphi_{FS}, E) - f(\varphi_{FD}, E)]. \qquad (1.10)$$

In general, the treatment of ballistic transport is that for which the carriers move over the barrier, so that we may take $T = 1$. We now rescale the energy through the introduction of the scaled k vectors as

$$k_x'^2 = \frac{\sqrt{m_x m_z}}{m_x} k_x^2, \quad k_z'^2 = \frac{\sqrt{m_x m_z}}{m_z} k_z^2, \qquad (1.11)$$

so that

$$E = \frac{\hbar^2}{2m^*}\left(k_x'^2 + k_z'^2\right), \quad m^* = \sqrt{m_x m_z}. \qquad (1.12)$$

With this transformation, we may change the variables in Equation 1.10 as

$$dk_x dk_z = \sqrt{\frac{m_x}{m^*}} dk_x' \sqrt{\frac{m_y}{m^*}} dk_z' = dk_x' dk_z' = k'dk'd\vartheta = \frac{m^*}{\hbar^2} dE d\vartheta. \qquad (1.13)$$

The angular integration can be carried out immediately, and Equation 1.10 becomes

$$I_{SD} = \frac{eWm^*}{\pi\hbar^2} \sum_{valleys} \sum_{n_y} \int v_x(k_x)[f(\varphi_{FS}, E) - f(\varphi_{FD}, E)]dE. \qquad (1.14)$$

The velocity can be assumed to be a thermal velocity, which is isotropic, so that

$$v_x \sim \sqrt{\frac{v^2}{2}} \sim \frac{\sqrt{2m_z}}{m^*} \sqrt{E}, \qquad (1.15)$$

where the scaled coordinates have been incorporated. If we now introduce the reduced coordinates

$$\eta = \frac{\varphi_{FS}}{k_B T}, \quad \chi_n = \frac{E_{n_y}}{k_B T}, \quad \varphi = \frac{eV_D}{k_B T}, \qquad (1.16)$$

the current can be written as

$$I = \frac{\sqrt{2}eW(k_B T)^{3/2}}{\pi \hbar^2} \sum_{valleys} \sum_{n_y} \sqrt{m_z} [F_{1/2}(\eta - \chi_n) - F_{1/2}(\eta - \chi_n - \phi)]. \qquad (1.17)$$

The functions $F_{1/2}$ are the Fermi–Dirac integrals of half-integer order [18].

However, there is a problem with Equation 1.17 and the development leading up to it. This problem lies in the fact that MOSFETs dissipate a significant amount of heat. If we use two thermal distribution functions at the lattice temperature, then these must be evaluated well into the reservoirs [19,20]. That is, we must use the distribution function in the metallic interconnects rather than in the drain region near the channel. If we want to use this latter region, which is the obvious point of discussion in the aforementioned derivations, then we must account for the higher electron temperature in this region. Each carrier that exits the channel into the drain brings with it an excess, directed energy of eV_D. This extra energy is rapidly thermalized by carrier–carrier scattering [21], which provides an elevated electron temperature $T_e > T$ in the drain. It is no simple task to determine this electron temperature and clearly gives a rationale for the use of detailed Monte Carlo simulations (classical) [22] or nonequilibrium Green's functions [14] in order to find the detailed distribution function that should be utilized in Equation 1.17. Moreover, the number of occupied subbands (in the y-direction) will be different in the drain end than in the source end. Hence, we should rewrite Equation 1.17, using primes to denote the expressions of Equation 1.16 evaluated with the electron temperature, as

$$I = \frac{\sqrt{2}eW(k_B T)^{3/2}}{\pi \hbar^2} \sum_{valleys} \left\{ \sum_{n_{y,S}} \sqrt{m_z} F_{1/2}(\eta - \chi_n) - \left(\frac{T_e}{T}\right)^{3/2} \sum_{n_{y,D}} \sqrt{m_z} F_{1/2}(\eta' - \chi_n' - \phi') \right\}.$$

$$(1.18)$$

It is clear that a good model for the electron temperature in the drain, near the channel, is necessary to really apply these ballistic formulas.

When the width of the device begins to get small as well, then quantization also occurs in this direction. While Natori [7] has mentioned this, it is relatively easy to incorporate this into Equation 1.17, leading to the Landauer formula, as is shown in Ferry [23]. We will not deal with this here, as the full quantum treatment is discussed in a later section.

1.3   GRANULARITY

By granularity, we refer to the failure of thermodynamic averaging in small devices. If we consider a silicon-on-insulator (SOI) MOSFET, with the silicon channel 10 nm thick, 20 nm wide, and 10 nm long, and doped to 10^{19} cm^{-3}, then there are only 20 dopant atoms in the channel. If the carrier density is 10^{13} cm^{-2}, then there are only

20 carriers in the channel at any one time. With such a small number of dopants and carriers, it is impossible to use average densities and statistics. Instead, the position of each impurity atom is quite important and device performance depends not only upon this number, but also upon the exact position of each of the impurities. Keyes [24] was the first to warn about threshold voltage fluctuations arising from variations in the number of dopant atoms in the channel, but did no simulations to evaluate the problem.

Perhaps the first to study the role of discrete dopants on transport were Boudville and McGill [25], who studied ohmic contacts to GaAs. Then, Joshi and Ferry [26] showed that, in heavily doped GaAs, an electron was typically interacting with three or more impurities at the same time. Wong and Taur [27] subsequently studied the role of discrete dopants in a Si MOSFET, and Zhou and Ferry [28–30] discussed the problem in MESFETs and HEMTs. Later, Vasileska et al. [31] and Asenov [32] reviewed MOSFET behavior, and the field has blossomed since then.

We can illustrate the problems inherent with the granularity, by looking, for example, at a simulation of a thin SOI MOSFET. In Figure 1.2, we plot the carrier density in an n-channel SOI MOSFET. The density is indicated by the gray scale of the plot, and we are looking down into the plane of the device. Figure 1.2a shows the case in which a purely classical simulation is incorporated, and it is quite clear that the variations in the carrier density are large. On the other hand, this device is small, and quantization should begin to occur. Figure 1.2b shows how the density fluctuations are reduced by introducing an effective potential (discussed in the next section) to account for quantum effects. While the density fluctuation has been reduced, it is still significant. Simulations such as these point out that each device, which will have a different number of actual donor and acceptor atoms and different configurations of these atoms, will have its own characteristic performance. While having millions of such devices on a chip can be viewed as an ensemble averaging process, it is important to note that the performance depends upon each individual device and not upon their average behavior. The variations in individual device behavior arise from the failure of thermodynamic averaging within the device, and we cannot invoke ensemble averaging when each device is important.

The importance of this small number of atoms in the active volume cannot be overestimated. A single charged impurity residing near the peak of the barrier, shown in Figure 1.1, has a much larger effect than one near the drain end [33]. Thus, the position of each and every impurity affects the resulting potential in slightly different ways, depending upon its location. As a result, the random device performance resulting from these distributions of small numbers of impurities can threaten the overall performance of the device. In many cases, it is now preferable to use undoped channels than those with only a few impurity atoms.

Dopant atoms are not the only problem that arises from the granularity of the device. Linton et al. [34,35] have pointed out that device variations can occur due to the line edge roughness of the gate polysilicon line. Variations in performance with top surface roughness (variations in thickness) for MOS structures [36] and for MOSFETs [37] have also been considered. Roughness at the oxide–semiconductor interface has usually been treated as a scattering process [38], but Brown et al. [39] have recently directly incorporated a model of the surface height

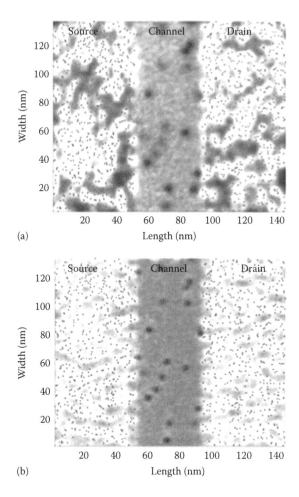

(a)

(b)

FIGURE 1.2 Electron density from a Monte Carlo simulation using MD for the carrier–carrier interaction. (a) Without the effective potential included to simulate quantum confinement, and with $V_G = 0.4$ V, $V_D = 0.1$ V. (b) With the effective potential included in the simulation, and with $V_G = 0.6$ V and $V_D = 0.1$ V. The higher gate voltage was used to get more electrons into the channel for image clarity. The lighter shades represent higher carrier densities, and the dots indicate the position of the impurities (donors in the source and drain, and acceptors in the channel). It is clear that the density tends to cluster around the impurities due to the lower potentials in this region.

variation to study the thickness variations in SOI MOSFETs. It is quite clear that a truly small semiconductor device can no longer be considered as a generic entity. It will have its own characteristic performance that will depend upon the configuration of the dopants, the variations of the oxide thickness and gate lines, and the variations in the "thickness" induced by roughness at the top and bottom (in SOI device) oxides. Limitation on the ultimate scalability may in the end depend upon the ability to control these fluctuations to a degree that allows the fabrication of billions of reasonably reliable devices.

1.4   QUANTUM BEHAVIOR IN THE DEVICE

As noted earlier, channel quantization in the direction normal to the oxide–semiconductor interface has been a fact of life for many years. This leads to important modifications that are readily seen in smaller devices. Two such effects are a shift in the threshold voltage, due to the rise of the lowest occupied subband above the conduction minimum, and a reduction in the gate capacitance, due to the setback of the maximum in the inversion density away from the interface. This latter produces a so-called quantum-induced capacitance, which is effectively in series with the normal gate capacitance [40]. If these are the major effects produced by the quantization, then they can be readily handled in a normal semiclassical theory by the introduction of a potential. There are a couple of popular methods of doing this. One is through the use of a so-called effective potential [41]. Another approach that has arisen in recent years is to include a *quantum potential* [42]. Both of these techniques appear as a correction to the solutions of the Poisson equation in self-consistent simulations. The effective potential arises from a smoothing of the self-consistent potential found from Poisson's equation. The quantum potential refers to an additive term to the self-consistent potential. Recently, the quantum potential has come to be called the "density-gradient" approach, since the quantum potential is often defined in terms of the second derivative of the square root of local density. Both of these approaches fail if the individual quantum levels in the inversion layer become resolved, or if the lateral quantization (in either width or thickness of an SOI layer) becomes important. Then, a full quantum mechanical model is required to handle the device. In the following, we first discuss the effective potential approach and the quantum potential approach. Then, we turn to the description of a full quantum mechanical simulation for ultrasmall SOI MOSFETs.

1.4.1   THE EFFECTIVE POTENTIAL

The first method we discuss involves smoothing the self-consistent potential, which introduces an *effective* potential. Here, the natural nonzero size of an electron wave packet in the quantized system [43] is used to introduce a smoothing of the local potential (found from Poisson's equation) [44]. This approach naturally incorporates the quantum potentials (discussed in the following), which are *approximations* to the effective potential. The introduction of an effective potential follows two trends that have been prominent in statistical physics during most of the twentieth century and into the current century. These are the nonzero size of an electron wave packet and the use of a modified potential to describe quantum effects within classical statistical mechanics. Here, we review these two approaches and show how they combine to give a form for the effective potential. We then show how the quantum potential derives from the effective potential as an approximation, and finally provide results from simulations to compare these approaches. We also estimate the problems in incorporating tunneling via this approach.

1.4.1.1   Effective Carrier Wave Packet

In order to describe the packet of a carrier in real space, one must account for the contributions to the wave packet from all occupied plane wave states [43]. That is, the states that exist in momentum space are the Fourier components of the real-space wave packet. If we want to estimate the size of this wave packet, we must utilize all Fourier components, not just a select few. (This approach is familiar from the definition of Wannier functions and their use to evaluate the size of a bound electron orbit near an impurity.) This is not the first attempt to define the nature of the quantum wave packet corresponding to a (semi)classical electron. Indeed, the study of the classical-quantum correspondence has really intensified over the past few decades, due in no small part to the rich nature of chaos in classical systems and the search for the quantum analog of this chaos. This has led to a number of studies of the manifestation of classical phase-space structure [45]. These have shown that meaningful sharp structure can exist in quantum phase-space representations, and these can profitably be used to explain (or to interpret) quantum dynamics: for example, to study the quantum effects that arise in otherwise classical simulations for semiconductor devices. The use of a Gaussian wave packet as a representation of the classical particle is the basis of the well-known coherent-state representation. However, if we have two such wave packets, there is a problem. When we take the two real-space wave packets and create a phase-space Wigner representation, then there is a superposition wave between the two phase space packets. This represents coherence between the two packets. We can approach the classical regime only by first destroying this decoherence [46]. Then, one can pass to the classical limit and the packets become discrete points in phase space. We return to this point shortly.

In the coherent state (Gaussian packet) approach, the phase-space representation of the quantum density localized at point $\mathbf{x}$ is given by [47,48]

$$\langle \mathbf{x} \mid \mathbf{p}, \mathbf{q} \rangle = \frac{1}{(\pi\sigma^2)^{N/4}} \exp\left[-\frac{(\mathbf{x}-\mathbf{q})^2}{2\sigma^2} + i\frac{\mathbf{p} \cdot (\mathbf{x}-\mathbf{q})}{2\hbar} \right]. \tag{1.19}$$

where
    $\mathbf{p}$ is the momentum of the wave packet
    $\mathbf{q}$ is the centroid position
    $\mathbf{x}$ is the general coordinate

As in most cases, the problem is to find the value of the spatial spread of the wave packet, which is defined by the parameter σ, which is related to the width of the wave packet. In this representation, the quantum particle has a phase space extent determined by the parameter σ, and this goes to zero as we pass to the classical limit. Hence, σ must be related to $\hbar$ in some manner. It was found earlier [43] that σ is given approximately by the thermal de Broglie wavelength.

For this approach to be valid, we must have wave packets that do not have coherence among the packets. This really means that the eigenvalue spectrum of the Schrödinger equation must be washed out by the thermal smearing. If this spectrum is distinguishable, then a single wave packet for each particle is not a valid approach, and our effective potential method will fail. When the approach is valid, we can then examine how the Gaussian wave packet leads to a smoothing of the classical potential. The scalar potential is related to the charge density through the static Lienard–Wiechert potential [49]

$$V(\mathbf{r}) = \frac{1}{4\pi\varepsilon} \int \frac{\rho(\mathbf{r}')}{|\mathbf{r} - \mathbf{r}'|} d^3\mathbf{r}'. \tag{1.20}$$

If we now introduce the discrete charge, this latter equation can be written as

$$V(\mathbf{r}) = \frac{1}{4\pi\varepsilon} \int \frac{e}{|\mathbf{r} - \mathbf{r}'|} \left\{ \sum_i \delta\left(\mathbf{r}' - \mathbf{r}_i^D\right) - \sum_j \langle \mathbf{r}' | \mathbf{p}_j, \mathbf{r}_j \rangle \right\} d^3\mathbf{r}'. \tag{1.21}$$

The first summation (index i) runs over the ionized donors, while the second summation (index j) runs over the free electrons. The coefficient of the second summation is the set of coherent states defined by Equation 1.19. The first summation provides a distinct contribution from the second, and we concentrate on the second, introducing a resolution of unity in terms of an integral over a delta function as

$$V_2(\mathbf{r}) = -\frac{e}{4\pi\varepsilon} \int \frac{1}{|\mathbf{r} - \mathbf{r}'|} \sum_j \int d^3\mathbf{r}'' \left| \langle \mathbf{r}' | \mathbf{p}_j, \mathbf{r}'' \rangle \right|^2 \delta(\mathbf{r}'' - \mathbf{r}_j) d^3\mathbf{r}'. \tag{1.22}$$

The squared magnitude Gaussian is independent of the momentum and is a function only of the difference (squared) of the two coordinate variables. Therefore, we can interchange these in this factor, at the same time changing the notation on the delta function accordingly, and then the integral can be rearranged to give

$$V_2(\mathbf{r}) = \int d^3\mathbf{r}'' V_{cl}(\mathbf{r}'') \left| \langle \mathbf{r}'' | 0, \mathbf{r} \rangle \right|^2, \tag{1.23}$$

where V_{cl} is the classical potential determined by the charges having only discrete points in phase space. An arbitrary treatment of the first term in Equation 1.21 in this fashion leads us to the result that the nonzero extent of the phase space wave packet of the carriers can be easily moved onto the potential, appearing as a smoothing of the potential by the Gaussian function [44]. However, we reiterate once more that this approach fails when the various eigenvalues of the quantization begin to be resolved. Nevertheless, comparisons with exact solutions of the Schrödinger–Poisson equations for an inversion layer show excellent agreement for those cases in which the approximations are valid [44,50].

1.4.1.2 Statistical Considerations

From the earliest days of quantum mechanics, there has been an interest in methods that allow the reduction of quantum calculations to classical ones, through the introduction of a suitable *effective potential*. In this regard, one would like to replace the potential in the partition function

$$n \sim \exp\left(-\frac{eV}{k_B T}\right), \qquad (1.24)$$

with a modified potential that will describe the density as determined by the quantum wave function. The earliest known approach was provided by Wigner [51], where he introduced an expansion of the classical potential in powers of $\hbar$ and $\beta = 1/k_B T$, which led to

$$V_{eff}(x) \sim V(x) + \frac{\hbar^2}{8mk_B T}\frac{\partial^2 V}{\partial x^2} + \cdots \qquad (1.25)$$

This series led to the well-known Wigner–Kirkwood expansion of the potential that is often used in solutions for the Wigner distribution function. However, the series has convergence problems below the Debye temperature and in cases with sharp potentials, such as the Si–SiO$_2$ interface. Feynman and Hibbs [52] found a similar result, but with the factor 8 replaced by 24. Feynman also introduced a different approach, in which an effective potential is introduced through the free energy: that is, through Equation 1.24 with the classical potential averaged over a Gaussian smoothing function, as in Equation 1.23. For the case of a free particle, he shows that an exact variational minimization leads to a Gaussian weighting of the potential around the classical path, and this automatically includes quantum effects into the trajectory. Indeed, Feynman found that the smoothing parameter σ should have the value

$$\sigma^2 = \frac{\hbar^2}{12mk_B T} = \frac{\lambda_D^2}{24\pi}, \qquad (1.26)$$

where λ_D is the thermal de Broglie wavelength.

Many people have extended the Feynman approach to the case of bound particles [53–57] and particles at interfaces [58]. The effective potential approach has been recently reviewed by Cuccoli et al. [59]. These approaches use the fact that the most likely trajectory in the path integral no longer follows the classical path when the electron is bound inside a potential well. The introduction of the effective potential and its effective Hamiltonian is closely connected to the return to a phase-space description, as discussed earlier. This can be done at present only for Hamiltonians containing a *kinetic* energy quadratic in the momenta and a coordinate-only dependence in the potential energy. That is, it is clear that some modifications will have to be made when nonparabolic energy bands, or a magnetic field, are present. However, the Gaussian approximation is well established as the method for incorporating the purely quantum fluctuations around the resulting path. The key new ingredient for bound states (such as in the potential well at the interface of a MOSFET) is the need to determine variationally the dominant path and hence the "correct" value for the parameter σ. For the case in which the bound states are well defined in the potential, both Feynman and Kleinert [54] and Cuccoli et al. [57] find

$$\sigma^2 = \frac{\hbar^2}{4mk_BT}\left[\frac{\coth(f)}{f} - \frac{1}{f^2}\right],$$ (1.27)

where

$$f = \frac{\hbar\omega_0}{2k_BT}$$ (1.28)

and $\hbar\omega_0$ is the spacing of the subbands. If we take the high-temperature limit, then we can expand for small f, and

$$\sigma^2 \sim \frac{\hbar^2}{12mk_BT},$$ (1.29)

to leading order, which agrees with Equation 1.26. In Si, this gives a value of 0.52 nm for the value to be used in the direction normal to the interface (at room temperature). A different mass would be used for transport along the channel, and this gives a value of 1.14 nm.

1.4.1.3 Using the Effective Potential in Device Simulation

We first used the effective potential to simulate a 50 nm gate length MOSFET [60]. Here, the transport was handled by an ensemble Monte Carlo technique and the effective potential was used to smooth the potential in the direction normal to the gate. This provided both the charge setback from the interface and a raised potential in the inversion layer to simulate the quantization effects. One problem with these early usages is the fact that only a single parameter is used in the effective potential at all points in the device, whereas the kinetic and potential energies of the carriers vary substantially.

Ahmed et al. [61] proposed a novel parameter-free effective potential scheme, in which the potential arises from a perturbation around the thermodynamic equilibrium. In this scheme, the size of each individual electron depends upon its unique kinetic energy, and hence, each individual electron has a different local effective potential. This scheme was applied to a 25 nm MOSFET, which included very high channel doping to avoid punch-through. This led to pronounced quantization in the channel potential, yet the effective potential gave excellent agreement with the observed threshold shift and current degradation.

More recently, Palestri et al. [62] have turned the axes for a multisubband simulation. As noted earlier, the effective potential fails to be very useful when the individual quantized levels can be discerned. These authors used the normal quantization in the direction normal to the gate to define the quantized levels of the inversion layer of a small SOI MOSFET. Then, they applied the effective potential approach to the potential *along the channel*. They still retain the single parameter approach, but they find that these Monte Carlo simulations give results that are in good agreement with a nonequilibrium Green's function approach.

1.4.2 THE QUANTUM POTENTIAL

As mentioned earlier, there is another approach to correct the classical potential to incorporate quantum effects when the subbands are not well formed. This arises from the hydrodynamic version of Schrödinger's equation. If it is assumed that the wave function can be written as [63,64]

$$\psi(\mathbf{r},t) = R(\mathbf{r},t)\exp\left[\frac{i}{\hbar}S(\mathbf{r},t)\right], \tag{1.30}$$

then Schrödinger's equation can be separated into two equations for the real and imaginary parts. This gives

$$\frac{\partial S}{\partial t} + \frac{1}{2m}(\nabla S)^2 + V - \frac{\hbar^2}{2Rm}\nabla^2 R = 0, \tag{1.31}$$

$$\frac{\partial R}{\partial t} + \nabla \cdot \left(\frac{R^2}{m}\nabla S\right) = 0. \tag{1.32}$$

The last equation is a form of the continuity equation, while the first is a form of the Euler equation, and in this equation, we identify the correction term as the *quantum potential*

$$V_Q(\mathbf{r},t) = -\frac{\hbar^2}{2mR(\mathbf{r},t)}\nabla^2 R(\mathbf{r},t). \tag{1.33}$$

Since the density is identified with the square magnitude of the wave function, Equation 1.33 has become known as the *density gradient* correction to the classical potential. The exact form of the quantum potential can take a variety of shapes, depending upon various approximations for the Wigner function, which have been discussed by Iafrate et al. [42], but Equation 1.33 represents the most common form that has been used in device modeling [44].

We first used the quantum potential in a full device simulation for our study on MESFETs [65,66]. This was then followed by its application to the HEMT [67]. In addition, we proposed a new approach that provided a combination of the Bohm and Wigner forms of the quantum potential [68]. Later, the equation was further developed by Gardner [69], and Grubin et al. explored its use for some other devices [70]. Its use in MOSFET simulation appears to have come somewhat later [71,72].

It is important to note that all of these various forms are related to one another. For example, the density-gradient potential is easily derived as a low-order expansion to the actual effective potential. We can expand the effective potential of Equation 1.23 when it is a slowly varying function of position. That is, we use a Taylor series expansion as

$$W(x) = \frac{1}{\sqrt{2\pi}\sigma}\int\limits_{-\infty}^{\infty} V(x+\xi)e^{-\xi^2/2\sigma^2}d\xi \cong \frac{1}{\sqrt{2\pi}\sigma}\int\limits_{-\infty}^{\infty}\left[V(x)+\xi\frac{\partial V}{\partial x}+\frac{\xi^2}{2}\frac{\partial^2 V}{\partial x^2}+\cdots\right]e^{-\xi^2/2\sigma^2}d\xi.$$

$$\tag{1.34}$$

The first term allows us to bring the potential outside the integral, while the second term vanishes due to the symmetry of the Gaussian. The third term becomes the leading correction term, which gives us

$$V_{eff}(x) = V(x) + \sigma^2 \frac{\partial^2 V}{\partial x^2} + \cdots \qquad (1.35)$$

We note that this result gives the Wigner form. A value for the smoothing parameter σ may be found if we compare with the results of Equation 4.7 to be

$$\sigma^2 = \frac{\hbar^2}{8mk_BT} = \frac{\lambda_D^2}{16\pi}, \qquad (1.36)$$

which is a factor of 1.5 larger than the Feynman result of Equation 1.29. Asenov et al. [73] have compared the density gradient approach and the effective potential approach and obtained similar results, which is to be expected. This is because the approximations on both begin to fail when the quantum corrections become comparable to the classical potential [44].

1.4.3   QUANTUM SIMULATIONS

There have been many suggestions for different full quantum methods to model ultrasmall semiconductor devices [74–76]. However, in each of these approaches, the length and the depth are modeled rigorously, while the third dimension (width) is usually included through the assumption that there is no interesting physics in this dimension (lateral homogeneity). Moreover, it is assumed often that the mode does not change shape as it propagates from the source of the device to the drain of the device. Other simulation proposals have simply assumed that only one subband in the orthogonal direction is occupied, therefore making higher-dimensional transport considerations unnecessary. These may not be valid assumptions, especially as we approach devices whose width is comparable to the channel length, both of which may be less than 10 nm.

It is important to consider all the modes that may be excited in the source (or drain) region, as this may be responsible for some of the interesting physics that we wish to capture. In the source, the modes that are excited are three dimensional (3D) in nature, even in a thin SOI device. These modes are then propagated from the source to the channel, and the coupling among the various modes will be dependent on the details of the total confining potential at each point along the channel. Moreover, as the doping and the Fermi level in short channel MOSFETs increases, we can no longer assume that there is only one occupied subband. In an effort to provide a more complete simulation method, we present a full 3D quantum simulation, based on the use of recursive scattering matrices, which is being used in our group to simulate short channel, fully depleted SOI MOSFET devices [77,78].

1.4.3.1   The Device Structure

The device under consideration is a fully depleted SOI MOSFET structure, shown schematically in Figure 1.3. We orient the x and z directions in order to correspond to the length and the height (thickness of the SOI layer) of the device, respectively. In the x direction, the source and the drain contact regions are 10 nm in length and 18 nm in width (lateral direction, the y axis). In an actual device, the length of the source and the drain of a MOSFET would be much longer, but this length captures the important energy relaxation length. We implement open boundary conditions at the ends of the structure and on the sidewalls. The gate length of this device is 11 nm corresponding to a dimension that will allow the gate to fully control that channel of the device. The actual channel length of the device used in these simulations is 9 nm. The channel itself is 9 nm in width, so that the Si layer is a wide-narrow-wide structure as shown in the figure. The entire structure is on a silicon layer that is taken to be only 6 nm thick, with a 10 nm buried oxide (BOX) layer below this layer. The gate oxide is taken to be 2 nm thick.

An important point relates to the crystal orientation of the device, as indicated in Figure 1.3. As is normal, we assume that the device is fabricated on a [100] surface of the Si crystal, and we then orient the channel so that the current will flow along the ⟨100⟩ direction. This direction is chosen so that all of the principle axes of the conduction band valleys line up with the coordinate axes. By this, we mean that the ⟨010⟩ direction lines up along the y direction and the ⟨001⟩ direction lines up with the z direction, and the six equivalent ellipsoids are oriented along the Cartesian coordinate axes. This is important so that the resulting quantization will split these ellipsoids into three pairs. Moreover, the choice of axes is most useful as the result-ing Hamiltonian matrix will be diagonal. In contrast, if we had chosen the ⟨110⟩ direction to lie along the channel, the six ellipsoids would have split into a twofold pair (those normal to the [100] plane) and a fourfold pair, but the Hamiltonian would not be diagonal, since the current axis makes an angle with each ellipsoid of the fourfold pair. Using our orientation complicates the wave function, as we will see, but allows for simplicity in terms of the amount of memory needed to store the

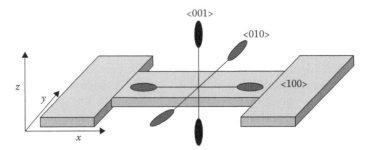

FIGURE 1.3 **(See color insert.)** A schematic diagram showing the crystal orientation of the SOI MOSFET for the quantum simulation (the directions are not to scale). The overlay shows how the six conduction band valleys of Si line up with the coordinate axes. This is discussed further in the text.

Hamiltonian and to construct the various scattering matrices (as well as the amount of computational time that is required).

1.4.3.2 The Wave Function and Technique

We can now write a total wave function, which is composed of three major parts, one for each of the three sets of valleys. That is, we can write the wave function as a vector

$$\Psi_T = \begin{bmatrix} \Psi^{(x)} \\ \Psi^{(y)} \\ \Psi^{(z)} \end{bmatrix}, \tag{1.37}$$

where the superscript refers to the coordinate axis along which the principal axis of the ellipsoid lies (the longitudinal mass direction). Thus, $\Psi^{(x)}$ refers to the two ellipsoids oriented along the x axis in Figure 1.3 (the $\langle 100 \rangle$ ellipsoids). Each of these three component wave functions is a complicated wave function on its own. Consider the Schrödinger equation for one of these sets of valleys (i corresponds to x, y, or z valleys):

$$\frac{-\hbar^2}{2} \left(\frac{1}{m_x} \frac{d^2}{dx^2} + \frac{1}{m_y} \frac{d^2}{dy^2} + \frac{1}{m_z} \frac{d^2}{dz^2} \right) \Psi^{(i)} + V(x,y,z)\Psi^{(i)} = E\Psi^{(i)}. \tag{1.38}$$

Here, it is assumed that the mass is constant, in order to simplify the equations (for nonparabolic bands, the reciprocal mass enters between the partial derivatives). We have labeled the mass corresponding to the principal coordinate axes, and these take on the values of m_L and m_T as appropriate. We then choose to implement this on a finite difference grid with uniform spacing a. Therefore, we replace the derivatives appearing in the discrete Schrödinger equation with finite difference representations of the derivatives. The Schrödinger equation then reads

$$-t_x(\psi_{i+1,j,k} + \psi_{i-1,j,k}) - t_y(\psi_{i,j+1,k} + \psi_{i,j-1,k}) - t_z(\psi_{i,j,k+1} + \psi_{i,j,k-1})$$

$$+ (V_{i,j,k} + 2t_x + 2t_y + 2t_z)\psi_{i,j,k} = E\psi_{i,j,k}, \tag{1.39}$$

where t_x, t_y, and t_z are the hopping energies

$$t_x = \frac{\hbar^2}{2m_x a^2},$$

$$t_y = \frac{\hbar^2}{2m_y a^2}, \tag{1.40}$$

$$t_z = \frac{\hbar^2}{2m_z a^2}.$$

Each hopping energy corresponds with a specific direction in the silicon crystal. The fact that we are now dealing with three sets of hopping energies is quite important.

There are other important points that relate to the hopping energy. The discretization of the Schrödinger equation introduces an artificial band structure, due to the periodicity that this discretization introduces. As a result, the band structure in any one direction has a cosinusoidal variation with momentum eigenvalue (or mode index), and the total width of this band is $4t$. Hence, if we are to properly simulate the real band behavior, which is quadratic in momentum, we need to keep the energies of interest below a value where the cosinusoidal variation deviates significantly from the parabolic behavior desired. For practical purposes, this means that $E_{max} < t$. The smallest value of t corresponds to the longitudinal mass, and if we desire energies of the order of the source–drain bias ~1 V, then we must have $a < 0.2$ nm. That is, we must take the grid size to be comparable to the Si lattice spacing!

With the discrete form of the Schrödinger equation defined, we now seek to obtain the transfer matrices relating adjacent slices in our solution space. For this, we will develop the method in terms of slices and follow a procedure first put forward by Ando [79] and by Usuki et al. [80,81] and used extensively by our group [74]. This is modified here by the two dimensions in the transverse plane. We begin first by noting that the transverse plane has $N_y \times N_z$ grid points. Normally, this would produce a second-rank tensor (matrix) for the wave function, and it would propagate via a fourth-rank tensor. However, we can reorder the coefficients into an $N_y N_z \times 1$ first-rank tensor (vector), so that the propagation is handled by a simpler matrix multiplication. Since the smaller dimension is the z direction, we use N_z for the expansion, and write the vector wave function as

$$\Psi^{(i)} = \begin{bmatrix} \psi^{(i)}_{1,N_y} \\ \psi^{(i)}_{2,N_y} \\ \vdots \\ \psi^{(i)}_{N_z,N_y} \end{bmatrix}. \tag{1.41}$$

Now, Equation 1.39 can be rewritten as a matrix equation as, with s an index of the distance along the x direction,

$$H^{(i)}\Psi^{(i)}(s) - T^{(i)}_x\Psi^{(i)}(s-1) - T^{(i)}_x\Psi^{(i)}(s+1) = EI\Psi^{(i)}(s), \tag{1.42}$$

where

I is the unit matrix
E is the energy to be found from the eigenvalue equation, and

$$H^{(i)} = \begin{bmatrix} H^{(i)}_0(\mathbf{r}) & \tilde{t}^{(i)}_z & \cdots & 0 \\ \tilde{t}^{(i)}_z & H^{(i)}_0(\mathbf{r}) & \cdots & \cdots \\ \cdots & \cdots & \cdots & \tilde{t}^{(i)}_z \\ 0 & \cdots & \tilde{t}^{(i)}_z & H^{(i)}_0(\mathbf{r}) \end{bmatrix}, \tag{1.43}$$

$$
T_x^{(i)} = \begin{bmatrix} \tilde{t}_x^{(i)} & 0 & \cdots & 0 \\ 0 & \tilde{t}_x^{(i)} & \cdots & 0 \\ \cdots & \cdots & \cdots & \cdots \\ 0 & 0 & \cdots & \tilde{t}_x^{(i)} \end{bmatrix}. \tag{1.44}
$$

The dimension of these two super-matrices is $N_z \times N_z$, while the basic Hamiltonian terms of Equation 1.43 have dimension of $N_y \times N_y$, so that the total dimension of the aforementioned two matrices is $N_y N_z \times N_y N_z$. In general, if we take k and j as indices along y, and η and v as indices along z, then

$$
\left(\tilde{t}_z^{(i)}\right)_{\eta v} = t_z^{(i)}\delta_{\eta v}, \quad \left(\tilde{t}_y^{(i)}\right)_{kj} = t_y^{(i)}\delta_{kj}, \quad \left(\tilde{t}_x^{(i)}\right)_{ss'} = t_z^{(i)}\delta_{ss'}, \tag{1.45}
$$

and

$$
H_0^{(i)}(\mathbf{r}) = \begin{bmatrix} V(s,1,\eta)+W & t_y^{(i)} & \cdots & 0 \\ t_y^{(i)} & V(s,2,\eta)+W & \cdots & 0 \\ \cdots & \cdots & \cdots & t_y^{(i)} \\ 0 & 0 & t_y^{(i)} & V(s,N_y,\eta)+W \end{bmatrix}. \tag{1.46}
$$

The quantity W is $2(t_x^{(i)} + t_y^{(i)} + t_z^{(i)})$ and is therefore independent of the valley index.

With this setup of the matrices, the general procedure follows the one laid out in the previous work [82]. One first solves the eigenvalue problem on slice 0 at the end of the source (away from the channel), which determines the propagating and evanescent modes for a given Fermi energy in this region. The wave function is thus written in a mode basis, but this is immediately transformed to the site basis, and one propagates from the drain end, using the scattering matrix iteration

$$
\begin{bmatrix} C_1^{(i)}(s+1) & C_2^{(i)}(s+1) \\ 0 & 1 \end{bmatrix} = \begin{bmatrix} 0 & 1 \\ -1 & \left(T_x^{(i)}\right)^{-1}(EI - H^{(i)}) \end{bmatrix}
$$

$$
\times \begin{bmatrix} C_1^{(i)}(s) & C_2^{(i)}(s) \\ 0 & 1 \end{bmatrix} \begin{bmatrix} 1 & 0 \\ P_1^{(i)}(s) & P_2^{(i)}(s) \end{bmatrix}. \tag{1.47}
$$

The dimension of these matrices is $2N_y N_z \times 2N_y N_z$, but the effective propagation is handled by submatrix computations, through the fact that the second row of this equation sets the iteration conditions

$$
C_2^{(i)}(s+1) = P_2^{(i)}(s) = \left[-C_2^{(i)}(s) + \left(T_x^{(i)}\right)^{-1}(EI - H^{(i)}) \right]^{-1}, \tag{1.48}
$$

$$
C_1^{(i)}(s+1) = P_1^{(i)}(s) = P_2^{(i)}(s)C_1^{(i)}(s).
$$

At the source end, $C_1(0) = 1$, and $C_2(0) = 0$ are used as the initial conditions. These are now propagated to the N_x slice, which is the end of the active region, and then onto the $N_z + 1$ slice. At this point, the inverse of the mode-to-site transformation matrix is applied to bring the solution back to the mode representation, so that the transmission coefficients of each mode can be computed. These are then summed to give the total transmission and this is used in a version of Equation 1.17 to compute the current through the device (there is no integration over the transverse modes, only over the longitudinal density of states and energy).

If we are to incorporate a self-consistent potential within the device, we must now solve Poisson's equation. Here, the density at each point in the device is determined from the wave function squared magnitude at that point, and this is used to drive Poisson's equation. Our solution for $C_1(N_x + 2)$ is the wave function at this point, and this is back-propagated using the recursion algorithm

$$\Phi_\xi^{(N_x+2,s,i)}(j,\eta) = P_1^{(i)}(s) + P_2^{(i)}\Phi_\xi^{(N_x+2,s+1,i)}(j,\eta). \tag{1.49}$$

Here, as before, the superscript "i" denotes the valley, while j and η denote the transverse position. Here, we are in the mode representation, and ξ is the mode index. The density at any site (s, j, η) is found by taking the sum over ξ of the occupied modes at that site, as

$$n(s,j,\eta) = \sum_\xi \left| \Phi_\xi^{(N_x+2,s,i)}(j,\eta) \right|^2. \tag{1.50}$$

1.4.3.3 Scattering in the Site Representation

It has often been argued that real scattering could not be put into the Landauer formula. But, Landauer's original paper had "scattering" in the title [12]. Indeed, so long as the transmission is computed *completely* through the structure, from one contact to another, this approach will work. The problem is in making sure that the total transmission is computed. Indeed, using a relaxation time will not work, since it does not conserve probability density. Rather, the correct approach must ensure a gain-loss formalism. What this means is that for each out-scattering process, there must be a corresponding in-scattering process treated within the formalism. Only with this complete treatment, probability density will be conserved throughout the simulation regime. We have found that this can be achieved with a self-energy formulation, so long as the actual self-energy matrix is transformed into the site representation used in the recursive scattering matrix [83].

Now, to be sure, this self-energy does not account for the self-consistent approach, which is common in strong scattering with Green's functions [84]. But, here, we are treating semiconductors, and scattering is usually quite weak in semiconductors. This is why they have high mobilities. Moreover, the dominant scattering is quite often impurity scattering, and this is treated in real space via the actual self-consistent potential [85,86]. While this approach was originally developed for classical simulations, it readily carries over to quantum simulations, since the potential is a local one-point function and is readily incorporated into the Hamiltonian in the recursive simulation.

Scattering, whether in the Green's function approach or in the present approach, is treated via the self-energy, as remarked earlier. The self-energy Σ has both real and imaginary parts, with the latter representing the dissipative interactions. The real part contributes an energy shift. In semiconductors, the scattering is weak, and is traditionally treated by first-order time-dependent perturbation theory, which yields the common Fermi golden rule for scattering rates. With such weak scattering, the real part of the self-energy can generally be ignored for the phonon interactions, and that part that arises from the carrier–carrier interactions is incorporated into the solutions of Poisson's equation by a local-density approximation, which approximately accounts for the Hartree–Fock corrections. In the many-body formulations of the self-energy, the latter is a two-site function in that it is written as

$$\Sigma(\mathbf{r}_1, \mathbf{r}_2). \tag{1.51}$$

In the case in which we are using transverse modes in the quantum wire of the transistor, this may be rewritten as

$$\Sigma(i, j; i', j', x_1, x_2). \tag{1.52}$$

Here, the scattering accounts for transitions from transverse mode i, j at position x_1 to i', j' at position x_2 (as in the previous section, x is taken as the longitudinal coordinate of, e.g., the nanowire). Generally, one then introduces a center-of-mass transformation

$$X = \frac{x_1 + x_2}{2}, \quad \xi = x_1 - x_2. \tag{1.53}$$

We then Fourier transform on the difference variable to give

$$\Sigma(i, j; i', j', X, k_x) = \frac{1}{2\pi} \int d\xi e^{i\xi k_x} \Sigma(i, j; i', j', X, \xi). \tag{1.54}$$

The center-of-mass position X remains in the problem as the mode structure may change as one moves along the channel. At this point, the left-hand side of Equation 1.54 is the self-energy computed by the normal scattering rates, such as is done in quantum wells and quantum wires previously [87]. However, these previous calculations usually used the Fermi golden rule, which is an evaluation of the bare self-energy. In many-body approaches, one normally does not use the energy-conserving delta function that is the central part of the Fermi golden rule. Rather, this function is broadened into the *spectral density*, through the use of the self-consistent Born approximation. In this way, off-shell effects are taken into account through the broadening of the relationship between momentum and energy. In semiconductors, however, we have already noted that the scattering is weak. It has been pointed out that these off-shell corrections are only important in fast processes where we are interested in femtosecond response and their neglect introduces only slight errors for times large compared to the collision duration [88]. Moreover, the broadening of the delta function will not be apparent when we reverse the Fourier transform of Equation 1.54, as the area under the spectral density remains normalized to unity. Since our recursion is in the site representation, rather than in a mode representation, we have to reverse the Fourier transform in Equation 1.54 to get the x-axis variation, and then do a mode-to-site unitary transformation to get the self-energy in the form necessary for the recursion. This is the subject of the rest of this section. Hence, we begin by seeking the imaginary part of the self-energy, which is related to the scattering rate via

$$\text{Im}\{\Sigma(i, j; i', j', X, k_x)\} = \hbar \left(\frac{1}{\tau}\right)_{i,j}^{i',j'}. \tag{1.55}$$

It is this latter scattering rate that we calculate. This result will be a function of the x-directed momentum (which is related, in turn, to the energy of the carrier) in the quantum wire. Finally, this scattering rate must be converted to the site representation with a unitary transformation

$$\Gamma_{ac} = \text{Im}\{\Sigma\} = U^+ \left(\frac{\hbar}{\tau}\right)_{i,j}^{i',j'} U, \tag{1.56}$$

where U is a unitary mode-to-site transformation matrix. The unitary matrix U^+ results from the eigenvalue solutions in the transverse slice, for example, in Section 1.4.3.1, and is composed of the various eigenfunctions in the site basis. Hence, it represents a mode-to-slice transformation. The actual scattering rates for quantum structures have been calculated earlier for nonpolar optical phonons and acoustic phonons that are appropriate for silicon [89].

1.4.3.4 Results

All of the aforementioned equations are written in the absence of a magnetic field, so that the Hamiltonian is symmetric. The various terms become more complicated if a magnetic field is present, as one may want for a study of spin transport, but this is beyond our present interest. Moreover, if there is no intervalley scattering (ballistic transport), then the equations for the three pairs of valleys are uncoupled. If intervalley scattering were to be present, then off-diagonal terms appear in the total Hamiltonian between valleys, and the iteration procedure of Equations 1.47 and 1.48 becomes much more difficult, with the matrices each being a factor of 3 increased in span. In the following, we will assume that *no* scattering is present, so that valleys that are unoccupied in the source will remain so throughout the device.

In Figure 1.4a, we plot the results of the transmission of incident modes as the Fermi energy is varied from 0 to 50 meV for a device at 300 K. Here, hard wall boundary conditions have been used (no self-consistent potential). In this method, we have taken into account the possibility of having the in-plane valleys contributing to the overall conductance of the device. Nevertheless, in the hardwall case, only the two surface normal ($\langle 001 \rangle$) valleys contribute to the conductance. This can be attributed to the fact that the surface normal valleys have the larger effective masse normal to the primary

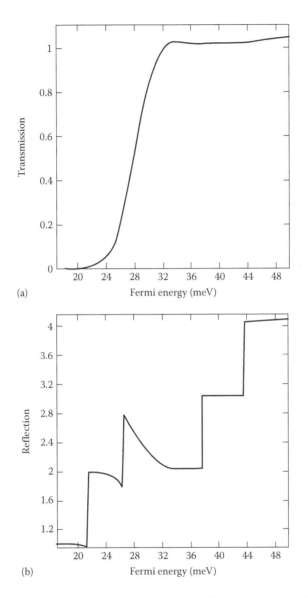

FIGURE 1.4 (a) Transmission and (b) reflection versus Fermi energy for a 9 nm channel length SOI MOSFET using hardwall potentials.

quantization direction (z direction) and, therefore, modes excited in these valleys will be the first to contribute. Further, we see that as the Fermi energy of the system is increased, the number of excited modes in the source of the device grows, but the transmission of these modes through the channel remains constant. This is confirmed in Figure 1.4b, where the reflection coefficient is plotted against increasing Fermi energy. Clearly, the number of modes increases, but the vast majority of these are reflected at the source channel interface. At approximately 24 meV, we see a decrease in the reflection coefficient followed by a sharp rise and subsequent decline. This behavior is expected as the onset of this decrease marks the point where the MOSFET begins to conduct. As we progress in energy, we see the sharp increase as another mode begins to propagate in the source of the device. This is followed by the exponential decrease back to 2 as the channel saturates with a full mode now propagating.

We now compare the hardwall results with results obtained using a self-consistent potential, found from solving Poisson's equation. The n^+ source and drain have been doped at 1×10^{20} cm^{-3} whereas the p-type channel of the device has been doped at 1×10^{18} cm^{-3}. In Figure 1.5, we plot the transmission resulting from varying the Fermi energy from 0 to 40 meV for all of the valleys, for a gate voltage of 1 V. In the case

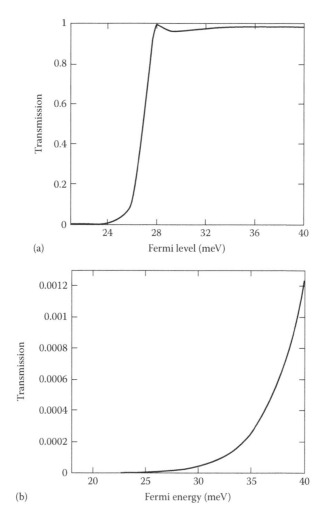

(a)

(b)

FIGURE 1.5   (a) Transmission for the surface-normal valleys versus Fermi energy for the 9 nm channel length SOI MOSFET using a self-consistent potential with a gate voltage of 1 V. (b) Transmission for the in-plane (upper) valleys versus Fermi energy for this device.

of the self-consistent potential, the final Fermi energy has been reduced to keep the energies within the artificial band structure. In Figure 1.5, we see that the turn-on energy for the transmission in both the in-plane and perpendicular valleys is very close to that of the hardwall case. In Figure 1.5a, we see that the self-consistent potentials reduce the contribution from the surface normal valleys. This is because the self-consistent potential squeezes the channel in the lateral y-direction. This greatly raises these valleys due to quantization in this direction, while two of the in-plane valleys are lowered in energy with respect to this first set. In Figure 1.5b, we also see that the upper valleys have begun to conduct. This can be attributed to the fact that, with the self-consistent potential, we see a softer variation in the potential. The potential allows for more leakage and higher-order contributions. The in-plane valleys now contribute to the current flow. Although there are more modes excited in the surface normal valleys, most of the modes are reflected at the source-channel constriction.

1.5   QUANTUM DOT SINGLE-ELECTRON DEVICES

Single-electron devices are of great interest, in particular for possible device application in integrated circuits [90]. The ability to control electron charging of a capacitive node by individual electrons makes these devices suitable candidates for memory applications [91,92]. As there are several chapters in this volume devoted to single-electron devices, we mention here only planar devices that are lithographically defined.

A major difficulty in fabricating planar single-electron transistors arises from the lithographic limits required in making small tunnel junctions in which the charging energy of the junction capacitor $e^2/2C \gg k_B T$. For room temperature operation, this requirement dictates lithographic control below 10 nm. In general, a single-electron transistor is made of a small "dot" isolated from the source and drain by two small tunnel junctions. For VLSI, it is preferable to work with devices fabricated in a semiconductor system, and quite novel ones have recently been fabricated using sidewall depletion gates [93]. Quantum confinement becomes relevant in silicon, and one may be able to observe quantum confinement effects and Coulomb blockade simultaneously in electrical measurements. Recently, single-electron dots have been created within an MOSFET [93–97]. In these structures, the dot is formed in the inversion layer created by a top gate (which is referred to as the inversion gate), with the lateral definition of the dots being provided by side gates (these gates provide the depletion of the dot, and are referred to as the depletion gates) embedded within the gate oxide. In essence, this is a multiple oxide system with stacked gates. The early work on this has recently been reviewed [96], and the recent work using sidewall depletion gates appears quite promising. The major issue at this point is technological—can the devices be fabricated with sufficiently small dimensions to operate at, or near to, room temperature.

1.6   MANY-BODY INTERACTIONS

In simulations of ultrasmall semiconductor devices, a number of important considerations have been either ignored or have been approximated in a manner, which is not representative of the actual physical interactions within the devices. Foremost of these is the study of the Coulomb interaction between the electrons and the impurities and between the individual electrons themselves. This Coulomb interaction has two parts: first, the nature of discrete impurities and how this affects device performance, and, secondly, how the Coulomb interaction affects the *transport* of the carriers through the device. The first of these has been discussed earlier. Here, we want to turn our attention to the carrier–carrier interaction.

Most ensemble Monte Carlo (EMC) simulation of small semiconductor devices does not include the details of the Coulomb interactions between the individual carriers, primarily because of the computational time and resources required. If carrier–carrier scattering is included, it is typically included through a k-space scattering process without much regard for the energy exchange in the process [98]. In such simulations, Ravaioli (U. Ravaioli, personal communication, 2000) has shown that the carriers will go several tens of nm into the drain before relaxing their energy and directed momentum. If this is a real effect, then actual device sizes will be significantly larger than the gate-related lengths in order to account for the actual hot carrier sizes [99]. Hence, it is important to know if the full Coulomb interaction, treated properly in real space (as opposed to approximations in terms of scattering processes), has a significant role in the transport of carriers in ultrasmall MOSFETs. We have previously discussed a full 3D model of an ultrasmall MOSFET, in which the transport is treated by a coupled EMC and molecular dynamics (MD) procedure to treat the Coulomb interaction in real space [21,100]. Impurities within the device, including the source and drain regions, are treated as discrete charges and are randomly sited according to the nominal doping density of each region. We find that the inclusion of the proper Coulomb interaction significantly affects both the energy and momentum relaxation processes, but also has a dramatic effect on the characteristic curves of the device. Relaxation occurs in the drain over a few nanometers, and the Coulomb "scattering" causes a significant shift in threshold voltage as well as a reduction in actual drain current. These effects are moderated somewhat in an SOI device due to the limited thickness of the Si layer and the small size of the drain [101].

The inherent real-space tracking of particle positions in the EMC allows us a more exact treatment of the Coulomb interaction between charged particles (particle–ion and particle–particle interactions). This is accomplished through the addition of an MD loop [29,102–104]. This coupled EMC-MD scheme has been shown to give simulation mobility results in excellent agreement with the experimental data for bulk samples with high substrate doping levels. It has also been corrected for both the degeneracy [105] and many-body exchange corrections to the ground state energy of the system [106]. Problems with this EMC-MD approach arise from the fact that both the e–e and e–i interactions are already included in the self-consistent potential via the solution of the Poisson equation (this is in the Hartree term). The magnitude of the resulting so-called mesh force depends upon the volume of the cell and, for commonly employed mesh sizes in device simulations, usually leads to "double counting" of the force if a separate Coulomb interaction is added to the EMC transport kernel [100]. Hence, careful treatment of the short range particle–particle interactions is needed to avoid the double counting of the force.

One brute force manner of overcoming this difficulty is to identify the correction terms necessary for the inelastic interaction between charge centers within an overall self-consistent particle-based device simulation, thus avoiding the problem of the double counting of the force. Briefly, we estimate the smoothed self-consistent potential on the grid points, as determined by the solutions to Poisson equation, and then determine the short-range-corrected Coulomb interaction to be used in our MD routine. This scheme has proven to be quite successful in explaining the doping dependence of the low-field electron mobility in highly doped resistors [21,100]. It also gives us confidence that this approach can be successfully used to accurately describe the fluctuations in various device parameters due to the atomistic nature and different distribution of the impurity atoms in the active region of the device. While *ad hoc*, it can be based on a more fundamental principle. Quite generally, we can replace the localized carrier by a function that is related strongly to the Gaussian wave packet discussed in Section 1.4.1.1. This divides the Coulomb potential into a short-range part and a long-range part. For example, we can write

$$\delta(r) \rightarrow \delta(r) + f(r) - f(r), \tag{1.57}$$

so that the Coulomb potential goes into something like

$$\frac{1}{r} \rightarrow [1 - \mathrm{erf}(r)]\frac{1}{r} + \mathrm{erf}(r)\frac{1}{r}. \tag{1.58}$$

The first term on the left-hand side is a short range function, which vanishes as the magnitude of r increases. On the other hand, the second term is a long range function, which vanishes at short distances. This is, of course, the principle of the potential splitting discussed earlier, in which the long range term is found from the solutions of Poisson's equation. Such a cutoff was introduced by Kelbg [107] in order to treat MD in plasma problems without incurring the very short range attraction between ions and electrons. This approach has been shown to be particularly useful in quantum many-body problems [108]. A similar split has been suggested by Kohn et al. [109] for electronic structure calculations, where the short range potential is kept within the density-functional approach and the long range potential is used for perturbation theory or configuration-interaction refinements of the results [110]. It is clear that the separation of the Coulomb potential into these short range and long range parts has a rich history and validates the splitting discussed earlier.

An alternative approach is to do the direct elastic Coulomb scattering by the traditional momentum space approach, but then add inelastic plasmon scattering [111,112]. This approach has also been used recently by Fischetti [113] to study the interaction of channel electrons with electrons in the gate. One advantage of this is the separation of the scattering from the role of the impurities in the self-consistent potential. On the other hand, the separation is rather artificial, and it is difficult to account for the density variation in the plasmon description. Moreover, one needs to take a nonequilibrium plasmon distribution function (the Bose–Einstein distribution must be taken at the electron temperature, which is quite difficult to evaluate, especially as a function of position). As a consequence, the best approach is the MD coupled Monte Carlo approach.

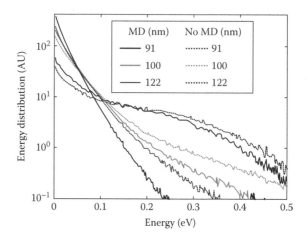

FIGURE 1.6 Energy distribution functions at various points in the drain. Here, the transport is simulated with an ensemble Monte Carlo technique using an MD routine to include the carrier–carrier scattering. The curves are with and without including the carrier–carrier scattering through MD. The drain metallurgical junction is at 92 nm, so the black curves are just before entering the drain. It is clear that losses are higher with the carrier–carrier interactions.

In Figure 1.6, we show the energy decay of the channel electrons as they move into the drain region [100] of a 50 nm gate length SOI MOSFET. This is indicated by the decay of the energy distribution functions as one moves further into the drain. It is clear that the inclusion of the electron–electron interaction causes a more rapid decay, which is indicative of plasmon emission being the major loss mechanism (while plasmons do not exist explicitly, this is the energy loss mechanism that explains such a rapid decay). This simulation is for a 4 nm thick SOI layer, with a drain extension region. For thinner SOI layers, or in the absence of the extension region, the number of electrons available as a whole really cuts down the effectiveness of this energy relaxation process. This could be a major problem in very small SOI devices in the future.

ACKNOWLEDGMENTS

The authors have enjoyed many helpful discussions with J.R. Barker, J.P. Bird, M.J. Gilbert, S.M. Goodnick, W. Gross, C. Jacoboni, I. Knezevic, S. Milicic, S. Ramey, and D. Vasileska, which have aided the flow of this work. The work itself was funded in part by the Office of Naval Research and the Semiconductor Research Corporation.

REFERENCES

1. Moore GE. Cramming more components onto integrated circuits. *Electronics* 1965;38(8):114–117.
2. Doyle BS, Doczy M, Hareland S, Kavalieros J, Linton T, Murthy A, Rios R, Chau R. High performance fully-depleted tri-gate CMOS transistors. *IEEE Electron Dev Lett* 2003;24(4):263–265.

3. Barker JR, Ferry DK. On the physics and modeling of small semiconductor devices—II: The very small device. *Solid-State Electron* 1980;23:531–544.
4. Fischetti M. Theory of electron transport in small semiconductor devices using the Pauli master equation. *J Appl Phys* 1998;83:270–291.
5. Likharev K. Sub-20 nm electron devices. In: Morkoç H, ed., *Advanced Semiconductor and Organic Nano-Technique*. New York: Academic Press, 2002.
6. Venugopal R, Ren Z, Datta S, Lundstrom MS, Jovanovic D. Simulating quantum transport in nanotransistors: Real versus mode-space approaches. *J Appl Phys* 2002;92:3730–3739.
7. Natori K. Ballistic metal-oxide-semiconductor field effect transistor. *J Appl Phys* 1994;76:4879–4890.
8. Gilbert MJ, Akis R, Ferry DK. Modeling of fully-depleted SOI MOSFETs in 3D using recursive scattering matrices. *J Comp Electron* 2003;2:329–334.
9. Lilienfeld JE. Method and apparatus for controlling electric currents. U.S. Patent 1,745,175, 1930.
10. Ferry DK, Hess H, Vogl P. Submicron IGFETs. II. In: Einspruch NG, ed., *VLSI Electronics: Microstructure Science*, Vol. 2. New York: Academic Press, 1981, pp. 67–103.
11. Ando T, Fowler A, Stern F. Electronic properties of two dimensional systems. *Rev Mod Phys* 1982;54:437–672.
12. Landauer R. Spatial variations of currents and fields due to localized scatterers in metallic conductors. *IBM J Res Develop* 1957;1:223–231.
13. Ferry DK, Bird JP. *Electronic Materials and Devices*. San Diego, CA: Academic Press, 2001.
14. Anantram MP, Svizhenko A. Role of scattering in nanotransistors. *IEEE Trans Electron Dev* 2003;50:1459–1466.
15. Akis R, Faralli N, Ferry DK, Goodnick SM, Phatak KA, Saraniti M. Ballistic transport in InP-based HEMTs. *IEEE Trans Electron Dev* 2009;56:2935–2943.
16. Natori K. Ballistic MOSFET reproduces current-voltage characteristics of an experimental device. *IEEE Electron Dev Lett* 2002;23:655–657.
17. Duke CB. *Tunneling in Solids*. New York: Academic Press, 1969.
18. Blakemore JS. *Semiconductor Statistics*. New York: Pergamon Press, 1962.
19. Landauer R. Electrical resistance of disordered one-dimensional lattice. *Philos Mag* 1970;21:863–875.
20. Landauer R. Electrical transport in open and closed systems. *Z Phys B* 1987;68:217–228.
21. Gross WJ, Vasileska D, Ferry DK. Ultra-small MOSFETs: The importance of the full Coulomb interaction on device characteristics. *IEEE Trans Electron Dev* 2000;47:1831–1837.
22. Ravaioli U, Ferry DF. MODFET ensemble Monte Carlo model including the quasi-two-dimensional electron gas. *IEEE Trans Electron Dev* 1986;33:677–680.
23. Ferry DK. *Quantum Mechanics*, 2nd edn. Bristol, U.K.: Institute of Physics Publishing, 2001.
24. Keyes RW. Physical limits in semiconductor devices. *Science* 1977;195:1230–1235.
25. Boudville WJ, McGill TC. Ohmic contacts to *n*-type GaAs. *J Vac Sci Technol B* 1985;3:1192–1196.
26. Joshi RP, Ferry DK. Effect of multi-ion screening on the electronic transport in doped semiconductors: A molecular dynamics analysis. *Phys Rev B* 1991;43:9734–9739.
27. Wong H-S, Taur Y. Three dimensional "atomistic" simulation of discrete dopant distribution effect in sub-0.1 μm MOSFETs. *1993 International Electron Device Meeting Technical Digest*. New York: IEEE Press, 1993, pp. 705–708.
28. Zhou J-R, Ferry DK. Three dimensional simulation of the effect of random impurity distributions on conductance for deep submicron devices. *Proceedings of the Third International Workshop on Computational Electronics*, Portland, OR, May 1994, pp. 74–77.

29. Zhou J-R, Ferry DK. 3D simulation of deep submicron devices: How impurity atoms affect conductance. *IEEE Comput Sci Eng* 1995;2(2):30–37.

30. Zhou J-R, Ferry DK. 3D discrete dopant effects on small semiconductor device physics. In: Hess K, Leburton JP, Ravioli U, eds., *Hot Carriers in Semiconductors*. New York: Plenum Press, 1996, pp. 491–496.

31. Vasileska D, Gross WJ, Ferry DK. Modeling of deep submicron MOSFETs: Random impurity effects, threshold voltage shifts, and gate capacitance attenuation. *Proceedings of the Sixth International Workshop on Computational Electronics*. New York: IEEE Press, 1998, pp. 259–262.

32. Asenov A. Efficient 3D "atomistic" simulation technique for studying of random dopant induced threshold voltage lowering and fluctuations in decanano MOSFETs. *Proceedings of the Sixth International Workshop on Computational Electronics*. New York: IEEE Press, 1998, pp. 263–266.

33. Vasileska D, Ahmed SS. Narrow-width SOI devices: The role of quantum-mechanical size quantization effect and unintentional doping on the device operation. *IEEE Trans Electron Dev* 2005;52:227–236.

34. Linton TD, Yu S, Shaheed R. Modeling 3D fluctuation effects in highly scaled VLSI devices. *VLSI Des* 2001;13:103–110.

35. Linton T, Shadhok M, Rice BJ, Schrom G. Determination of the line edge roughness specification for 34 nm devices. *2002 International Electron Device Meeting Technical Digest*. New York: IEEE, 2002, pp. 303–306.

36. Rack MJ, Vasileska D, Ferry DK, Sidorov M. Surface roughness of SiO_2 from a remote microwave plasma enhanced chemical vapor deposition process. *J Vac Sci Technol B* 1998;16:2165–2170.

37. Asenov A, Kaya S, Davies JH. Intrinsic threshold voltage fluctuations in decanano MOSFETs due to local oxide thickness variations. *IEEE Trans Electron Dev* 2002;49:112–119.

38. Goodnick SM, Ferry DK, Wilmsen CW, Lilienthal Z, Fathy D, Krivanek OL. Surface roughness at the $Si(100)$-SiO_2 interface. *Phys Rev B* 1985;32:8171–8186.

39. Brown AR, Lema FA, Asenov A. Intrinsic parameter fluctuations in nanometer scale thin body SOI devices induced by interface roughness. *Superlatt Microstruct* 2003;34:283–291.

40. Vasileska D, Schroder D, Ferry DK. Scaled silicon MOSFETs: Degradation of the total gate capacitance. *IEEE Trans Electron Dev* 1997;44:584–587.

41. Ferry DK, Shifren L, Ramey S, Akis R. The effective potential in device modeling: The good, the bad, and the ugly. *J Comput Electron* 2002;1:59–65.

42. Iafrate GJ, Grubin HL, Ferry DK. Utilization of quantum distribution functions for ultra-submicron device transport. *J Phys* 1981;41:C7-307–C7-312.

43. Ferry DK, Grubin HL. Electrons in semiconductors: How big are they? *Proceedings of the International Workshop on Computational Electronics*, Osaka, Japan, 1998, pp. 84–87. doi: 10.1109/IWCE.1998.742716.

44. Ferry DK. The onset of quantization in ultra-submicron semiconductor devices. *Superlatt Microstruct* 2000;27:61–66.

45. Skodje RT, Rohrs HW, van Buskirk J. Flux analysis, the correspondence principle, and the structure of quantum phase space. *Phys Rev A* 1989;40:2894–2916, and references therein.

46. Zurek WH. Decoherence, einselection, and the quantum origins of the classical. *Rev Mod Phys* 2003;75:715–775.

47. Glauber RJ. Coherent and incoherent states of the radiation field. *Phys Rev* 1963;131:2766–2788.

48. Klauder JR, Sudarshan ECG. *Fundamentals of Quantum Optics*. New York: Benjamin, 1968.

49. Reitz JR, Milford FJ. *Foundations of Electromagnetic Theory*. Reading, MA: Addison-Wesley, 1960.

50. Ahmed SS, Akis R, Vasileska D. Quantum effects in SOI devices. *Technical Proceedings of the Fifth International Conference on Modeling and Simulation of Microsystems*. Boston, MA: Computational Publications, 2002, pp. 518–521.

51. Wigner E. On the quantum correction for thermodynamic equilibrium. *Phys Rev* 1932;40:749–759.

52. Feynman RP, Hibbs AR. *Quantum Mechanics and Path Integrals*. New York: McGraw-Hill, 1965.

53. Giachetti R, Tognetti V. Variational approach to quantum statistical mechanics of nonlinear systems with applications to Sine-Gordon chains. *Phys Rev Lett* 1985;55:912–915.

54. Feynman RP, Kleinert H. Effective classical partition functions. *Phys Rev A* 1986;34:5080–5084.

55. Cao J, Berne BJ. Low temperature variational approximation for the Feynman quantum propagator and its application to the simulation of quantum systems. *J Chem Phys* 1990;92:7531–7539.

56. Voth GA. On the use of Feynman-Hibbs effective potentials to calculate quantum mechanical free energies of activation. *J Chem Phys* 1991;94:4095–4096.

57. Cuccoli A, Macchi A, Neumann M, Tognetti V, Vaia R. Quantum thermodynamics of solids by means of an effective potential. *Phys Rev B* 1992;45:2088–2096.

58. Kriman A, Ferry DK. Statistical properties of hard-wall potentials. *Phys Lett A* 1989;138:8–12.

59. Cuccoli A, Giachetti R, Tognetti V, Vaia R, Verrucchi P. The effective potential and effective Hamiltonian in quantum statistical mechanics. *J Phys Condens Matter* 1995;7:7891–7938.

60. Ferry DK, Akis R, Vasileska D. Quantum effects in MOSFETs: Use of an effective potential in 3D Monte Carlo simulation of ultra-short channel devices. *International Electron Devices Meeting*, Washington, DC, 2000, pp. 287–290. doi: 10.1109/IEDM.2000.904313.

61. Ahmed SS, Ringhofer C, Vasileska D. Parameter-free effective potential method for use in particle-based device simulations. *IEEE Trans Nanotechnol* 2005;4:465–471.

62. Palestri P, Lucci L, Dei Tos S, Esseni D, Selmi L. An improved empirical approach to introduce quantization effects in the transport direction in multi-subband Monte Carlo simulations. *Semicond Sci Technol* 2010;25:055011.

63. Madelung E. Quantentheorie in hydrodynamischer form. *Z Phys* 1926;40:322–328.

64. Bohm D. A suggested interpretation of the quantum theory in terms of "hidden" variables. *Phys Rev* 1952;85:166–179.

65. Zhou JR, Ferry DK. Simulation of ultra-small GaAs MESFET using quantum moment equations. *IEEE Trans Electron Dev* 1992;39:473–478.

66. Zhou JR, Ferry DK. Simulation of ultra-small GaAs MESFET using quantum moment equations—II. Velocity overshoot. *IEEE Trans Electron Dev* 1992;39:1793–1796.

67. Zhou JR, Ferry DK. Modeling of quantum effects in ultrasmall HEMT devices. *IEEE Trans Electron Dev* 1993;40:421–426.

68. Ferry DK, Zhou JR. Form of the quantum potential for use in hydrodynamic equations for semiconductor device modeling. *Phys Rev B* 1993;48:7944–7950.

69. Gardner CL. The quantum hydrodynamic model for semiconductor devices. *SIAM J Appl Math* 1994;54:409–427.

70. Grubin HL, Kreskovsky JP, Govindan TR, Ferry DK. Uses of the quantum potential in modeling hot-carrier semiconductor devices. *Semicond Sci Technol* 1994;9:855–858.

71. Ancona MG, Yu Z, Dutton RW, Voorde PV, Cao M, Vook D. Density-gradient analysis of MOS tunneling. *IEEE Trans Electron Dev* 2000;47:2310–2319.

72. Wettstein A, Schenk A, Fichtner W. Quantum device-simulation with the density-gradient model on unstructured grids. *IEEE Trans Electron Dev* 2001;48:279–284.
73. Asenov A, Watling JR, Brown AR, Ferry DK. The use of quantum potentials for confinement and tunneling in semiconductor devices. *J Comput Electron* 2002;1:503–513.
74. Pikus FG, Likharev KK. Nanoscale field-effect transistors: An ultimate size analysis. *Appl Phys Lett* 1997;71:3661–3663.
75. Datta S. Nanoscale device modeling: The Green's function method. *Superlatt Microstruct* 2000;28:253–278.
76. Knoch J, Lengeler B, Appenzeller J. Quantum simulations of an ultrashort channel single-gated n-MOSFET on SOI. *IEEE Trans Electron Dev* 2002;49:1212–1218.
77. Gilbert MJ, Akis R, Ferry DK. Modeling fully depleted SOI MOSFETs in 3D using recursive scattering matrices. *J Comput Electron* 2003;2:329–334.
78. Gilbert MJ, Ferry DK. Efficient quantum three-dimensional modeling of fully depleted ballistic silicon-on-insulator metal-oxide-semiconductor field-effect-transistors. *J Appl Phys* 2004;95:7954–7960.
79. Ando T. Quantum point contacts in magnetic fields. *Phys Rev B* 1991;44:8017–8027.
80. Usuki T, Takatsu M, Kiehl RA, Yokoyama N. Numerical analysis of electron-wave detection by a wedge-shaped point contact. *Phys Rev B* 1994;50:7615–7625.
81. Usuki T, Saito M, Takatsu M, Kiehl RA, Yokoyama N. Numerical analysis of ballistic-electron transport in magnetic fields by using a quantum point contact and a quantum wire. *Phys Rev B* 1995;52:8244–8255.
82. Akis R, Ferry DK, Bird JP. Magnetotransport fluctuations in regular semiconductor ballistic quantum dots. *Phys Rev B* 1996;54:17705–17715.
83. Akis R, Gilbert M, Ferry DK. Fully quantum mechanical simulations of gated silicon quantum wire structures: Investigating the effects of changing cross-section on transport. *J Phys Conf Ser* 2006;38:87–90.
84. Fetter AL, Walecka JD. *Quantum Theory of Many-Particle Systems*. New York: McGraw-Hill, 1971.
85. Joshi RP, Ferry DK. Effect of multi-ion screening on the electronic transport in doped semiconductors: A molecular dynamics approach. *Phys Rev B* 1991;41:9734–9739.
86. Gross WJ, Vasileska D, Ferry DK. A novel approach for introducing the electron-electron and electron-impurity interactions in particle-based simulations. *IEEE Trans Electron Dev* 1999;20:563–565.
87. Ferry DK, Goodnick SM, Bird JP. *Transport in Nanostructures*, 2nd edn. Cambridge, U.K.: Cambridge University Press, 2009.
88. Ferry DK., Kriman AM, Hida H, Yamaguchi S. Collision retardation and its role in femtosecond-laser excitation of semiconductor plasmas. *Phys Rev Lett* 1991;67:633–635.
89. Gilbert MJ, Akis R, Ferry DK. Phonon-assisted ballistic to diffusive crossover in silicon nanowire transistors. *J Appl Phys* 2005;98:094303.
90. Grabert H, Devoret MH, eds. *Single Charge Tunneling*. New York: Plenum Press, 1991.
91. Yano K, Ishii T, Hashimoto T, Kobayashi T, Murai F, Seki K. Room-temperature single-electron memory. *IEEE Trans Electron Dev* 1994;41:1628–1638.
92. Welser JJ, Tiwari S, Rishton S, Lee KY, Lee Y. Room temperature operation of a quantum-dot flash memory. *IEEE Electron Dev Lett* 1997;18:278–280.
93. Kim DH, Sung S-K, Kim KR, Lee JD, Park B-G, Choi BH, Hwang SW, Ahn D. Silicon single-electron transistors with sidewall depletion gates and their application to dynamic single-electron transistor logic. *IEEE Trans Electron Dev* 2002;49:627–635.
94. Khoury M, Gunther A, Pivin Jr DP, Rack MJ, Ferry DK. Silicon quantum dot in a metal-oxide-semiconductor field effect transistor (MOSFET) structure. *Jpn J Appl Phys* 1999;38:469–472.
95. Khoury M, Rack MJ, Gunther A, Ferry DK. Spectroscopy of a Si quantum dot. *Appl Phys Lett* 1999;74:1576–1578.

96. Khoury M, Gunther A, Milicic S, Rack MJ, Goodnick SM, Vasileska D, Thornton TJ, Ferry DK. Single-electron quantum dots in silicon MOS structures. *Appl Phys A* 2000;71:415–421.

97. Simmel F, Abusch-Magder D, Wharam DA, Kastner MA, Kotthaus JP. Statistics of the Coulomb-blockade peak spacings of a silicon quantum dot. *Phys Rev B* 1999;59:R10441–R10444.

98. Takenaka N, Inoue M, Inuishi Y. Influence of inter-carrier scattering on hot electron distribution function in GaAs. *J Phys Soc Jpn* 1979;47:861–868.

99. Ferry DK, Barker JR. Issues in general quantum transport with complex potentials. *Appl Phys Lett* 1999;74:582–584.

100. Gross WJ, Vasileska D, Ferry DK. A novel approach for introducing the electron-electron and electron-impurity interactions in particle-based simulations. *IEEE Electron Dev Lett* 1999;20:463–465.

101. Ramey SM, Ferry DK. A new model for including discrete Dopant ions into Monte Carlo simulations. *IEEE Trans Nanotechnol* 2003;2:193–197.

102. Jacoboni C. Recent developments in the hot electron problem. *Proceedings of the International Conference on Physics on Semiconductors* Rome, Italy: Tipograf. Marves, 1974, pp. 1195–1199.

103. Bosi S, Jacoboni C. Monte Carlo high-field transport in degenerate GaAs. *J Phys C: Sol State Phys* 1976;9:315–321.

104. Lugli P, Ferry DK. Dynamical screening of hot carriers in semiconductors from a coupled molecular-dynamics and ensemble Monte Carlo simulation. *Phys Rev Lett* 1986;56:1295–1297.

105. Lugli P, Ferry DK. Degeneracy in the ensemble Monte Carlo method for high-field transport in semiconductors. *IEEE Trans Electron Dev* 1985;32:2431–2437.

106. Kriman AM, Kann MJ, Ferry DK, Joshi R. Role of the exchange interaction in the short-time relaxation of a high-density electron plasma. *Phys Rev Lett* 1990;65:1619–1622.

107. Kelbg G. Zur theorie des mikrofeldes im plasma. *Ann Phys (Berlin)* 1964;13:385–394.

108. Morawetz K. Relation between classical and quantum particle systems. *Phys Rev E* 2002;66:022103-1–022103-4.

109. Kohn W, Meir Y, Makarov DE. Van der Waals energies in density functional theory. *Phys Rev Lett* 1998;80:4153–4156.

110. Pollet R, Savin A, Leininger T, Stoll H. Combining multideterminantal wave functions with density functionals to handle near degeneracy in atoms and molecules. *J Chem Phys* 2002;116:1250–1258.

111. Lugli L, Ferry DF. Electron-electron interaction and high field transport in Si. *Appl Phys Lett* 1985;46:594–596.

112. Lugli L, Ferry DF. Investigation of plasmon-induced losses in ballistic transport. *IEEE Electron Dev Lett* 1985;6:25–27.

113. Fischetti MV. Long-range Coulomb interactions in small Si devices. Part II. Effective electron mobility in thin oxide structures. *J Appl Phys* 2001;89:1232–1250.

2 Tri-Gate Transistors

Suman Datta

CONTENTS

Abstract 37
2.1   Introduction 37
2.2   Fully Depleted Tri-Gate Transistors 38
2.3   Electrostatics of Tri-Gate Transistors 39
2.4   Transport in Tri-Gate Transistors 41
    2.4.1   Long Channel Transport 41
    2.4.2   Short Channel Transport 42
2.5   Variation In Tri-Gate Transistors 46
2.6   Beyond Silicon Tri-Gate Transistors 48
References 50

ABSTRACT

One of the foremost challenges of sustaining Moore's law of doubling CMOS transistor density on an integrated circuit every 24 months is the rapid deterioration of short channel effects with decreasing channel length of the transistors. This has led to a rapid flurry of device research and technology development of fully depleted (FD) transistor architecture since the turn of the century, which are, in principle, more immune to short channel effects than their nondepleted counterparts. There are two classes of FD channel transistors: (a) single-gate planar silicon-on-insulator (SOI) transistors and (b) multigate Tri-gate or FinFET transistors on either bulk or SOI substrate. While each architecture has its pros and cons, recent trends suggest that Tri-gate transistors on bulk silicon substrate are increasingly becoming the leading transistor option for advanced logic manufacturers. This chapter reviews the essential physics of operation of Tri-gate transistors.

2.1   INTRODUCTION

According to Moore's Law, the number of transistors per integrated circuit doubles every 2 years, and has been the guiding principle for the semiconductor industry for over four decades. The sustenance of the Moore's law requires continued transistor channel length scaling and performance improvement. The physical gate length, L_G, of the silicon transistors introduced in 2009 by Intel Corporation in the 32 nanometer (nm) logic technology node is approximately 30 nm [1]. The planar single-gate transistors were fabricated on bulk silicon substrate and featured second generation of high-k and metal gate stack with an equivalent oxide thickness (EOT) of 0.9 nm, highly strained silicon channels for mobility enhancement and epitaxially regrown source and drain regions to improve external resistance. It was projected at that time that the transistor L_G may reach approximately 20 nm by 2015. Further improvement in control of short channel effects in the 22 nm node and beyond CMOS transistors would likely not come from reduction in the EOT of the high-k dielectric and the metal gate stack due to reliability constraints. Moreover, due to explosive growth in the mobile computing market stemming from the widespread adoption of the smartphones and tablets, the industry felt there is an urgent need for next-generation nanoscale transistors to be "energy efficient," which implies that they should operate with the lowest switching power (also called the dynamic power) and the lowest off-state leakage power (also called the stand-by power). Having already harnessed the benefits of key transistor innovations such as the highly strained-Si channels [2,3] for mobility enhancement and the high-k/metal-gate stacks [4,5] for higher drive current and lower gate leakage, the advanced transistor research community accelerated the research on the nonplanar multiple gate CMOS transistor architecture such as Tri-gate transistors [6,7]. The fundamental motivation behind adoption of Tri-gate transistors is that a combination of multiple gates and thin body channel would enable FD transistor operation at reduced channel lengths and maintain well-tempered transistor behavior with well-controlled short channel effects at the 22 nm node and beyond.

2.2   FULLY DEPLETED TRI-GATE TRANSISTORS

The single most important knob in the control of short channel effects in MOSFETs in the post-EOT scaling era is the body thickness. Figure 2.1 shows the schematic of the single-gate FDSOI and Tri-gate transistors. In the case of single-gate FDSOI devices, the silicon body thickness (T_{Si}) needs to be about a third to a half of the electrical gate length in order to maintain full substrate depletion under gate control. Scaling this device to 20 nm gate length (say) dimensions, for example, requires a

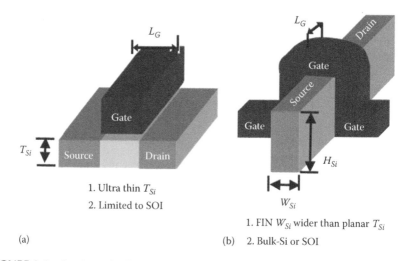

1. Ultra thin T_{Si}
2. Limited to SOI

(a)

1. FIN W_{Si} wider than planar T_{Si}
2. Bulk-Si or SOI

(b)

FIGURE 2.1   A schematic diagram of (a) single-gate FDSOI planar transistor and (b) Tri-gate nonplanar transistor showing the body/fin thickness requirement for each to maintain acceptable short channel effect performance.

3σ body thickness uniformity of 0.7 nm on a silicon film thickness of 7 nm, which is quite challenging to achieve in commercial production settings. For nonplanar double-gate FinFET or Tri-gate devices, the fin thickness (W_{Si}) requirement for the silicon between the two gates could be relaxed to approximately two-thirds of the physical gate length, since each gate controls half the body thickness. This relaxes the requirement of a 20 nm gate length Tri-gate transistor, for example, to a fin width thickness of 14 nm with a corresponding 3σ body thickness uniformity control of 1.4 nm, which is more practical to achieve in a commercial production fabrication facility. However, despite the relaxation of the fin width requirement in the case of Tri-gate transistors, the fin dimension is still smaller than the physical gate length, the most critical lithography step in printing transistor features, and will require either next-generation patterning technique (such as extreme UV lithography, direct write electron beam pattering) or sidewall image transfer and patterning technique. In the jargon of the semiconductor industry, the latter is referred to as the self-aligned double patterning (SADP) technique and is the preferred commercial fin patterning technique.

2.3 ELECTROSTATICS OF TRI-GATE TRANSISTORS

In addition to the fin width, W_{Si}, the electrostatic robustness of the Tri-gate transistor depends on various other factors. Two of the most important device design parameters that control the short channel effects in Tri-gate transistors are the shape of the fin as well as the impurity doping concentration of the fin. The short channel effects in Tri-gate transistors, like their planar counterparts, are quantified by two parameters: (a) the subthreshold slope under high drain bias or saturated drain bias (SS SAT), and (b) the drain induced barrier lowering (DIBL) parameter, which is essentially the threshold voltage difference between the saturated and linear drain bias conditions normalized to the drain bias difference. Figure 2.2 illustrates the dependence of the SS SAT and DIBL values

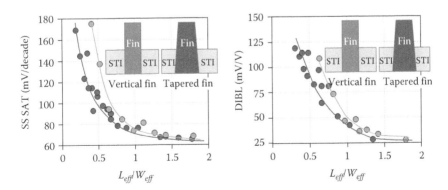

FIGURE 2.2 Dependence of the Tri-gate transistor short channel effect parameters such as subthreshold slope under saturated drain bias conditions, SS SAT, and DIBL on the fin profile as a function of electrical channel length thickness, L_{eff}, normalized to the electrical fin width thickness, W_{eff}.

in Tri-gate transistors with two different fin profiles as a function of the electrical channel length normalized to the electrical fin width. The electrical channel length, L_{eff}, is typically defined as the physical gate length minus the overlap length of the source and drain extension regions, while the electrical fin width, W_{eff}, is defined as

$$W_{eff} = W_{Si} + 2\left(\frac{\varepsilon_{Si}}{\varepsilon_{ox}}\right) T_{ox}$$

It is evident from Figure 2.2 that the Tri-gate transistor vertical fin profile has stronger electrostatics compared to the one with tapered fin profile. This can be explained considering the fact, that, for a given channel length, the transistor with the tapered fin profile has an effectively wider W_{eff} near the bottom of the fin, thereby making itself more susceptible to worse short channel effect. Another important parameter that determines the short channel performance of Tri-gate transistors is the doping concentration of the fin. Figure 2.3 shows the SS SAT and DIBL values of the Tri-gate transistor with three different fin doping concentrations. It is evident that, for a given fin width thickness, W_{Si}, the short channel performance of the Tri-gate transistor improves with increasing fin doping concentration. It is to be noted that, in practical Tri-gate transistors, the fin doping concentration is typically targeted between 1×10^{17} and 5×10^{17} cm^{-3} that is approximately an order of magnitude lower than that in planar single-gate transistors fabricated on bulk silicon substrate. The lower channel doping concentration provides several benefits in Tri-gate transistors including lower impurity scattering in the channel, leading to mobility enhancement as well as lower random dopant fluctuation (RDF) effect, leading to reduction in device-to-device variation.

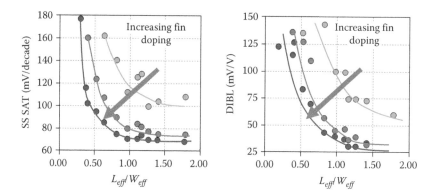

FIGURE 2.3 Dependence of the Tri-gate transistor short channel effect parameters such as subthreshold slope under saturated drain bias conditions, SS SAT, and DIBL on fin doping concentration as a function of electrical channel length thickness, L_{eff}, normalized to the electrical fin width thickness, W_{eff}.

2.4 TRANSPORT IN TRI-GATE TRANSISTORS

2.4.1 LONG CHANNEL TRANSPORT

The performance of Tri-gate transistors depends not only on the electrostatics but
also on the transport of electrons and holes in the fin along the channel direction.
It is critical to carefully examine the impact of the fin sidewalls on the transport
of carriers in the Tri-gate transistors. Figure 2.4 illustrates the electron field effect
drift mobility in long channel Tri-gate NMOSFETs benchmarked against planar
single-gate transistors fabricated on bulk silicon substrates with high and low
channel impurity concentrations. Both the Tri-gate transistors and the planar bulk
silicon transistors are fabricated with high-k dielectric and metal gate electrodes.
The universal electron mobility curve for silicon NMOSFETs with SiO_2 gate
dielectric is included for reference as well. Figure 2.4 suggests that, particularly
at low transverse gate electric field, the electron mobility in Tri-gate transistors is
higher than the planar NMOSFETs. This is because of lower doping concentration in the fin as well as due to the phenomenon of "volume inversion" in Tri-gate
transistors. The volume inversion in Tri-gate transistors occurs at low gate electric
field, when the electrostatic coupling between the depletion regions associated
with the gate electrodes on the fin sidewalls leads to the "flattening" of the electric
potential in the cross-section of the fin, leading to less confinement of carriers to
the sidewall surface. This results in more electronic conduction throughout the
volume of the fin, which, in turn, reduces the rate of scattering. It is to be noted
that, for both the Tri-gate and planar MOSFETs with high-k dielectrics, the electron mobility is overall lower than the universal mobility curve established for
the Si/SiO_2 interface supporting an inversion layer. This is due to the additional
remote optical phonon scattering arising from the soft phonons in the polariz-
able high-k dielectric. Further, one should note that the (110) crystallographic

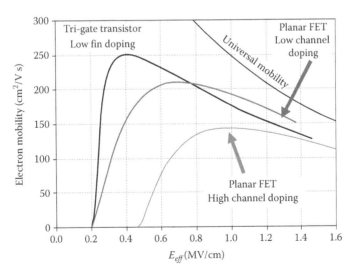

FIGURE 2.4   Room temperature electron drift mobility in n-channel Tri-gate transistors.

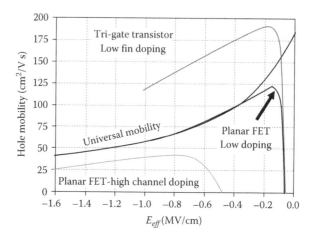

FIGURE 2.5   Room temperature hole drift mobility in n-channel Tri-gate transistors.

orientation of the fin sidewall does not lead to a significant degradation in the electron mobility in Tri-gate transistors compared to their planar counterpart. On the other hand, the hole transport in Tri-gate transistors benefits significantly from the conduction along the (110) oriented fin sidewalls. Figure 2.5 plots the field effect hole mobility in Tri-gate transistors and benchmarks it against the planar silicon NMOSFETs with (100) surface orientation. The hole mobility in Tri-gate transistors with the high-k dielectric and metal gate electrode gets a significant boost from both the <110> crystal orientation effect as well as lower fin doping and exceeds the universal hole mobility curve for SiO_2-gated planar PMOSFETs. This implies that the difference between electron and hole mobilities is going to be markedly reduced as the industry transitions from planar CMOS to the Tri-gate CMOS transistor technology.

2.4.2 SHORT CHANNEL TRANSPORT

Modern short channel planar NMOS and PMOS transistors benefit from enhanced electron and hole transport properties due to incorporation of strain in the channel. In the case of NMOS transistors, the uniaxial tensile strain along the <110> channel direction increases electron mobility on the (100) channel surface. The mobility enhancement arises from the lifting of the degeneracy of the energy levels between the four in-plane and the two out-of-plane ellipsoidal valleys that constitute the Brillouin space. This results in repopulation of the electrons within the two out-of-plane valleys, which has lower longitudinal transport mass and higher transverse confinement mass. In the case of n-channel Tri-gate transistors, the situation is more complicated as there are three surfaces that contribute to current conduction: two sidewall surfaces with (110) crystal orientation, and single top surface with (100) orientation. The source to drain current is oriented along <110> direction and stays the same in Tri-gate transistors. Further, the three-dimensional (3D)

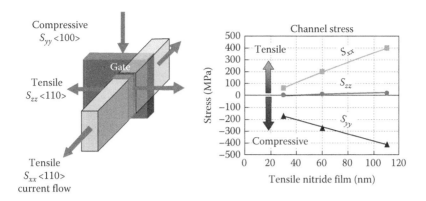

FIGURE 2.6 Dependence of the stress direction and stress magnitude inside the fin of the Tri-gate transistors when a tensile strained silicon nitride film is deposited conformally on top of the Tri-gate fin. (From Kavalieros, J. et al., Tri-gate transistor architecture with high-κ gate dielectrics, metal gates and strain engineering, *VLSI Symposium Technical Digest*, 2006, pp. 62–63.)

configuration of the Tri-gate transistor adds complexity to the direction and the magnitude of strain that can be incorporated in the channel. Figure 2.6 depicts the sign and the magnitude of the mechanical stress that can be induced within the 3D fin structure when a conformal tensile strained silicon nitride film is deposited on top of the Tri-gate fin [8,14]. It is evident that a tensile strained silicon nitride film with a thickness of 100 nm can impart an average tensile stress of 350 MPa along the channel direction, an average tensile stress of about 10 MPa on the fin sidewalls, and, finally, an average compressive stress of 400 MPa on the top surface of the fin. It is important to analyze the piezoresistive response of the fin sidewalls and the fin top surface to this stress profile [15].

Figure 2.7a plots the percent mobility gain for the fin top surface, which shows that a maximum electron mobility gain of approximately 18%–20% is achievable primarily from the compressive stress on the fin top. Figure 2.7b plots the percent mobility gain for the fin sidewalls, which shows that a maximum electron mobility gain of 22% is possible again primarily from the compressive in-plane stress and the tensile stress along the current direction. It almost seems that simply pressing on the fin from above results in electron mobility enhancement for both top and sidewall conduction planes in Tri-gate NMOS.

Strain-engineered short channel Tri-gate PMOSFET is also a topic of significant interest [9]. Uniaxial strain in short channel Tri-gate PMOSFETs is achieved by embedding larger lattice constant materials (e.g., Si_xGe_{1-x}) as source/drain (S/D) stressor regions [11,12]. However, in this case, there is always a possibility of strain relaxation of the embedded (eS/D) regions through their free surfaces due to the absence of shallow trench isolation (STI) in Tri-gate transistors, as the fin stands above the plane of the STI.

A nested device layout, which is a common strategy in the physical design of circuits [10], has to be adopted to minimize S/D relaxation associated with free surfaces

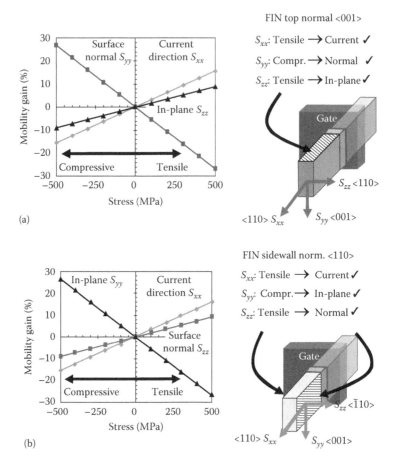

FIGURE 2.7 (a) Electron mobility for the current conducting fin top surface as a function of the stress direction and stress magnitude. (b) Electron mobility for the current conducting fin sidewalls as a function of the stress direction and stress magnitude. (From Kavalieros, J. et al., Tri-gate transistor architecture with high-κ gate dielectrics, metal gates and strain engineering, *VLSI Symposium Technical Digest*, 2006, pp. 62–63.)

of the eS/D regions and thus maximize the channel strain. Thus, the average uniaxial strain retention for top plane and sidewall in Tri-gate PMOSFETs will depend on the following factors: (1) the average distance between the source and the drain, (2) the embedded source drain (eS/D) Ge content, (3) the number of nested gates, (4) the number of nested fins, (5) the shape of the eS/D regions, (6) the eS/D etch depth, and (7) the contacted gate pitch.

The stress-inducing mechanism of the eS/D, the shape of the eS/D regions, and the presence of free surfaces result in a varying stress profile along the channel length, as shown in Figure 2.8. Hence, an average stress needs to be extracted for both the (100)/<110> and the (110)/<110> channels to estimate the achievable

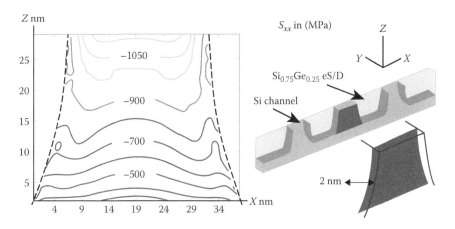

FIGURE 2.8   Sidewall [(110)/<110>] stress profiles (S_{xx} [stress along channel length] in MPa) a short channel Tri-gate PMOSFET with one active gate and two dummy gates on each side. The dummy gates are included to retain the compressive strain in the channel. The Tri-gate PMOSFET has an embedded SiGe source drain regions with 25% germanium content. (From Mujumdar, S. et al., *IEEE Trans. Electr. Dev.*, 59, 72, 2012.)

mobility enhancement. The average stress is taken as the average value of the stress profiles for the respective conduction planes, where the stress profiles are obtained for a channel depth of 2 nm into the fin. Figure 2.9 shows the average stress values for the (100)/<110> channel of the Tri-gate transistor for three different Ge contents in the eS/D regions. The channel stress for the planar pMOSFET case with the corresponding eS/D Ge contents [1] is also shown for comparison. The percentage reduction in the average channel stress compared to the planar case is shown in the inset. The absence of STI causes significant reduction in the channel stress of the Tri-gate structures compared to the planar case. The nested fin (two dummy fins) structure shows the highest average (100)/<110> channel stress reduction, indicating significant relaxation of the S/D regions through their free surfaces. The nested gate (two dummy gates) structure shows an approximately 23% reduction in the average channel stress for the (100)/<110> channel orientation, stressing the importance of gate nesting in improving the channel stress. The double-nested structure displays an even higher average (100)/<110> channel stress. The free eS/D sidewalls (parallel to the channel length) of the nested gate (two dummy gates) are eliminated in the double-nested structure post merger of the eS/D regions, resulting in increased average channel stress for the double-nested structure. The nested gate (four dummy gates) structure shows least degradation in the channel stress compared to the planar case among all four Tri-gate structures. It is evident that a hole mobility enhancement of over 300% for the top plane and 65% for the sidewall plane can be achieved in Tri-gate PMOSFETs with 40% Ge content eS/D regions and with a suitable transistor layout scheme.

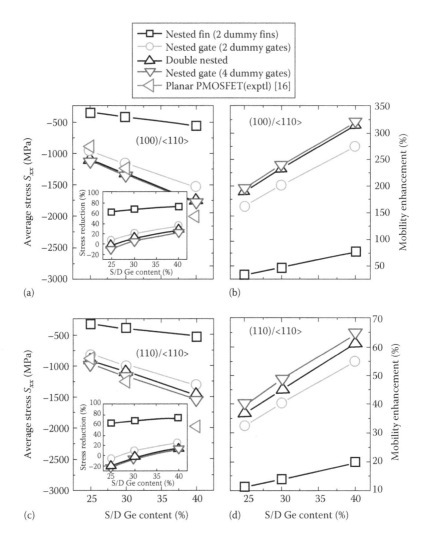

FIGURE 2.9 Average channel stress and corresponding hole mobility enhancement for the (100)/<110> and (110)/<110> channels of the Tri-gate PMOSFETs. Mobility enhancements are obtained from the hole mobility versus stress plots. (From Mujumdar, S. et al., *IEEE Trans. Electr. Dev.*, 59, 72, 2012; Irie, H. et al., In-plane mobility anisotropy and universality under uni-axial strains in n- and p-MOS inversion layers on (100), (110), and (111) Si, *International Electron Device Conference Meeting Tech. Dig. (IEDM)*, pp. 225–228, 2004.)

2.5 VARIATION IN TRI-GATE TRANSISTORS

At the sub-22 nm technology node, where a majority of semiconductor companies are expected to introduce the Tri-gate transistor technology for both energy efficiency as well as performance, variation could be a significant issue in the determination of the successful deployment of the technology. Due to the small feature sizes, the Tri-gate transistors will be highly sensitive to the process variations.

In addition to usual parameter variation such as gate length, L_G, oxide thickness, T_{ox}, and RDF parameters, the Tri-gate transistor deals with new parameter variation such as fin width, W_{FIN}, fin height, H_{FIN}, and the fin tapering angle, θ variation parameters unlike planar devices. The fin lithography and etching steps of Tri-gate results in sidewall roughness (SWR) of the fin. Due to continued scaling, the critical dimensions of the Tri-gate transistor such as the fin width or body thickness, W_{FIN}, are now becoming comparable with the parameters of SWR/SR viz. root-mean-square (rms) amplitude (Δ) and correlation length (Λ) [7–9].

With increased sources of variation and higher impact of SWR/SR, it is essential to quantitatively understand the sources of variation in Tri-gate transistors and their impact on low voltage circuit operation. Figure 2.10 shows the 3D simulation structure of a silicon Tri-gate fabricated on bulk (with tapered pro-file) silicon substrate calibrated to experimental data [16]. The calibrated Si Tri-gate transistor has a gate length, L_G, of 26 nm; effective oxide thickness, EOT, of 0.9 nm with 0.5 nm of SiO_2 acting as a interfacial layer between the HfO_2 and the fin; channel doping, N_{ch}, of 1×10^{16} cm^{-3}; and S/D doping, N_{sd}, of 1×10^{20} cm^{-3} operating at 0.8 V supply voltage. The fin width at the middle part of the tapered Tri-gate is 8 nm with tapering angle of 84°. Figure 2.10 plots the variance of the threshold voltage, V_T, due to all process parameter variations except SWR/SR. For a 1 nm change in W_{FIN}, H_{FIN}, and L_G parameters and 0.5 nm change in T_{ox}, the variance of V_T is shown. The estimation of ΔV_T due to RDF is determined using the impedance field method [17]. Though there is an increase in the number of variation sources in Tri-gate structures, they show lower overall V_T variation than the planar transistors on both bulk and SOI substrates. Tri-gate transistors show high sensitivity of V_T variation to fin height and fin width variation as expected. However, they enjoy the advantage of markedly reduced variation from RDF compared to planar bulk transistors. They have less sensitivity

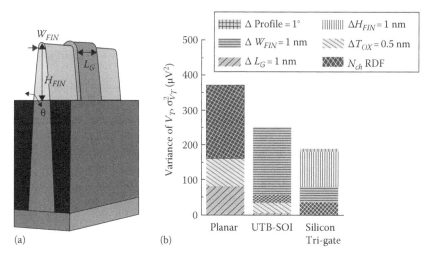

FIGURE 2.10 (a) Three-dimensional numerical simulation model of a Tri-gate transistor on bulk silicon substrate. (b) Variance of threshold voltage, V_T, of planar MOSFET nonbulk silicon substrate, ultrathin body planar MOSFET on SOI substrate, and nonplanar silicon Tri-gate transistor on bulk silicon substrate. (From Agrawal, N. et al., *IEEE Trans. Electr. Dev.*, 60, 3298, 2013.)

(a)

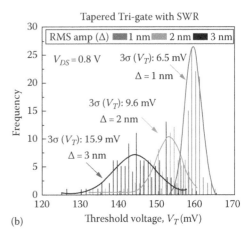

(b)

FIGURE 2.11 **(See color insert.)** (a) Three-dimensional numerical simulation model of a Tri-gate transistor on bulk silicon substrate incorporating the SWR. (b) Monte Carlo simulation of SWR-induced V_T variation in tapered Si Tri-gate NMOSFETs. Reducing the rms amplitude reduces 3σ (V_T). (From Agrawal, N. et al., *IEEE Trans. Electr. Dev.*, 60, 3298, 2013.)

to body thickness variation than their planar ultrathin body transistors on SOI substrate. The overall improvement in variation makes Tri-gate transistors suitable for next-generation low-power logic products. However, it is imperative to study in detail the effect of the sidewall roughness on parametric fluctuation in Tri-gate transistors.

To study the SWR/SR effect, one can implement a 1D Fourier synthesis of the Gaussian autocorrelation function that is used to generate random roughness on the sidewalls of Tri-gate Fin [18]. A total of 100 3D device samples with randomly generated SWR on Si Tri-gate transistors (with tapered fins) were simulated. For the SWR parameters, the state-of-the-art numbers reported by commercial fabrication facilities are the following: rms amplitude (Δ) of 1, 2, and 3 nm and autocorrelation length (Λ) of 20 nm [19]. Figure 2.11 shows the histogram of V_T variation of these ensembles of Tri-gate devices.

As the rms amplitude of SWR increases, 3σ of V_T increases, since the local variation in W_{FIN} also increases and the mean value of V_T ($\mu(V_T)$) decreases. With thinner W_{FIN}, V_T of the transistor typically increases due to the increase in confinement. With SWR in a device, there will always be regions that are thicker or thinner than the nominal W_{FIN} in the channel. In 3D Tri-gate with SWR, the thicker regions dominate the subthreshold characteristics by forming a parallel path of conduction from source to drain and thus reducing the mean of V_T. It is expected that Tri-gate transistors with more rectangular fin profiles will have a narrower V_T distribution function.

2.6 BEYOND SILICON TRI-GATE TRANSISTORS

Beyond silicon Tri-gate transistors, it is expected that germanium (Ge) [21] and/or compound semiconductor materials (III–V) such as indium gallium arsenide ($In_{0.70}Ga_{0.30}As$) [22] and InSb [23], due to their superior carrier transport properties, will replace the Si channel materials within the Tri-gate transistor configuration for future digital logic applications. While planar quantum–well FETs (QWFETs) in Ge and III–V material system have already demonstrated their superior performance compared to the state-of-the-art planar Si MOSFETs, particularly for below 0.5 V supply voltage logic applications, albeit at longer channel lengths, a key research challenge remains in addressing the scalability of both Ge and III–V based quantum-well FETs to sub-10 nm technology node applications while still maintaining their superior transport advantage over their Si counterpart. Recently, there have been experimental demonstrations of nonplanar, multigate, modulation doped, strained $In_{0.7}Ga_{0.3}As$ quantum well FET (MuQFET) [23], combining the advantages of Tri-gate configuration and high mobility quantum well channel. Figure 2.12a and b shows the schematic of a nonplanar lattice matched $In_{0.53}Ga_{0.47}As$ FINFET and a nonplanar modulation-doped pseudomorphic $In_{0.70}Ga_{0.30}As$ MuQFET, respectively. The pseudomorphic structure with higher indium content (70%) shows much higher electron mobility than its corresponding lattice-matched composition with indium content of 53%. Recently, Radosavljevic et al. [22] demonstrated an $In_{0.53}Ga_{0.47}As$ FinFET device suitable for low-power logic applications. Figure 2.12c shows the measured

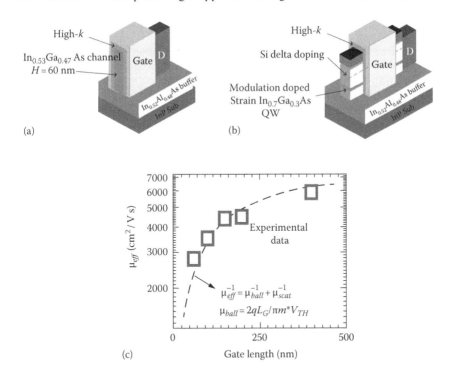

FIGURE 2.12 **(See color insert.)** A schematic diagram of compound semiconductor-based nonplanar transistor architectures; experimental electron field effect mobility as a function of gate length for $In_{0.7}Ga_{0.3}As$ multi-gate quantum-well FET (MuQFET). (From Lu, L. et al., Device circuit co-design using classical and non-classical III–V multi-gate quantum-well FETs (MuQFETs), *International Electron Devices Meeting Technical Digest (IEDM)*, 2011, pp. 4.5.1–4.5.4.)

electron mobility in $In_{0.7}Ga_{0.3}As$ MuQFETs with higher indium percentage by Lu et al. [24] as a function of gate length. The extracted mobility sometimes referred to as the apparent mobility exhibits an interesting trend of mobility reduction with channel length scaling, and captures the behavior of the short channel MuQFET accurately as it operates in the quasiballistic transport regime. In the quasiballistic regime of operation, the apparent mobility of the MuQFET is a combination of the ballistic mobility (given by $2qL_G/\pi m^* v_{TH}$, where m^* is the effective mass, v_{TH} is the thermal velocity) and the scattering limited diffusive mobility. This suggests that, in the future, we will likely witness a transformative evolution of silicon Tri-gate transistors into compound semiconductor-based Tri-gate transistors.

REFERENCES

1. P. Packan, S. S. Akbar, M. Armstrong, D. Bergstrom, M. Brazier, H. Deshpande, K. Dev et al. (2009). High performance 32 nm logic technology featuring 2nd generation high-k+ metal gate transistors. *International Electron Device Conference Meeting Technical Digest (IEDM)*, pp. 1–4.
2. S. Thompson, M. Armstrong, C. Auth, S. Cea, R. Chau, G. Glass, T. Hoffman et al. (2004). A logic nanotechnology featuring strained-silicon. *IEEE Electr. Dev. Lett.* **25**, 191–193.
3. S. Datta, G. Dewey, M. Doczy, B. S. Doyle, B. Jin, J. Kavalieros, R. Kotlyar, M. Metz, N. Zelick, and R. Chau (2003). High mobility Si/SiGe strained channel MOS transistors with HfO_2/TiN gate stacks. *International Electron Device Conference Meeting Technical Digest (IEDM)*, pp. 28.1.1–28.1.4.
4. R. Chau, S. Datta, M. Doczy, J. Kavalieros, and M. Metz (2003). Gate dielectric scaling for high-performance CMOS: From SiO_2 to high-K. *Extended Abstracts of International Workshop on Gate Insulator (IWGI)*, pp. 124–126.
5. R. Chau, S. Datta, M. Doczy, B. Doyle, J. Kavalieros, and M. Metz (2004). High-K/metal-gate stack and its MOSFET characteristics. *IEEE Electr. Dev. Lett.* **25**, 408–410.
6. R. Chau, B. Doyle, J. Kavalieros, D. Barlage, A. Murthy, M. Doczy, R. Arghavani, and S. Datta (2002). Advanced depleted-substrate transistors: Single-gate, double-gate and tri-gate. *Extended Abstracts of International Conference on Solid State Devices and Materials (SSDM)*, pp. 68–69.
7. B. Doyle, S. Datta, M. Doczy, S. Hareland, B. Jin, J. Kavalieros, T. Linton, A. Murthy, R. Rios, and R. Chau (2003). High performance fully-depleted tri-gate CMOS transistors. *IEEE Electron Dev. Lett.* **24**, 263–265.
8. J. Kavalieros, B. S. Doyle, S. Datta, G. Dewey, and R. Chau (2006). Tri-gate transistor architecture with high-κ gate dielectrics, metal gates and strain engineering. *Digest of Technical Papers VLSI Technology Symposium*, pp. 62–63.
9. S. Mujumdar, K. Maitra, and S. Datta (2012). Layout dependent strain optimization for p-channel non-planar tri-gate transistors. *IEEE Trans. Electr. Dev.* **59**, 72–78.
10. M. Alioto (2009). Analysis and evolution of layout density of FinFET logic gates. *Proceedings of the ICM*, pp. 106–109.
11. J. J. Wortman and R. A. Evans (January 1965). Young's modulus, shear modulus, and Poisson's ratio in Si and Germanium. *J. Appl. Phys.* **36**(1), 153–156.
12. N. Serra, F. Conzatti, D. Esseni, M. De Michielis, P. Palestri, L. Selmi, S. Thomas et al. (2009). Experimental and physics-based modeling assessment of strain induced mobility enhancement in FinFETs. *International Electron Device Conference Meeting Technical Digest (IEDM)*, pp. 1–4.

13. H. Irie, K. Kita, K. Kyuno, and A. Toriumi (2004). In-plane mobility anisotropy and universality under uni-axial strains in n- and p-MOS inversion layers on (100), (110), and (111) Si. *International Electron Device Conference Meeting Technical Digest (IEDM)*, pp. 225–228.

14. S. Ito, H. Namba, K. Yamaguchi, T. Hirata, and K. Ando (2009). Mechanical stress effect of etch-stop nitride and its impact on deep submicron transistor design. *International Electron Device Conference Meeting Technical Digest (IEDM)*, pp. 247–250.

15. M. Chu, Y. Sun, U. Aghoram, and S. E. Thompson (2009). Strain: A solution for higher carrier mobility in nanoscale MOSFETs. *Annu. Rev. Mater. Res.* **39**, 203–229.

16. C. Auth, C. Allen, A. Blattner, D. Bergstrom, M. Brazier, M. Bost, M. Buehler et al. (2012). A 22 nm high per-formance and low-power CMOS technology featuring fully-depleted tri-gate transistors, self-aligned contacts and high density MIM capacitors. *Proceedings of the VLSI Symposium on Technology*, pp. 131–132.

17. A. Asenov, S. Kaya, and A. Brown (May 2003). Intrinsic parameter fluctuations in decananometer MOSFETs introduced by gate line edge roughness. *IEEE Trans. Electr. Dev.* **50**(5), 1254–1260.

18. *Sentaurus Device User Guide* (2011). Synopsys, Inc., Mountain View, CA.

19. S. Jin, M. Fischetti, and T.-W. Tang (2007). Modeling of surface-roughness scattering in ultrathin-body SOI MOSFETs. *IEEE Trans. Electr. Dev.* **54**, 2191–2203.

20. N. Agrawal, Y, Kimura, R. Arghavani, and S. Datta (2013). Impact of transistor architecture (bulk planar, trigate on bulk, ultrathin-body planar SOI) and material (silicon or III semiconductor) on variation for logic and SRAM applications. *IEEE Trans. Electr. Dev.* **60**, 3298–3304.

21. A. Agrawal, M. Barth, G. B. Rayner Jr., V. T. Arun, C. Eichfeld, G. Lavallee, S-Y. Yu et al. (2014). Enhancement mode strained (1.3%) germanium quantum well FinFET (W_{Fin} = 20 nm) with high mobility (hole = 700 cm^2/V s), low EOT (~0.7 nm) on bulk silicon substrate. *IEEE International Electron Device Meeting (IEDM) Technical Digest*, pp. 414–417.

22. M. Radosavljevic, G. Dewey, D. Basu, J. Boardman, B. C. Kung, J. Fastenau, S. Kabehie et al. (2011). Electrostatics improvement in 3-D tri-gate over ultra-thin body planar InGaAs quantum well field effect transistors with high-K gate dielectric and scaled gate-to-drain/gate-to-source separation. *IEEE International Electron Device Meeting (IEDM) Technical Digest*, pp. 1–4.

23. S. Datta, T. Ashley, J. Brask, L. Buckle, M. Doczy, M. Emeny, D. Hayes et al. (2005). 85 nm gate length enhancement and depletion mode InSb quantum well transistors for ultra high speed and very low power digital logic applications. *International Electron Devices Meeting Technical Digest (IEDM)*, pp. 763–766.

24. L. Lu, V. Saripalli, V. Narayanan, and S. Datta (2011). Device circuit co-design using classical and non-classical III–V multi-gate quantum-well FETs (MuQFETs). *International Electron Devices Meeting Technical Digest (IEDM)*, pp. 4.5.1–4.5.4.

3 Variability in Scaled MOSFETs

Toshiro Hiramoto

CONTENTS

3.1 Introduction 54
3.2 Variability in 65 nm Transistors 54
    3.2.1 Classification of Transistor Variability 54
    3.2.2 Dependence on the Number of Transistors 56
    3.2.3 Origin of Random Variability 58
    3.2.4 Size Dependence of Variability 59
    3.2.5 Drain Current Variability 60
    3.2.6 Origin of COV Variability 62
3.3 Variability of 11G Transistors 63
    3.3.1 Measurement of 11G Transistors 63
    3.3.2 V_{thc} Variability of 11G Transistors 64
    3.3.3 V_{thex} and I_{on} Variability of 11G Transistors 66
    3.3.4 SS Variability of 11G Transistors 67
3.4 Stability of SRAM Cells..68
    3.4.1 Variability of Static Noise Margin 68
    3.4.2 V_{dd} Dependence of SNM 69
    3.4.3 DIBL Dependence of SNM 71
3.5 Intrinsic Channel FD SOI Transistors 72
    3.5.1 V_{th} and Drain Current Variability...72
    3.5.2 Stability in FD SOTB SRAM...74
3.6 Self-Suppression of Variability..75
    3.6.1 Mechanism of Self-Improvement 75
    3.6.2 Measurements of $|V_{th}|$ Shift by High Voltage Stress........................77
    3.6.3 Measurements of Self-Improvement of SRAM Stability 78
3.7 Conclusions..79
Acknowledgments..79
References 80

3.1 INTRODUCTION

For the past 40 years, the size of metal-oxide-semiconductor field-effect-transistors (MOSFETs) has been scaled down in order to attain higher performance, lower power dissipation, and higher integration in large-scale integrated circuits (VLSI). It is well known that the miniaturized size of MOSFETs has brought about various technical issues including short channel effects, degraded reliability due to a high electric field, and performance degradation due to parasitic resistance and capacitance. Among them, one of the most significant technical problems is the variability of transistor characteristics [1–4]. Owing to the transistor variability, the circuits do not function correctly even though each individual transistor behaves correctly, or the margin in the circuit operation is reduced causing manufacturing yield to drop precipitously. This problem of the variability in characteristics may impose a limit on the transistor scaling, performance improvement, and power dissipation reduction. Therefore, it is an urgent issue to understand the root causes and find solutions. However, the origins that lead to variability in characteristics cover a very wide range from atomic-level impurity distribution to manufacturing equipment. The understanding of the physics behind the variability problems and the development of effective measures to the problems are essential.

In this chapter, the present status of the transistor variability is reviewed. In particular, the random variability and its impact on static random access memory (SRAM) are described. As a solution to the variability issue, a fully-depleted (FD) silicon-on-insulator (SOI) transistor with intrinsic channel is described. A new concept for the suppression of random variability in SRAM is also introduced.

3.2 VARIABILITY IN 65 nm TRANSISTORS

In this section, the status and basic behaviors of transistor variability are illustrated by taking the 65 nm planar bulk technology as an example.

3.2.1 CLASSIFICATION OF TRANSISTOR VARIABILITY

There are so many types of transistor variability in VLSI. Figure 3.1 schematically shows the classification of transistor variability. The transistor characteristics differ from one lot to another. This is lot-to-lot variability (or interlot variability). Within the same lot, the transistor characteristics differ from one wafer to another (wafer-to-wafer variability or interwafer variability). Within the same wafer, transistor characteristics differ from one chip to another (chip-to-chip variability or interchip variability). Even in the same chip, the transistor characteristics differ from one transistor to another (intrachip variability).

Figure 3.2 shows an example of intrawafer variability and intrachip variability of transistors fabricated by the 65 nm bulk planar technology [5]. A large number of transistors were measured using a device-matrix-array (DMA) test-element group (TEG). Each chip has one million (1M) transistors with identical gate length (L) and gate width (W). In order to show the overall and systematic variability within the wafer and chip, each datum point in Figure 3.2 represents the average threshold voltage (V_{th}) of 1k transistors (each wafer has 1k data points). Apparently, some chips

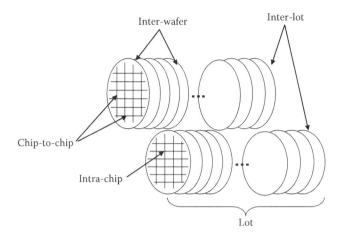

FIGURE 3.1   A schematic diagram of interlot, interwafer, interchip, and intrachip variabilities.

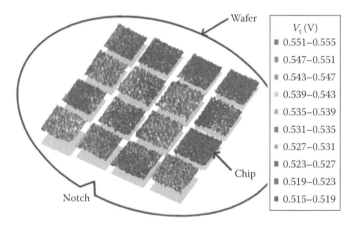

FIGURE 3.2   Measured V_{th} variability of nFETs in a wafer and in a chip. Each chip has 1M nFETs. Average V_{th}'s of 1k transistors are shown (each wafer has 1k data points). (From Tsunomura, T. et al., *Jpn. J. Appl. Phys.*, 48, 124505, Copyright 2009 The Japan Society of Applied Physics.)

have higher V_{th}, and others have lower V_{th}. This is the chip-to-chip variability. There is also a V_{th} variation within a chip. It is found that the magnitude of the intrawafer variability is larger than that of the overall within-chip variability.

Figure 3.3a shows detailed intrachip variability of the 65 nm transistors, where each datum point corresponds to V_{th} of each transistor [5]. Surprisingly, V_{th} differs largely from one transistor to another. The measured V_{th} values are mathematically separated into a random component and systematic component in Figure 3.3b and c [5]. It is found that the random component is much larger than the systematic component. These results indicate that as long as the transistor layout patterns are regular and identical, the random component is dominant and the systematic component is negligible in the planar bulk transistors.

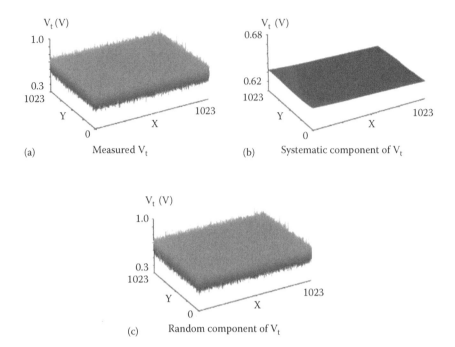

FIGURE 3.3   (a) Measured V_{th} variability of 1M nFETs within a chip. (b) Separated systematic component. (c) Separated random component. (From Tsunomura, T. et al., *Jpn. J. Appl. Phys.*, 48, 124505, Copyright 2009, The Japan Society of Applied Physics.)

3.2.2   DEPENDENCE ON THE NUMBER OF TRANSISTORS

The range of the random variability depends on the number of transistors. Figure 3.4a shows I–V characteristics of 100 n-type transistors (nFETs) within a chip. Gate length L is 60 nm and gate width W is 120 nm. Clear V_{th} variability is found, but the range of V_{th} is not so wide. The cumulative plot of V_{th} is also shown. The V_{th} data lie on an almost straight line and V_{th} ranges within $\pm 2.6\sigma$, where σ is the standard deviation and approximately 43 mV in this wafer. When the number of transistors increases by 100 times (10,000 transistors), I–V characteristics are varied as shown in Figure 3.4b. The V_{th} range increases to $\pm 3.8\sigma$.

When the number of transistors increases further by 100 times (1M transistors), I–V characteristics are shown in Figure 3.4c [5], where the V_{th} range further increases to $\pm 5\sigma$. It is also clearly shown that since V_{th} data lie on a straight line in the cumulative plot, V_{th} follows the normal distribution up to $\pm 5\sigma$, and therefore, the variability is random. V_{th} ranges over a wide range from −0.28 to 0.73 V, which may cause severe yield loss and margin degradation in circuit operation.

1M nFETs in other chips in the same wafers and other wafers were also measured. Although the average V_{th}'s are slightly different from one chip to another, V_{th} also follows the normal distribution up to $\pm 5\sigma$ in all chips and σ is almost the same (not shown). From these results, the transistor variability in lots, wafers,

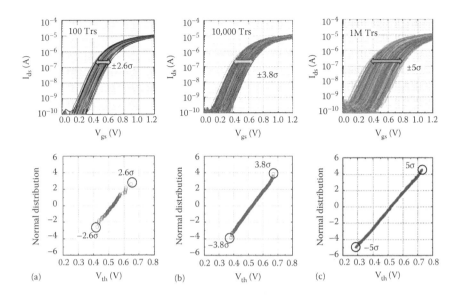

FIGURE 3.4 Measured I–V characteristics and cumulative V_{th} distribution of (a) 100 nFETs, (b) 10,000 nFETs, (c) 1M nFETs. $L = 60$ nm and $W = 120$ nm. (From Tsunomura, T. et al., *Jpn. J. Appl. Phys.*, 48, 124505, Copyright 2009, The Japan Society of Applied Physics.)

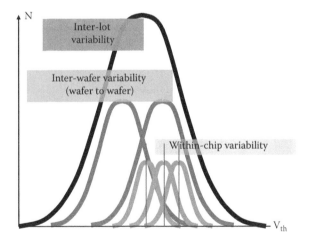

FIGURE 3.5 A schematic distribution showing the relationship among interlot, interwafer, interchip, and within-chip variabilities.

and chips is schematically shown in Figure 3.5. The intrawafer variability is smaller than the interlot variability. Each wafer has different average V_{th} and each chip also has different average V_{th}. The within-chip variability is defined by the average V_{th} and σ. It is also found that V_{th} follows the normal distribution up to ±5σ in pFETs and the variability in nFETs is larger than that in pFETs [5,6] (not shown).

3.2.3 Origin of Random Variability

In order to examine the origins of the aforementioned huge random variability of measured V_{th}, the transistors that exhibit extreme V_{th} values are directly observed by the transmission electron microscope (TEM) [6]. Figure 3.6 shows a top view and the cross-sectional TEM images of −5σ, median, and +5σ nFETs. Observed gate oxide thickness (t_{ox}), measured strain, and images of gate polysilicon grains are also shown. It is surprising to find that transistor sizes (L, W, and t_{ox}) of −5σ FET and +5σ FET are almost identical even though V_{th}'s are so different. This result clearly suggests that the large random variability is not caused by the variation of transistor dimensions.

It is now well recognized that the origin of the random variability of transistor characteristics is the random dopant fluctuation (RDF). V_{th} of a transistor is primarily determined by dopant concentration in the depletion layer in the transistor channel. However, the number of dopants in the depletion layer varies from one transistor to another, causing the V_{th} variability. It is well known that when the dopants are randomly distributed, the number of dopants follows the Poisson distribution. When the number of dopants increases, the Poisson distribution is well approximated by the normal distribution. This is why the measured V_{th} follows the normal distribution.

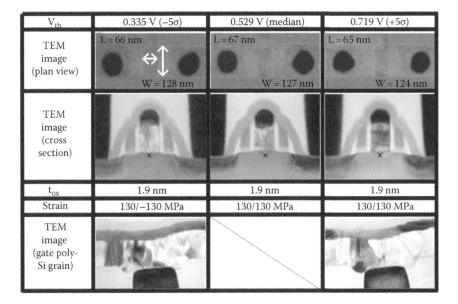

V_{th}	0.335 V (−5σ)	0.529 V (median)	0.719 V (+5σ)
TEM image (plan view)	L = 66 nm W = 128 nm	L = 67 nm W = 127 nm	L = 65 nm W = 124 nm
TEM image (cross section)			
t_{ox}	1.9 nm	1.9 nm	1.9 nm
Strain	130/−130 MPa	130/130 MPa	130/130 MPa
TEM image (gate poly-Si grain)			

FIGURE 3.6 Observations of nFETs that have extremely low V_{th} (−5σ), median V_{th}, and extremely high V_{th} (+5σ). Top view and cross-sectional TEM images, measured t_{ox}, measured strain, and observed poly-Si grains are shown. (Modified from Tsunomura, T. et al., Analyses of 5σ V_{th} fluctuation in 65 nm-MOSFETs using Takeuchi plot, *Symposium on VLSI Technology*, Honolulu, HI, 2008, pp. 156–157.)

3.2.4 SIZE DEPENDENCE OF VARIABILITY

When the random variability is dominant in the transistor variability, the standard deviation σ of transistor parameters has transistor size dependence. This phenomenon can be understood by the nature of the Poisson distribution. In a Poisson distribution, σ is given by $\sqrt{\mu}$, where μ is the average number, and the normalized variability (σ/μ) is given by $1/\sqrt{\mu}$. In a smaller transistor, the number of average dopant atoms included in the depletion region is smaller, and hence, the normalized variability (σ/μ) of the dopant number is larger. When the transistor size increases, the variability of dopant number is averaged out and becomes smaller. Therefore, the variability of a transistor parameter increases as the transistor size is scaled down.

The average number μ of dopant atoms is proportional to LW, where L is the gate length and W is the gate width. Then, the variability is given by a simple function of $1/\sqrt{LW}$ [3]. Figure 3.7a shows σV_{th} as a function of $1/\sqrt{LW}$ [7]. When t_{ox} is fixed and only transistor size is varied, measured data lie on the same straight line. The slope of this line is often called Pelgrom coefficient (A_{vt}) and this figure is called Pelgrom plot [3]. Then, σV_{th} is given by $\sigma V_{th} = A_{vt}/\sqrt{LW}$. This is a very useful equation. When A_{vt} of a semiconductor process is known and the transistor size is fixed, σV_{th} is easily derived.

It is known that A_{vt} depends on gate oxide thickness t_{ox} and N_A. In order to develop a more universal relationship between σV_{th} and transistor parameters, it is proposed to use the Takeuchi plot [8]. Figure 3.7b shows a Takeuchi plot, where the data in Figure 3.7a is replotted [7]. The horizontal axis is $\sqrt{T_{inv}(V_{th} + V_0)/LW}$, where T_{inv} is the electrical gate oxide thickness at inversion and V_0 is given by $-(V_{FB} + 2\varphi_F)$, where V_{FB} is the flat band voltage and φ_F is the Fermi energy. Even though t_{ox} and

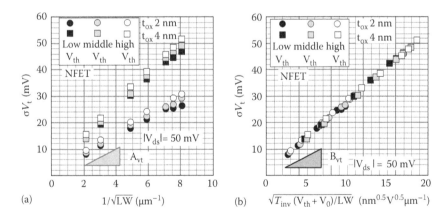

(a) $1/\sqrt{LW}$ (μm⁻¹)   (b) $\sqrt{T_{inv}(V_{th} + V_0)/LW}$ (nm⁰·⁵V⁰·⁵μm⁻¹)

FIGURE 3.7 (a) Pelgrom plot of measured σV_{th} of nFETs with various sizes and t_{ox}. The slope is defined as A_{vt}. (b) Takeuchi plot of measured σV_{th} of nFETs with various sizes and t_{ox}. The slope is defined as B_{vt}. (Tsunomura, T. et al., Process condition dependence of random V_{th} variability in NFETs and PFETs, *International Conference on Solid State Devices and Materials* (*SSDM*), Sendai, Japan, pp. 1010–1011, Copyright 2009 The Japan Society of Applied Physics.)

N_A are different, all data lie on the same straight line. The slope of Takeuchi plot is defined as B_{vt}. This Takeuchi plot indicates that σV_{th} increases as V_{th} becomes higher. The Takeuchi plot is useful when σV_{th} should be obtained in different V_{th}.

3.2.5 DRAIN CURRENT VARIABILITY

Drain current variability, as well as V_{th} variability, is a major concern in VLSI, because it directly causes huge variations in memory and logic circuit performances. Figure 3.8a shows I–V characteristics of 1M nFETs fabricated by 65 nm bulk technology. Large drain current variability is observed [9]. Figure 3.8b shows the cumulative plot of the on-current (I_{on}), which shows that I_{on} also follows the normal distribution up to $\pm 5\sigma$. Obviously, the origins of drain current variability are V_{th} variability and G_m variability. However, it has been found that there is a third origin of drain current variability [9].

Generally, there are two definitions of threshold voltage. One is the threshold voltage defined by subthreshold constant current ($I_0 = 10^{-7} \times W/L$). This threshold voltage is called V_{thc} in this study. The other is the threshold voltage determined by extrapolating drain current (V_{gs} intercept of tangent line with largest slope in I_{ds}–V_{gs} characteristics). This is called V_{thex}.

Figure 3.9 shows I–V characteristics of two nFETs, which have identical V_{thc} and G_m [9]. I_{on} differs significantly even though V_{thc} and G_m are the same. Please note that the onset point of drain current in the linear scale plot is different, and therefore, V_{thex} is quite different in these two nFETs. Here, we define "current onset voltage (COV)" as COV = $V_{thex} - V_{thc}$ [9,10]. The drain current rises rapidly and I_{on} is high when COV is small, while the drain current rise is slower when COV is large. Therefore, the COV variability contributes to the I_{on} variability.

It has been found that V_{thc}, G_m, and COV are almost mutually independent [9]. Then, I_{on} variability can be separated into three components of V_{thc}, G_m, and COV. Figure 3.10 shows the decomposition of measured I_{on} variability [9]. I_{on} variability

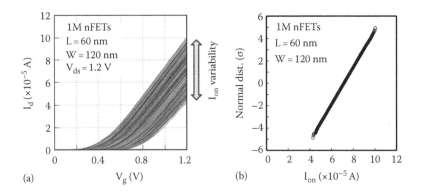

(a)

(b)

FIGURE 3.8 (a) Measured I–V characteristics of 1M nFETs. (b) Cumulative plot of the on-current (I_{on}) of 1M nFETs. (Modified from Tsunomura, T. et al., Analysis and prospect of local variability of drain current in scaled MOSFETs by a new decomposition method, *VLSI Symposium on Technology*, Honolulu, HI, 2010, pp. 97–98.)

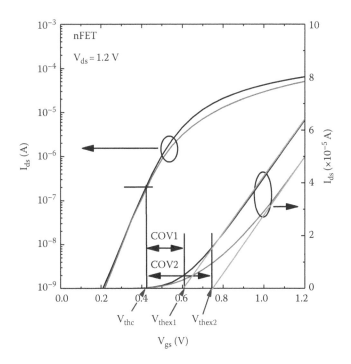

FIGURE 3.9 Measured I–V characteristics of two nFETs with identical V_{thc} and G_m, showing the difference of COV. (Modified from Tsunomura, T. et al., Analysis and prospect of local variability of drain current in scaled MOSFETs by a new decomposition method, *VLSI Symposium on Technology*, Honolulu, HI, 2010, pp. 97–98.)

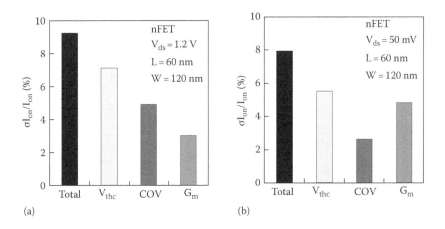

FIGURE 3.10 Decomposition of measured I_{on} variability of nFETs into V_{thc}, G_m, and COV components. (a) Saturation region at $V_{ds} = 1.2$ V. (b) Linear region at $V_{ds} = 0.05$ V. (Modified from Tsunomura, T. et al., Analysis and prospect of local variability of drain current in scaled MOSFETs by a new decomposition method, *VLSI Symposium on Technology*, Honolulu, HI, 2010, pp. 97–98.)

(σI_{on}) is normalized to median I_{on}, and the decomposition of $\sigma I_{on}/I_{on}$ is shown. It is found that V_{thc} is the major component of I_{on} variability, and COV is the second biggest component in the saturation region ($V_{ds} = 1.2$ V) and is larger than the G_m component. To suppress the I_{on} variability, the understanding and suppression of COV variability is indispensable.

3.2.6 ORIGIN OF COV VARIABILITY

Why is COV not constant and fluctuates from one transistor to another? This is because the current path in the subthreshold region and strong inversion region is different. In a planar bulk MOS transistor, the channel potential fluctuates due to RDF. The subthreshold current flows in the potential valley, which is often called the percolation path. Therefore, V_{thc}, which reflects the subthreshold current, is determined by how deep the potential valley is. In the strong inversion region, on the other hand, the potential fluctuation by RDF is screened by inversion charges. Therefore, V_{thex}, which reflects the strong inversion current, is determined by the average potential of the channel.

In order to examine the aforementioned model of COV, 3D device simulation is performed. Assuming RDF, I–V characteristics of 200 transistors are simulated. Among them, two transistors, which have the smallest COV and the largest COV, are selected. Figure 3.11 shows simulated potential of the "potential dividing line"

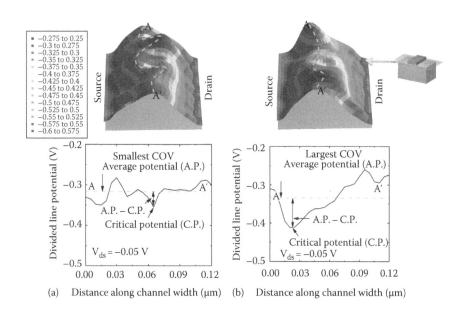

FIGURE 3.11 (**See color insert.**) Simulated potential distribution in the transistor channel and the potential on the "dividing line" along the channel width direction. (a) The transistor with the smallest COV among 200 pFETs. (b) The transistor with the largest COV among 200 pFETs. (Modified from Kumar, A. et al., Origin of "current-onset voltage" variability in scaled MOSFETs, *IEEE Silicon Nanoelectronics Workshop*, Honolulu, HI, 2010, pp. 7–8.)

along the channel width direction [10]. It is clearly shown that the transistor with the smallest COV has only shallow potential valleys, while the transistor with the largest COV has a very deep potential valley. The potential depth (the difference between average and minimum channel potential) is also simulated. It is found that while V_{thc} and V_{thex} have poor correlation with the potential depth, COV has a strong correlation with the potential depth [10] (not shown). These results confirm the model that the COV variability is caused by the potential fluctuation due to RDF.

3.3 VARIABILITY OF 11G TRANSISTORS

Generally, the ideal normal distribution of V_{th} is assumed to predict the yield of large-scale integrated circuits and memories in the circuit simulation. In the previous sections, it is shown that V_{th} follows the normal distribution up to $\pm 5\sigma$ in both nFETs and pFETs [5,6]. However, the state-of-the-art VLSI chips contain more than 1G (one billion) transistors in a chip. To our best knowledge, V_{th} distributions of 1G-level transistors have not been reported, mainly because the measurement takes an overwhelmingly long time. Therefore, the 6σ distribution is still unknown. In this section, a special DMA-TEG for ultrafast V_{th} monitoring is designed and fabricated, and 10G-level transistor variability is measured and the distribution is analyzed [11,12].

3.3.1 MEASUREMENT OF 11G TRANSISTORS

Figure 3.12 shows a schematic of the fast V_{th} monitoring circuit using the amp mode [11]. The device under test (DUT) is selected by a decoder and the current, I_{DUT}, is compared with the constant current, I_{REF}. The gate force voltage, V_{GF}, is scanned at an interval of 25 mV, and V_{GF} at which I_{DUT} exceeds I_{REF} is defined as measured V_{th}.

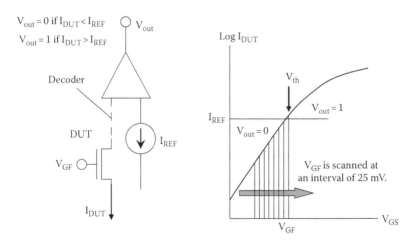

FIGURE 3.12 A schematic diagram of fast V_{th} monitoring circuit for measuring 11G transistors. (Modified from Mizutani, T. et al., Measuring threshold voltage variability of 10G transistors, *International Electron Devices Meeting (IEDM)*, Washington, DC, 2011, pp. 563–566.)

Please note that the measured V_{th} is V_{thc}, instead of V_{thex}. The designed DMA TEG has 256M transistors in a chip, and the measurement time is approximately 3 h per chip.

DMA TEG chips were fabricated by the 65 nm technology. Gate length L is 60 nm and gate width W is 120 nm. I_{REF} is set to 100 nA. V_{th}'s of 44 chips were measured, and therefore, the total number of transistors is 11G (11 billion), which corresponds to 6.5σ. After V_{th}'s of all transistors were measured and V_{th} distribution of 256M transistors was determined in each chip, I–V characteristics were also measured only for transistors with extremely high V_{th} or low V_{th} (beyond $\pm5.0\sigma$) and median V_{th} in order to investigate the origin of nonnormal distribution of V_{th}.

3.3.2 V_{thc} VARIABILITY OF 11G TRANSISTORS

Figure 3.13 shows measured cumulative plots of V_{th} of 11G nFETs and pFETs at $|V_{ds}| = 50$ mV [11]. It has been found that nFETs show good normality up to more than $\pm6.5\sigma$, although slight deviation is observed in the low V_{th} region below -4σ. This is the first observation of V_{th} distribution of 10G-level transistors. On the other hand, pFETs also have an almost normal distribution in the high V_{th} region up to more than 6.5σ. However, apparent distribution "tail" is observed in the low V_{th} region below -4σ. This large "tail" may affect the yield of large-scale logic circuits and SRAM, and determining the cause is urgent.

Figure 3.14 shows measured I–V characteristics at $|V_{ds}|$ of 50 mV and 1.2 V of some of the transistors with extremely high V_{th} and low V_{th} (beyond $\pm5.0\sigma$) [11]. For transistors with extremely low V_{th}, apparently larger drain-induced barrier lowering (DIBL) is observed in both nFETs and pFETs. In particular in pFETs, degraded subthreshold slope (SS) is observed. In order to examine the DIBL degradation in

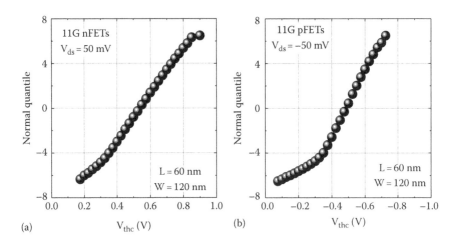

(a)

(b)

FIGURE 3.13 Measured cumulative distributions of V_{th} of 11G transistors at $|V_{ds}| = 50$ mV. (a) nFETs and (b) pFETs. (Modified from Mizutani, T. et al., Measuring threshold voltage variability of 10G transistors, *International Electron Devices Meeting (IEDM)*, Washington, DC, 2011, pp. 563–566.)

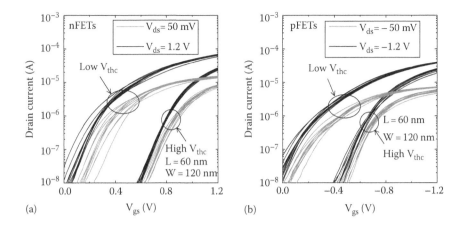

(a)

(b)

FIGURE 3.14 Examples of measured I–V characteristics at $|V_{ds}|$ of 50 mV and 1.2 V with extremely high V_{th} and low V_{th} (beyond ±5.0σ). (a) nFETs and (b) pFETs. (Modified from Mizutani, T. et al., Measuring threshold voltage variability of 10G transistors, *International Electron Devices Meeting (IEDM)*, Washington, DC, 2011, pp. 563–566.)

more detail, DIBL distribution of 1000 transistors in the extremely low V_{th} region (below −5.0σ), medium V_{th}, and extremely high V_{th} region (above 5.0σ) is measured [11] (not shown). Clearly, transistors in low V_{th} region have anomalously large DIBL in both nFETs and pFETs. However, a clear difference between nFET and pFET is not found in DIBL.

We found pronounced differences of characteristics in COV between nFET and pFET. Figure 3.15 shows distributions of measured COV in extremely low V_{th} region (below −5.0σ), medium V_{th}, and extremely high V_{th} region (above 5.0σ) [11].

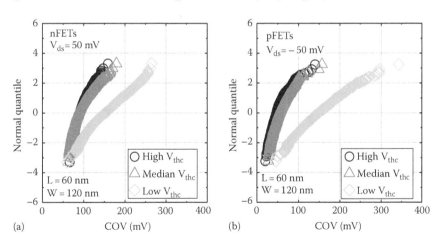

(a)

(b)

FIGURE 3.15 Distributions of measured COV in extremely low V_{th} region (below −5.0σ), medium V_{th}, and extremely high V_{th} region (above 5.0σ). (a) nFETs and (b) pFETs. (Modified from Mizutani, T. et al., Measuring threshold voltage variability of 10G transistors, *International Electron Devices Meeting (IEDM)*, Washington, DC, 2011, pp. 563–566.)

Again, transistors in low V_{th} region have anomalously large COV in both nFET and pFET. Especially in pFETs, COV of low V_{th} transistors is abnormally large. This result indicates that the distribution tail observed in pFET is closely related with degraded COV and may be caused by local percolation paths that are formed in the regions with extremely small number of dopants due to RDF, which will be discussed again later.

3.3.3 V_{thex} AND I_{on} VARIABILITY OF 11G TRANSISTORS

It is impossible to measure V_{thex} of all 11G transistors, because I–V curve measurements are necessary to derive V_{thex}, which takes an extremely long time. By measuring I–V characteristics of low V_{thc} region and high V_{thc} region with additional measurements of I–V characteristics of another 4k transistor TEG (which corresponds to the distribution center), the distribution of V_{thex} is derived [12]. Figure 3.16 shows cumulative plots of measured V_{thex} of 11G nFETs and pFETs at $|V_{ds}| = 50$ mV [12]. It is newly found that V_{thex} does not have a long distribution tail, contrary to V_{thc}, even in pFETs. Although V_{thc} has an impact on the standby power, circuit operations and performance are mainly affected by V_{thex} rather than V_{thc}. Therefore, one expects that the impact of the long V_{thc} distribution tail on circuit design is minimal.

Using a similar method, the V_{thc} and I_{on} distributions at $V_{ds} = 1.2$ V are determined [12]. As expected, V_{thc} has a long tail in low V_{thc} region at $|V_{ds}| = 1.2$ V even in nFETs (not shown), because DIBL is very large in low V_{thc} region. Figure 3.17 shows cumulative plots of measured I_{on} of 11G nFETs and pFETs at $|V_{ds}| = 1.2$ V [12]. nFETs have much larger I_{on} variability than pFETs, because V_{th} variability in nFETs is larger than that in pFETs. Moreover, I_{on} deviates from the normal distribution in the low I_{on} region in both nFETs and pFETs. Some transistors have

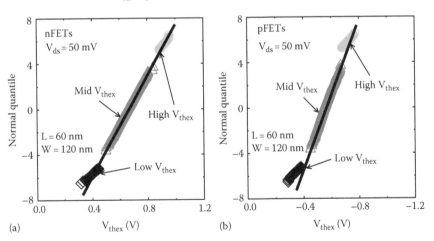

(a) (b)

FIGURE 3.16 Cumulative distributions of measured V_{thex} of 11G transistors at $|V_{ds}| = 50$ mV. (a) nFETs and (b) pFETs. (Modified from Mizutani, T. et al., Analysis of transistor characteristics in distribution tails beyond ±5.4σ of 11 billion transistors, *International Electron Devices Meeting (IEDM)*, Washington, DC, 2013, pp. 826–829.)

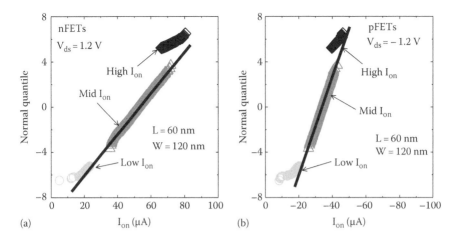

FIGURE 3.17 Cumulative distributions of measured I_{on} of 11G transistors at $|V_{ds}| = 1.2$ V. (a) nFETs and (b) pFETs. (Modified from Mizutani, T. et al., Analysis of transistor characteristics in distribution tails beyond $\pm 5.4\sigma$ of 11 billion transistors, *International Electron Devices Meeting (IEDM)*, Washington, DC, 2013, pp. 826–829.)

abnormally low I_{on}, which may have a greater impact on circuit performance and yield. It is found that $|V_{thc}|$ of the FETs with lowest I_{on} is not necessarily the highest in both nFET and pFET [12] (not shown), indicating that the origin of abnormally low I_{on} is not high V_{thc} but some other mechanisms such as extraordinarily high contact resistance.

3.3.4 SS VARIABILITY OF 11G TRANSISTORS

Subthreshold swing (SS) is one of the most important parameters, which determine the off-current of transistors. In spite of the importance, few experimental data [13] have been reported on the variability of SS. Looking at I–V curve in Figure 3.14 carefully, SS in high subthreshold current (SS7′, at $3 \times 10^{-7} \times (W/L)$ A) is apparently degraded and fluctuates significantly in low V_{thc} region in pFET, although SS in deep subthreshold region (SS8, at $1 \times 10^{-8} \times (W/L)A$) is not degraded. It is also found that SS7′ has very poor correlation with DIBL [12] (not shown), indicating that the degradation of SS7′ is not caused by the short channel effect. Moreover, degraded SS7′ has very good correlation with degraded COV [12] (not shown).

In order to examine the reason that SS7′ is degraded but SS8 is not degraded in low V_{thc} region, 3D device simulation was performed assuming two types of deep potential valleys in the transistor channel as shown in Figure 3.18a [12]. It is found that when the valley is narrow enough, punch-through is prevented. Although SS in high subthreshold current region is degraded, better SS in deep subthreshold region is obtained, as shown in Figure 3.18b [12]. These results indicate that the narrow potential valleys caused by RDF are responsible for degraded SS and degraded COV in high subthreshold current region.

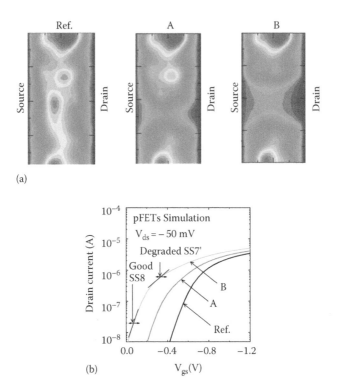

FIGURE 3.18 (a) Potential distribution of the three transistor channels assumed in the simulation. (b) Simulated I–V characteristics of the three transistors. (Modified from Mizutani, T. et al., Analysis of transistor characteristics in distribution tails beyond ±5.4σ of 11 billion transistors, *International Electron Devices Meeting (IEDM)*, Washington, DC, 2013, pp. 826–829.)

3.4 STABILITY OF SRAM CELLS

The instability in SRAM cells due to the variability of individual transistors in the cells is known as a crucial problem that will prevent further device integration and V_{dd} lowering [14]. The yield and the minimum operating voltage (V_{min}) are mainly determined by SRAM cells in recent VLSI. Therefore, the analysis of cell imbalances at the transistor level is essential for better understanding of SRAM stability at low V_{dd}. In this section, static noise margin (SNM) of SRAM cells and V_{th} of six individual transistors in the cells are directly measured and their variability is intensively analyzed using a DMA TEG of 16 kbit SRAM cells.

3.4.1 VARIABILITY OF STATIC NOISE MARGIN

Figure 3.19 shows a schematic of the 16 kbit SRAM DMA-TEG [15]. The TEG is based on the transistor DMA TEG that we have developed [5,6]. DUT is connected to six switch transistors, so that DUT is electrically isolated from other devices. DUTs are arrayed in a matrix manner and each DUT can be accessed by using decoder

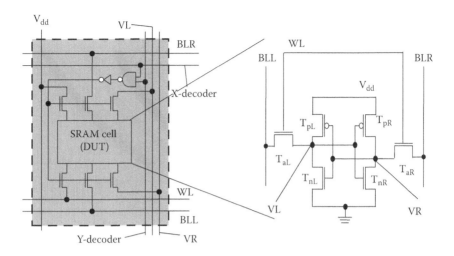

FIGURE 3.19 A schematic diagram of SRAM DMA-TEG and the circuit of six-transistor SRAM. (Modified from Hiramoto, T. et al., *IEEE Trans. Electr. Dev.*, 58, 2249, 2011.)

circuits. In the present SRAM DMA TEG, DUT is replaced by an SRAM mini array of 6 × 8 cells. Terminals of V_{dd}, word line (WL), left bit line (BLL), right bit line (BLR) as well as two internal storage nodes (VL, VR) at the center cell of the SRAM miniarray are connected to the six switched transistors and can be accessed, and the rest of SRAM cells are dummy cells. Since the internal storage nodes are accessible, noise margins as well as characteristics of the 6 individual transistors can be directly measured. The 16 kbit SRAM DMA TEG was fabricated with 65 nm technology.

I–V characteristics of 1k access NMOS transistors (T_a), drive NMOS transistors (T_n), and PMOS transistors (T_p) were measured [15] (not shown). The transistor characteristics vary significantly. The normal distributions of V_{th}'s of T_a, T_n, and T_p are confirmed. It is also confirmed that T_p has a smaller V_{th} variability than T_a and T_n.

Figure 3.20a shows measured butterfly curves of 1 kbit SRAM cells at V_{dd} of 1.2 V [15]. The WL is set to V_{dd}. There are huge variations of the butterfly curves. Here, one-side SNM is defined: SNM(L) is the square of the left eye of the butterfly curve, and SNM(R) is the square of the right eye. Please note that the SNM is defined as the smaller square of two eyes of the butterfly curve. Figure 3.20b shows cumulative distribution of measured 16 kbit SNM at several values of V_{dd} [15]. It is found that, while the one-sided SNM follows the normal distribution up to ±4σ even when V_{dd} is lowered from 1.2 to 0.4 V [15] (not shown), SNM does not follow the normal distribution. It is also shown that SNM is degraded when V_{dd} decreases and some cells fail at V_{dd} of 0.4 V.

3.4.2 V_{dd} Dependence of SNM

Figure 3.21 shows measured SNM as a function of V_{dd} [15]. Among 16 kbit SRAM cells, cells that have SNM values between 0.20 and 0.21 V at V_{dd} of 1.2 V are selected, and their V_{dd} dependences are shown. It is very interesting to note that,

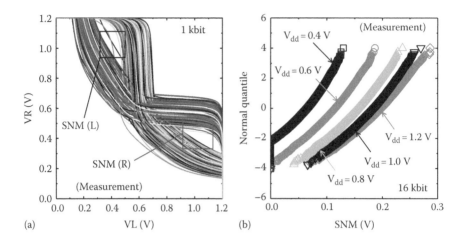

FIGURE 3.20 (a) Measured butterfly curves of 1 kbit SRAM cells at V_{dd} = 1.2 V. (b) Cumulative distributions of measured 16 kbit SNM at several values of V_{dd}. (Modified from Hiramoto, T. et al., *IEEE Trans. Electr. Dev.*, 58, 2249, 2011.)

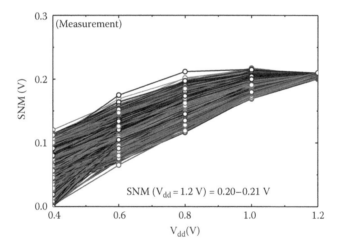

FIGURE 3.21 Measured SNM as a function of V_{dd} in SRAM cells whose SNM values are between 0.20 and 0.21 V at V_{dd} = 1.2 V. (Modified from Hiramoto, T. et al., *IEEE Trans. Electr. Dev.*, 58, 2249, 2011.)

even though SNM at 1.2 V is very similar, the V_{dd} dependence is very different depending on the cell: some cells show improvement of SNM when V_{dd} is lowered down to 0.8 V, and some cells show very severe degradation of SNM when V_{dd} decreases. At V_{dd} of 0.4 V, SNM of some cells remains at a high value above 0.1 V, but some cells fail (SNM is 0 V).

However, it has been found that this peculiar V_{dd} dependence is not simply explained by V_{th} variability alone [15]. Actually, when taking the variability of cell transistors into account in the circuit simulation, only V_{th} variability is generally considered. The circuit simulation results of SNM or V_{min} are not necessarily consistent with the measured data [15]. Therefore, the effects of transistor parameters other than V_{th} should be examined.

3.4.3   DIBL DEPENDENCE OF SNM

In order to investigate the impact of device parameters on SNM, a unique method using a half-cell is employed [16]. The 16 kbit SRAM has 32 kbit half-cells. Among 32 kbit half-cells, half-cells in which V_{th}'s of three transistors (T_n, T_p, and T_a) are within the median value ±10 mV are selected. As a result, 183 half-cells are selected. Then, the selected 183 half-cells have almost identical V_{th}'s for the three transistors and the effects of V_{th} variability are eliminated. It is found that there is no clear correlation between measured SNM of the selected half-cells and measured G_m [16] (not shown), indicating that G_m variability has no clear effect on SNM. It is also found that the body factor has no clear effect on SNM.

Figure 3.22 shows measured SNM of the selected half-cells as a function of measured DIBL of T_n and T_p [16]. Here, DIBL is defined by $V_{th}(V_{ds} = 50 \text{ mV}) - V_{th}(V_{ds} = 0.6 \text{ V})$. Apparently, SNM has a negative correlation with DIBL of T_n and T_p, and SNM is more degraded when DIBLs of T_n and T_p are larger. There is clear experimental evidence that DIBL variability degrades SRAM stability. This negative correlation was not found in T_a [16] (not shown). These results show that the DIBL variability should be taken into account to explain SNM variability and its V_{dd} dependence.

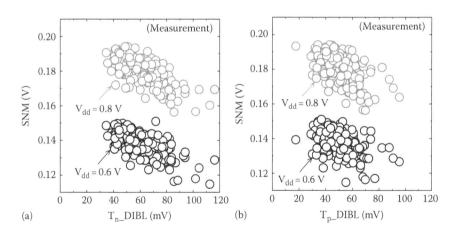

(a)   (b)

FIGURE 3.22   Measured SNM of the selected half-cells as a function of measured DIBL of T_n and T_p. (Modified from Song, X. et al., Impact of DIBL variability on SRAM static noise margin analyzed by DMA SRAM TEG, *International Electron Devices Meeting (IEDM)*, 2010, San Francisco, CA, pp. 62–65.)

3.5 INTRINSIC CHANNEL FD SOI TRANSISTORS

The major origin of random V_{th} variability is RDF. It is reported that DIBL and COV variabilities are also caused by channel potential fluctuations due to RDF [9,10]. These results suggest that, if the dopant atoms are removed from the channel, not only V_{th} variability but DIBL variability and COV variability will be suppressed. In this section, intrinsic channel FD SOI MOSFETs were fabricated and their variability was compared with that of conventional bulk MOSFETs using DMA TEG.

3.5.1 V_{th} AND DRAIN CURRENT VARIABILITY

Figure 3.23 shows a schematic of intrinsic channel FD SOI nFET and pFET [17,18]. The channels are not intentionally doped. The SOI is very thin (t_{SOI} = 12 nm in this study) to suppress the short channel effect. The buried oxide (BOX) is also very thin (t_{BOX} = 10 nm in this study), which enables us to control V_{th} by back bias. This device is also called a silicon-on-thin-BOX (SOTB) transistor [17–19]. FD SOTB transistors were fabricated with 65 nm technology. For comparison, conventional bulk transistors, where the channels are doped (2×10^{18} cm^{-3}), were also fabricated for reference. A poly-Si gate was used in both FD SOTB and bulk MOSFETs, while high-k/SiON gate dielectric was used to adjust V_{th} of FD SOTB transistors. T_{inv} is almost the same (approximately 2.6 nm). The gate length is 60 nm and gate width is 120 nm. The characteristics of both bulk and intrinsic channel FD SOTB transistors were measured using DMA TEG.

Figure 3.24 compares I_d–V_{gs} characteristics of 1k transistors of bulk nFETs and intrinsic channel FD SOTB nFETs [20]. Apparently, FD SOTB transistors have a smaller variability. Figure 3.25 shows cumulative plots of V_{thc} in bulk and SOTB nFETs [20]. V_{thc} shows a normal distribution in both bulk and SOTB, and V_{thc} variability is suppressed in SOTB nFETs (σ = 17.8 mV in linear region) compared with bulk (σ = 37.5 mV). This is because the intrinsic channel FD SOTB transistors have a very small number of dopants in the channel.

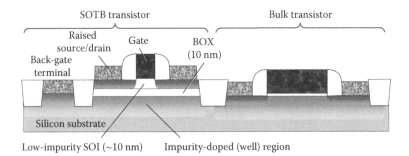

FIGURE 3.23 A schematic diagram of an intrinsic channel FD SOI FET. The device is also called an STOB transistor. A bulk FET cointegrated with an SOTB FET is also shown in this figure. (From Sugii, N. et al., Ultralow-voltage operation SOTB technology toward energy efficient electronics, *International Conference on Solid State Devices and Materials* (*SSDM*), Fukuoka, Japan, pp. 736–737, Copyright 2013 The Japan Society of Applied Physics.)

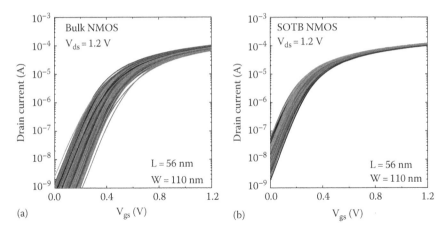

FIGURE 3.24 Measured I_d–V_{gs} characteristics of 1k transistors of (a) bulk nFETs and (b) intrinsic channel FD SOTB nFETs. (Modified from Mizutani, T. et al., Reduced drain current variability in fully depleted silicon-on-thin-BOX (SOTB) MOSFETs, *IEEE Silicon Nanoelectronics Workshop*, Honolulu, HI, 2012, pp. 71–72.)

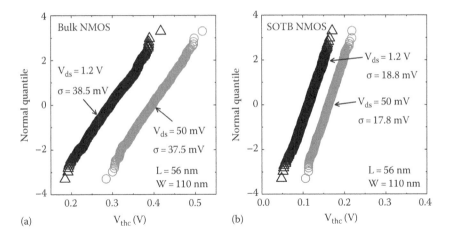

FIGURE 3.25 Cumulative distributions of measured V_{thc} at $V_{ds} = 1.2$ V in (a) bulk nFETs and (b) SOTB nFETs. (Modified from Mizutani, T. et al., Reduced drain current variability in fully depleted silicon-on-thin-BOX (SOTB) MOSFETs, *IEEE Silicon Nanoelectronics Workshop*, Honolulu, HI, 2012, pp. 71–72.)

Figure 3.26 shows cumulative plots of on-current (I_{on}) in bulk and SOTB nFETs [20]. The I_{on} current is normalized to median I_{on} and $\sigma I_{on}/I_{on}$ is shown in the figure. I_{on} variability is also reduced in the FD SOTB transistors. It is found that not only V_{thc} variability suppression but also COV variability suppression contributes to the I_{on} variability reduction in FD SOTB transistors (not shown). Figure 3.27 compares simulated potential fluctuations of transistor channels in bulk and FD SOTB MOSFETs [21]. Thanks to the intrinsic channel, the potential of an FD SOTB transistor channel is very smooth, leading to the reduction of V_{th} and I_{on} variabilities as well as DIBL and COV variabilities.

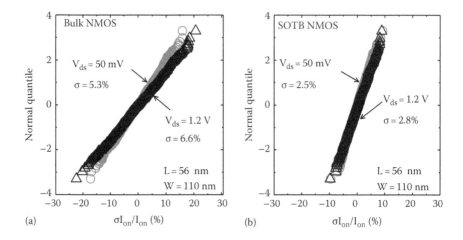

FIGURE 3.26 Cumulative distributions of measured $\sigma I_{on}/I_{on}$ at $V_{ds} = 0.05$ V and $V_{ds} = 1.2$ V in (a) bulk nFETs and (b) SOTB nFETs. (Modified from Mizutani, T. et al., Reduced drain current variability in fully depleted silicon-on-thin-BOX (SOTB) MOSFETs, *IEEE Silicon Nanoelectronics Workshop*, Honolulu, HI, 2012, pp. 71–72.)

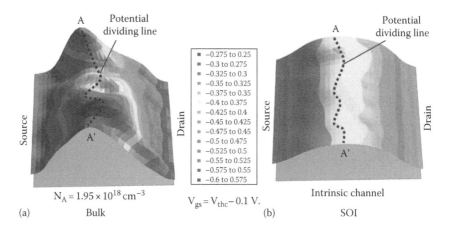

FIGURE 3.27 **(See color insert.)** Simulated potential distributions of transistor channels in (a) bulk nFETs and (b) SOTB nFETs. (Modified from Hiramoto, T. et al., Suppression of DIBL and current-onset voltage variability in intrinsic channel fully depleted SOI MOSFETs, *IEEE International SOI Conference*, San Diego, CA, 2010, pp. 170–171.)

3.5.2 Stability in FD SOTB SRAM

Small variability in FD SOTB transistors allows a drastically reduction in the operating voltage of FD SOTB SRAMs. Figure 3.28 compares the measured butterfly curves of 1 kbit bulk and FD SOTB SRAMs at V_{dd} of 0.4 V [22]. Some cells fail at 0.4 V in bulk SRAM, while all 1 kbit cells operate even at 0.4 V in FD SOTB SRAM. The minimum operation voltage (V_{min}) of 48 kbit SRAM is reduced from 0.542 V in

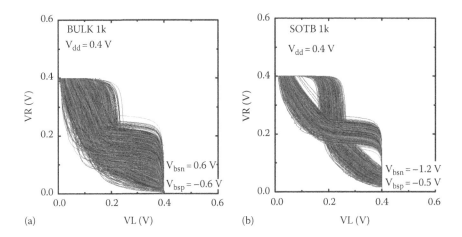

FIGURE 3.28 Measured butterfly curves of 1 kbit cells at $V_{dd} = 0.4$ V: (a) bulk SRAM and (b) FD SOTB SRAM. (From Mizutani, T. et al., *Jpn. J. Appl. Phys.*, 53, 04EC18, Copyright 2014 The Japan Society of Applied Physics.)

bulk to 0.290 V in FD SOTB (not shown) [22]. The intrinsic channel is essential to achieve ultralow voltage operation below 0.4 V in large-scale SRAM cell array.

3.6 SELF-SUPPRESSION OF VARIABILITY

Even in the intrinsic channel FD SOI MOSFETs, variability remains, and the variability seems to be enhanced in the next generations as the transistor size shrinks. Therefore, a new concept to cope with the variability problem is strongly required. In this section, a postfabrication scheme for self-suppression of SRAM variability (or self-improvement of SRAM stability) is proposed [23] and experimentally demonstrated [24]. This concept requires techniques of nonvolatile V_{th} shift of transistors and V_{th} shifts with high voltage stress were utilized in the experiments.

3.6.1 MECHANISM OF SELF-IMPROVEMENT

Figure 3.29 shows a schematic of a six-transistor SRAM cell [25]. The two storage nodes in the cell are named as VL and VR. In the self-improvement technique, the stress voltage is applied to the V_{dd} terminal of SRAM cell array. V_{dd} is raised from 0 V to the stress voltage (3.2 V in this study), keeping the word line (WL) at 0 V. The scan time (stress time) is only several seconds. Since the V_{th} shift of nFETs was small enough while the $|V_{th}|$ shift of pFETs is much larger at a stress voltage of 3.2 V, the self-improvement technique in the retention operation is explained based on the pFET $|V_{th}|$ shift in the following.

The SRAM cell is bistable in the retention condition when V_{dd} is high enough and can store one bit per cell. The storage node VR can be either "high" or "low" in the bistable operation. However, when V_{dd} is very low (e.g., 0.1 V), each SRAM cell is not bistable because of the unbalance of the four transistors that compose two inverters.

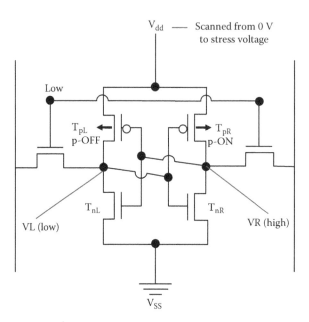

FIGURE 3.29 A schematic diagram of a six-transistor SRAM cell. It is assumed that VR is fixed to "high" at very low V_{dd}. (From Hiramoto, T. et al., *IEICE Trans. Electron.*, E96-C, 759, Copyright 2013 IEICE. With permission.)

This is caused by the V_{th} variability. Then, VR is fixed to only "high" in almost half of the cells at very low V_{dd} and VR is fixed to only "low" in the other cells. In this technique, V_{dd} is just scanned from 0 to 3.2 V.

At the beginning of V_{dd} scan (0 V), VR is fixed to "high" in almost half-cells. As an example, let us assume that a cell whose VR is fixed to "high" at the beginning of V_{dd} scan, as shown in Figure 3.29. This means that the strength to pull up VR is stronger than the strength to pull up VL in this cell. Therefore, T_{pR} or T_{nL} may be stronger (lower $|V_{th}|$) than T_{pL} or T_{nR}. Here, we call a pFET connected to the high node as "p-ON" and a pFET connected to the low node as "p-OFF", because the former is at the *on* state and the latter is at the *off* state. Similarly, "n-ON" and "n-OFF" are defined. It should be noted that, if p-ON is weakened, the cell stability is certainly improved, because the strength to pull up VR is weakened. Therefore, it is shown in the following that p-ON is the stronger pFET (that should be weakened for cell stability improvement) and p-OFF is the weaker pFET (that should be strengthened).

Next, let us consider the situation where V_{dd} is raised to 3.2 V. The negative gate bias is automatically applied to only p-ON that is stronger, because this transistor is at the *on* state. This bias condition is just the same as that of negative bias temperature instability (NBTI) stress, as shown in Figure 3.30a. Positively charged interface traps are generated. Then, $|V_{th}|$ of p-ON is selectively raised and this transistor is weakened. On the other hand, the bias condition of p-OFF, which is weaker, is shown in Figure 3.30b. It is known that $|V_{th}|$ is raised by the *off* state due to negative charge generation in oxide near the drain, and this transistor is strengthened. As a result, the cell stability is improved [25].

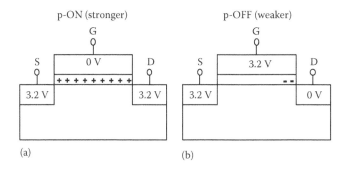

FIGURE 3.30 Bias conditions of pFETs when high voltage is applied at the V_{dd} terminal: (a) p-ON at the ON-state and (b) p-OFF at the OFF-state. (From Hiramoto, T. et al., *IEICE Trans. Electron.*, E96-C, 759, Copyright 2013 IEICE. With permission.)

3.6.2 Measurements of $|V_{th}|$ Shift by High Voltage Stress

SRAM DMA TEG was fabricated with 40 nm bulk technology and the self-improvement technique was applied to 4 kbit SRAM cells. We pay attention to the $|V_{th}|$ shift of p-ON and p-OFF. By checking "high" or "low" of VL and VR, it is easy to determine which pFET is p-ON or p-OFF. Figure 3.31a shows measured $|V_{th}|$ shift of p-ON, which was originally stronger and should be weakened for the self-improvement [25]. It is found that a majority of p-ON transistors show positive $|V_{th}|$ shift, indicating that p-ON is weakened. This positive $|V_{th}|$ shift is caused by the NBTI stress.

Figure 3.31b shows measured $|V_{th}|$ shift of p-OFF, which is weaker [25]. Almost all p-OFF transistors show negative $|V_{th}|$ shift, indicating that p-OFF which was originally weak is strengthened by the high voltage stress. The shift is even larger than that of p-ON. The self-improvement mechanism works.

Figure 3.31 shows that not all cells exhibit the favorable $|V_{th}|$ shift, that is, some cells exhibit negative $|V_{th}|$ shift of p-ON and positive $|V_{th}|$ shift of p-OFF,

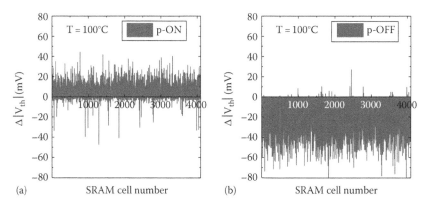

FIGURE 3.31 Measured $|V_{th}|$ shifts of pFETs in 4 kbit SRAM cells. (From Hiramoto, T. et al., *IEICE Trans. Electron.*, E96-C, 759, Copyright 2013 IEICE. With permission.)

which is the opposite direction to the self-improvement. It is found that cells that show the opposite $|V_{th}|$ shift are originally stable cells [25] (not shown). It is thought that these cells are so stable that "high" or "low" of the storage nodes is not determined at the beginning of V_{dd} scan. As a result, p-ON and p-OFF are interchanged resulting in the opposite $|V_{th}|$ shift. However, since these cells are still stable enough after the opposite $|V_{th}|$ shift, this phenomenon does not result in yield loss of SRAM.

3.6.3 Measurements of Self-Improvement of SRAM Stability

Figure 3.32 shows examples of measured butterfly curves in the retention condition (WL is 0 V) before and after applying high voltage stress [25]. Here, RetNM(L) is defined as the square of the left eye of the butterfly curve and RetNM(R) is the square of the right eye. RetNM is the smaller square of two eyes of the butterfly curve. The change of butterfly curves by the stress is explained in the following.

As mentioned in Section 3.6.1, VR is fixed to "low" or "high" at the beginning of the V_{dd} scan. In a cell in Figure 3.32a, "VR = low" is stable at low V_{dd}, and hence RetNM(R) is larger and RetNM(L) is smaller. The right pFET (T_{pR}) is the p-OFF. When high voltage stress is applied, $|V_{th}|$ of p-OFF is lowered, and the inverter curve by T_{pR} and T_{nR} moves in the right direction, as shown in Figure 3.32a. Similarly, raised $|V_{th}|$ of p-ON moves the inverter curve by T_{pL} and T_{nL} down. In this way, both p-ON and p-OFF contribute to enlarge RetNM(L), resulting in the cell stability improvement.

In a cell in Figure 3.32b, on the other hand, "VR = high" is stable, and hence, RetNM(R) is smaller. In this specific cell, only negative $\Delta|V_{th}|$ of p-OFF

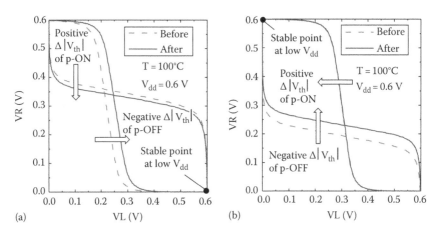

FIGURE 3.32 Measured butterfly curves in the retention condition before and after applying high voltage: (a) a cell where "VR = low" is stable at low V_{dd} and (b) a cell where "VR = high" is stable at low V_{dd}. (From Hiramoto, T. et al., *IEICE Trans. Electron.*, E96-C, 759, Copyright 2013 IEICE. With permission.)

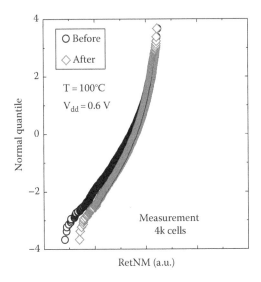

FIGURE 3.33 Cumulative distributions of measured RetNM of 4 kbit SRAM cells before and after applying high voltage. (From Hiramoto, T. et al., *IEICE Trans. Electron.*, E96-C, 759, Copyright 2013 IEICE. With permission.)

contributes, but no $\Delta|V_{th}|$ of p-ON is observed. The cell stability of this cell is also improved, because RetNM(R) is enlarged.

Figure 3.33 shows measured RetNM distributions before and after the high voltage stress to V_{dd} in 4k SRAM cells [25]. Clear improvement of RetNM is observed, particularly in the worst cell that have originally the smallest RetNM. Since V_{min} is determined by the worst cell, this self-improvement technique largely contributes to the yield enhancement of SRAM cells.

3.7 CONCLUSIONS

The present status of the variability in scaled transistors is reviewed. The variability of a large number of transistors is extensively measured, and it is shown that the main origin of random variability in bulk transistors is RDF. The relationship between cell transistor variability and cell stability in SRAM was analyzed. As an approach to variability suppression, two methods are described: (1) utilization of intrinsic channel FD MOSFET to avoid RDF and (2) novel self-improvement technique of SRAM stability. These results will largely contribute to further device scaling and further minimization of energy consumption in future VLSI.

ACKNOWLEDGMENTS

A part of this work was performed under the MIRAI project, LEAP project, and ELP project supported by NEDO.

80   Nanoscale Silicon Devices

REFERENCES

1. Hoeneisen B and Mead CA. 1972. Fundamental limitations in microelectronics—I. MOS technology. *Solid-State Electron.* 15: 819–829.
2. Keyes RW. 1975. The effect of randomness in the distribution of impurity atoms on FET thresholds. *Appl. Phys.* 8: 251–259.
3. Pelgrom MJM, Duinmaijer ACJ, and Welbers APG. 1989. Matching properties of MOS transistors. *IEEE J. Solid-State Circuits* 24: 1433–1440.
4. Kuhn K, Giles MD, Becher D et al. 2011. Process technology variation. *IEEE Trans. Electr. Dev.* 58: 2197–2208.
5. Tsunomura T, Nishida A, and Hiramoto T. 2009. Verification of threshold voltage variation properties in scaled transistors with ultra large-scale device matrix array test element group. *Jpn. J. Appl. Phys.* 48: 124505.
6. Tsunomura T, Nishida A, Yano F et al. 2008. Analyses of 5σ V_{th} fluctuation in 65 nm-MOSFETs using Takeuchi plot. *Symposium on VLSI Technology*, Honolulu, HI, pp. 156–157.
7. Tsunomura T, Nishida A, Takeuchi K et al. 2009. Process condition dependence of random V_{th} variability in NFETs and PFETs. *International Conference on Solid State Devices and Materials (SSDM)*, Sendai, Japan, pp. 1010–1011.
8. Takeuchi K, Fukai T, Tsunomura T et al. 2007. Understanding random threshold voltage fluctuation by comparing multiple fabs and technologies. *International Electron Devices Meeting (IEDM)*, Washington DC, pp. 467–470.
9. Tsunomura T, Kumar A, Mizutani T et al. 2010. Analysis and prospect of local variability of drain current in scaled MOSFETs by a new decomposition method. *VLSI Symposium on Technology*, Honolulu, HI, pp. 97–98.
10. Kumar A, Mizutani T, Shimizu K et al. 2010. Origin of "current-onset voltage" variability in scaled MOSFETs. *IEEE Silicon Nanoelectronics Workshop*, Honolulu, HI, pp. 7–8.
11. Mizutani T, Kumar A, and Hiramoto T. 2011. Measuring threshold voltage variability of 10G transistors. *International Electron Devices Meeting (IEDM)*, Washington DC, pp. 563–566.
12. Mizutani T, Kumar A, and Hiramoto T. 2013. Analysis of transistor characteristics in distribution tails beyond $\pm 5.4\sigma$ of 11 billion transistors. *International Electron Devices Meeting (IEDM)*, Washington DC, pp. 826–829.
13. Mizutani T, Yamamoto Y, Makiyama H et al. 2013. Statistical analysis of subthreshold swing in fully depleted silicon-on-thin-buried-oxide and bulk metal-oxide-semiconductor field effect transistors. *Jpn. J. Appl. Phys.* 52: 04CC02.
14. Bhavnagarwala AJ, Tang X, and Meindl JD. 2001. The impact of intrinsic device fluctuations on CMOS SRAM cell stability. *IEEE J. Solid-State Circuits* 36: 658–665.
15. Hiramoto T, Suzuki M, Song X et al. 2011. Direct measurement of correlation between SRAM noise margin and individual cell transistor variability by using device matrix array. *IEEE Trans. Electr. Dev.* 58: 2249–2256.
16. Song X, Suzuki M, Saraya T et al. 2010. Impact of DIBL variability on SRAM static noise margin analyzed by DMA SRAM TEG. *International Electron Devices Meeting (IEDM)*, San Francisco, CA, pp. 62–65.
17. Tsuchiya R, Horiuchi M, Kimura S et al. 2004. Silicon on tin BOX: A new paradigm of the CMOSFET for low-power and high-performance application featuring wide-range back-bias control. *International Electron Devices Meeting (IEDM)*, San Francisco, CA, pp. 631–634.
18. Sugii N, Iwamatsu T, Yamamoto Y et al. 2013. Ultralow-voltage operation SOTB technology toward energy efficient electronics. *International Conference on Solid State Devices and Materials (SSDM)*, Fukuoka, Japan, pp. 736–737.

19. Yamamoto Y, Makiyama H, Shinohara H et al. 2013. Ultralow-voltage operation of silicon-on-thin-BOX (SOTB) 2 Mbit SRAM down to 0.37 V utilizing adaptive back bias. *VLSI Symposium on Technology*, Kyoto, Japan, pp. T212–T213.

20. Mizutani T, Yamamoto Y, Makiyama H et al. 2012. Reduced drain current variability in fully depleted silicon-on-thin-BOX (SOTB) MOSFETs. *IEEE Silicon Nanoelectronics Workshop*, Honolulu, HI, pp. 71–72.

21. Hiramoto T, Mizutani T, Kumar A et al. 2010. Suppression of DIBL and current-onset voltage variability in intrinsic channel fully depleted SOI MOSFETs. *IEEE International SOI Conference*, San Diego, CA, pp. 170–171.

22. Mizutani T, Yamamoto Y, Makiyama H et al. 2014. Comparison and distribution of minimum operation voltage in fully depleted silicon-on-thin-buried-oxide and bulk static random access memory cells. *Jpn. J. Appl. Phys.* 53: 04EC18.

23. Suzuki M, Saraya T, Shimizu K et al. 2009. Post-fabrication self-convergence scheme for suppressing variability in SRAM cells and logic transistors. *Symposium on VLSI Technology*, Kyoto, Japan, pp. 148–149.

24. Suzuki M, Saraya T, Shimizu K et al. 2010. Direct measurements, analysis, and post-fabrication improvement of noise margins in SRAM cells utilizing DMA SRAM TEG. *Symposium on VLSI Technology*, Honolulu, HI, pp. 191–192.

25. Hiramoto T, Kumar A, Saraya T et al. 2013. Experimental demonstration of post-fabrication self-improvement of SRAM cell stability by high-voltage stress. *IEICE Trans. Electron.* E96-C: 759–765.

4 Self-Heating Effects in Nanoscale 3D MOSFETs

Tsunaki Takahashi and Ken Uchida

CONTENTS

4.1   Introduction 83
4.2   Bulk and SOI FinFET Structures 84
4.3   Parameters for Thermal Analysis of Nanoscale Devices 85
      4.3.1   Thermal Conductivity 85
      4.3.2   Interface Thermal Resistance 86
4.4   Thermal Analysis of FinFETs 87
      4.4.1   DC Operation... 87
      4.4.2   Analog Operation 90
4.5   Thermal Modeling 91
      4.5.1   Thermal Modeling of Devices... 92
      4.5.2   Thermal Modeling of Interconnects and Vias 93
      4.5.3   Thermal Modeling of Circuits 97
      4.5.4   Thermal Advantage of Advanced Interconnect Material 98
References 101

4.1 INTRODUCTION

The size of metal-oxide-semiconductor field-effect transistors (MOSFETs) has been continually decreased for the higher performance of large-scale integrations (LSIs). The device scaling results in smaller delay time and higher device density in a chip without the increase in power consumption. In order to suppress off state leakage current in nanoscale devices, silicon-on-insulator (SOI) and/or three-dimensional (3D) device structures have been intensively investigated. Because of their excellent short channel effect immunity, a 3D structure MOSFET is applied for the mass production at the 22 nm technology node[1]: fin-type field-effect transistors (FinFETs).

However, it has been reported that the thermal properties of 3D FETs become worse than those of conventional planar bulk MOSFETs.[2,3] Although the 3D and SOI structures are effective to enhance gate controllability of the channel potential, these structures simultaneously weaken the thermal coupling between the channel and the substrate, because the thermal conductivity of silicon dioxide (~1.4 W m^{-1} K^{-1}) is much smaller than that of silicon (~148 W m^{-1} K^{-1}). Figure 4.1 schematically shows the dominant heat dissipation paths of conventional planar bulk MOSFET and planar SOI MOSFET. During circuit operations, joule heat $P = I_d V_d$ (I_d: drain current,

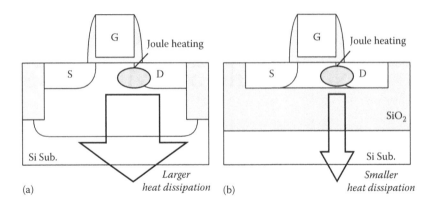

FIGURE 4.1 Schematics of dominant heat dissipation paths of (a) conventional planar bulk and (b) planar SOI MOSFET.

V_d: drain voltage) is generated at the channel of each operated transistor. In conventional planar bulk MOSFETs, the joule heat immediately dissipates to the substrate because of the higher thermal conductivity of silicon. On the other hand, in the SOI and 3D MOSFETs, the smaller heat dissipation causes a lattice temperature increase at the channel from the substrate. Thus, the local device temperatures of these devices are generally higher than chip (or die) temperature: self-heating effects. Since the channel temperature increase results in the degradation in carrier mobility, the power supply voltage increases in order to satisfy the requirement of *on* current. Furthermore, self-heating effects cause not only performance degradation but also reliability issues. Therefore, it is crucial to suppress the self-heating effects for nanoscale devices.

Self-heating effects of MOSFETs have been studied since 1990s for planar SOI MOSFETs and 3D MOSFETs. Experimental evaluations of self-heating effects have been carried out typically using the four-terminal gate resistance technique[4–6] or AC conductance method.[7–11] Although operation temperatures of transistors are directly obtained only by the four-terminal gate resistance technique, the technique requires a device structure designed specifically for this purpose. Therefore, the AC conductance method has been widely utilized. The thermal properties of MOSFETs have been also investigated by simulations in particular for nanoscale 3D devices.[2,3,12,13] In the thermal analysis of nanoscale devices, it is important that the thermal conductivities of nanoscale or heavily doped silicon region are accurately modified from that of pure bulk silicon.

4.2 BULK AND SOI FinFET STRUCTURES

Figure 4.2 shows the device structures of FinFETs. The first experimental demonstration of FinFETs was reported by Hisamoto et al.[14] and the device was called "DELTA." In the DELTA structure, an electrically isolated fin-shaped channel region was fabricated on bulk silicon wafers by applying the low thermal oxidation rate of silicon nitride. From the 1990s, the DELTA-like structure was fabricated using SOI wafers, and these devices are called SOI FinFETs (Figure 4.2c).

Another FinFET structure was proposed by Okano et al.[15] The structure was fabricated on bulk wafers using shallow trench isolation (STI) technology and called

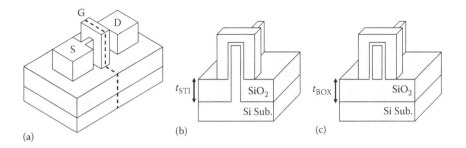

FIGURE 4.2   (a) Birds-eye view of FinFETs. Cross-sectional views of (b) bulk and (c) SOI FinFETs.

bulk FinFETs (Figure 4.2b). In bulk FinFETs, a fin-shaped silicon channel region is electrically connected to the substrate. Although the fin region under the channel must be heavily doped in order to suppress short channel effects, bulk FinFETs have an advantage in production cost. From 2012, bulk FinFETs were applied to mass production of the 22 nm technology node.[1]

In both the bulk and SOI FinFETs, a silicon dioxide layer exists between the channel and substrate: STI for bulk and BOX for SOI FinFETs, respectively. Compared to planar devices, the cross-sectional area of heat dissipation to the substrate per effective channel width is smaller in FinFETs due to their vertical channel structure. Because of these oxide layers and the smaller heat dissipation area, the self-heating effects of FinFETs are generally more severe than that of planar SOI MOSFETs.

4.3   PARAMETERS FOR THERMAL ANALYSIS OF NANOSCALE DEVICES

In nanoscale electron devices, electrical properties, such as carrier density and mobility, change from those of bulk material due to quantum size effects. In a similar way, thermal properties also change. Therefore, physical parameters of thermal properties must be modified in the thermal analysis of nanoscale devices. In particular, thermal conductivity and interface thermal resistance are significant in nanoscale MOSFETs.

4.3.1   THERMAL CONDUCTIVITY

Generally, thermal conductivities in semiconductors are determined by the mean free path of phonon transport. It is known that phonons are scattered by dopant ions, lattice imperfections, material boundary, etc., as shown in Figure 4.3a. Therefore, if the operation temperatures of nanoscale devices are calculated using the thermal conductivity of pure bulk silicon, the temperatures will be underestimated. Figure 4.3b shows the calculated temperature dependence of the thermal conductivities of pure bulk,[16] heavily phosphorus doped,[17] and 7 nm-thick silicon layers.[16] Thermal conductivity of silicon is greatly reduced in the heavily doped or thin silicon layer owing to a decrease in the mean free path of phonons. Because of these large thermal conductivity differences

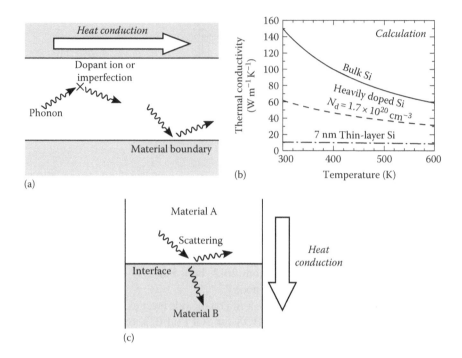

FIGURE 4.3 (a) A schematic diagram of phonon transport in a material. (b) Simulated temperature dependence of thermal conductivities of pure bulk (From Liu, W. et al., *IEEE Trans. Electron Dev.*, 53(8), 1868, 2006.), heavily phosphorus-doped (From Asheghi, M. et al., *J. Appl. Phys.*, 91(8), 5079, 2002.), and 7 nm thin-layer (From Liu, W. et al., *IEEE Trans. Electron Dev.*, 53(8), 1868, 2006.) silicon. (c) Schematic of phonon transport between two materials.

from the bulk value, it is important to incorporate the reduced thermal conductivities in thermal analysis of nanoscale devices. Namely, the thermal conductivity in heavily doped and thin layer silicon must be applied in source/drain and fin (or channel) regions of FinFETs for accurate evaluations of self-heating effects.

4.3.2 INTERFACE THERMAL RESISTANCE

It is known that an additional phonon scattering exists at the interface between different materials as illustrated in Figure 4.3c, which is called the interface thermal resistance.[18] The interface thermal resistance is determined by the combination of two materials and regardless of film thickness. Therefore, the interface thermal resistance exerts serious effects on the thermal properties of extremely thin gate dielectrics, whose thermal resistance itself is negligibly small. Since the gate insulator in nanoscale transistors consists of extremely thin multidielectric films including silicon dioxide as an interfacial layer, the interface thermal resistance between silicon dioxide and high-k materials is important in the thermal analysis of deeply scaled devices.

4.4 THERMAL ANALYSIS OF FinFETs

4.4.1 DC OPERATION

The magnitude of self-heating effects is evaluated using thermal resistance, R_{th}, which corresponds to a temperature increase per amount of heat. For a solid material, thermal resistance is defined as

$$R_{th} \equiv \frac{L}{A} \frac{1}{\lambda}, \tag{4.1}$$

where A, L, and λ are cross-sectional areas of heat flow, length, and the thermal conductivity of material, respectively (Figure 4.4a). For devices, thermal resistance is generally determined using input power P and an increase in device temperature from the substrate (Figure 4.4b):

$$R_{th} = \frac{T_{dev} - T_{sub}}{P} = \frac{T_{dev} - T_{sub}}{I_d V_d}. \tag{4.2}$$

A larger thermal resistance corresponds to severe self-heating effects, because the same input power causes a higher temperature increase in larger R_{th} devices.

Generally, self-heating effects in SOI FinFETs are more serious than that in bulk FinFETs when BOX and STI thicknesses are the same, because the silicon fin region of bulk FinFETs is connected to the substrate. However, BOX thickness of SOI FinFETs can be decreased, because the BOX thickness hardly affects short channel effects immunity.[19] In thin-BOX structures, thermal resistance might be decreased, because the heat dissipation to the substrate is increased. Therefore, the thermal characteristics of thin-BOX SOI FinFETs should be compared with that of bulk FinFETs.

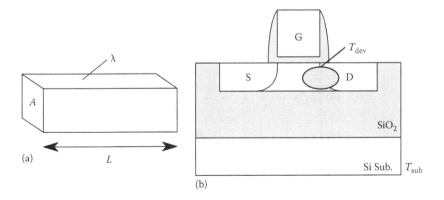

FIGURE 4.4 (a) Solid material and (b) device parameters for determination of thermal resistance.

Figure 4.5 shows the relationship between simulated thermal resistance of the SOI FinFETs and BOX thickness at 14 nm technology node.[19] In Figure 4.5, the thermal resistance of the bulk FinFET with STI thickness of 100 nm is also shown for comparison. As BOX thickness is decreased, the thermal resistance of SOI FinFETs is greatly reduced, thanks to the greater heat dissipation to the substrate. When BOX thickness is less than 50 nm, the thermal resistance of SOI FinFETs is smaller than that of the bulk FinFET with the STI thickness of 100 nm, implying weaker self-heating effects of thin-BOX SOI FinFETs. Here, it should be noted that the STI thickness of bulk FinFETs cannot be decreased to less than 100 nm, because the off-leakage current of thin-STI bulk FinFETs exceeds the requirement of 14 nm technology node.[19]

The weaker self-heating effects in thin-BOX SOI FinFETs result in higher DC performance because of higher carrier mobility at lower temperatures. Figure 4.6 shows I_d–V_d characteristics for the bulk and 5 nm thin-BOX SOI FinFETs.[19] The drain current of the 5 nm thin-BOX SOI FinFET is much higher than that of the bulk FinFET in the higher drain and gate voltage region, because self-heating effects become prominent at a higher input power.

However, the bias voltages of practical circuit operations are sufficiently lower than the bias region of the power supply voltage. In order to accurately evaluate the self-heating effects under practical conditions, the lattice temperature under an analog operation bias (drain voltage of 0.3 V and gate voltage of 0.5 V) is shown in Figure 4.7. The maximum temperature of the 5 nm thin-BOX SOI FinFET is approximately 50 K lower than that of the bulk FinFET, indicating the thermal advantage of extremely thin-BOX SOI FinFETs under the analog operation.

Because of the difficulties in fabricating nanoscale fin structures, the fin width might unintentionally fluctuate. In device simulations, it was clarified that the thermal resistance of SOI FinFETs is almost independent of fin width, whereas

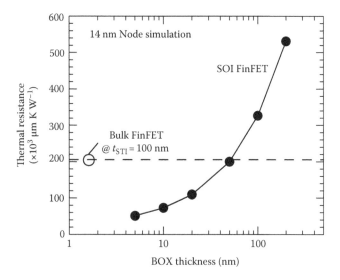

FIGURE 4.5 Simulated thermal resistance of 14 nm node SOI FinFETs versus BOX thickness.[19] Thermal resistance of bulk FinFETs with STI thickness of 100 nm is also shown by dashed line.

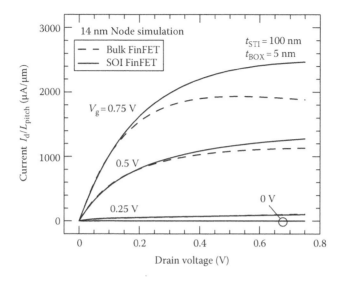

FIGURE 4.6 Simulated drain current versus drain voltage characteristics for 14 nm node bulk FinFET with STI thickness of 100 and 5 nm thin-BOX SOI FinFET. (From Takahashi, T. et al., *Jpn. J. Appl. Phys.*, 52, 04CC03, 2013.) Drain current is normalized by fin pitch length, L_{pitch}, which is twice of gate length.

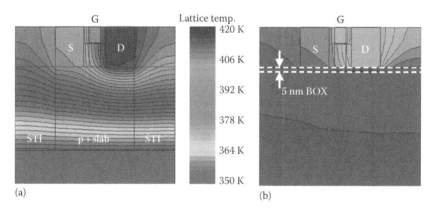

FIGURE 4.7 (See color insert.) Simulated contour plots of lattice temperatures for 14 nm node (a) bulk and (b) 5 nm thin-BOX SOI FinFETs under analog operation bias (drain voltage of 0.3 V and gate voltage of 0.5 V). (From Takahashi, T. et al., *Jpn. J. Appl. Phys.*, 52, 04CC03, 2013.)

that of bulk FinFETs greatly increases as fin width decreases.[19] The difference in the fin width dependence originates from the heat flow paths under the devices. In bulk FinFETs, the dominant heat flow path from the channel to the substrate is through the p+ doped silicon slab region, whose width is defined as fin width (see Figures 4.2b and 4.7a). Therefore, in bulk FinFETs, the fin width fluctuations result in the fluctuations in operation temperature. Thus, thin-BOX SOI FinFETs show not only the lower analog operation temperature but also smaller thermal variability.

4.4.2 ANALOG OPERATION

So far, the thermal characteristics of bulk and SOI FinFETs have been investigated in terms of thermal resistance and lattice temperatures. Since a device temperature increase is proportional to its input power, self-heating effects have a serious effect on a circuit operation, which requires larger input power per transistor, such as analog operations. In this section, the impacts of self-heating effects on analog performances are described.

Analog performance is investigated by calculating the cut-off frequency (f_T) and the maximum oscillation frequency (f_{max}). f_T and f_{max} are extracted from the frequency dependence of the $|h_{21}|$ parameter and Mason's unilateral power gain ($|U|$).[20,21] f_T and f_{max} are defined as the frequency where $|h_{21}|$ and $|U|$ are equal to 0 dB, respectively.

In the simulations for 14 nm node FinFETs,[19] f_T of the 5 nm BOX SOI FinFET is slightly higher than that of the bulk FinFET, and the difference increases as V_d increases. On the other hand, f_{max} values are almost the same between the bulk and 5 nm BOX SOI FinFETs in spite of the lower analog operation temperature of the 5 nm BOX SOI FinFET (Figure 4.7).

Although the thin-BOX structure is effective for the suppression of operation temperature, it also has a negative effect on circuit performance from the viewpoint of the electrical property. As BOX thickness decreases, the parasitic capacitance between the device and substrate increases. Since the parasitic capacitance corresponds to the delay time in high-frequency operations, the extremely thin-BOX structure may degrade the analog performance. Figure 4.8 shows the simulated BOX thickness dependences of f_T and f_{max} for 14 nm node SOI FinFETs.[19] Both f_T and f_{max}

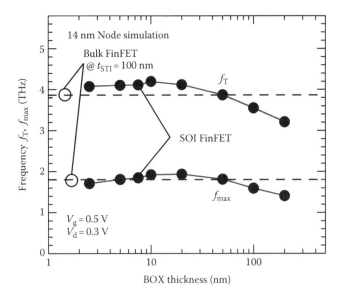

FIGURE 4.8 Simulated cut-off frequency, f_T, and maximum oscillation frequency, f_{max}, versus BOX thickness characteristics of 14 nm node SOI FinFETs[19]. f_T and f_{max} of bulk FinFET are also shown by the dashed lines for comparison.

take maximum values at BOX thickness of approximately 10 nm. In the thicker BOX region, f_T and f_{max} degrade owing to the increase in temperature (self-heating effects). On the other hand, in the extremely thin-BOX region, f_T and f_{max} degrade owing to the increase in the parasitic capacitance. Because of both the self-heating effects and parasitic capacitance, the analog performance of SOI FinFETs is maximized by optimizing the BOX thickness.

4.5 THERMAL MODELING

Typically, thermal properties of devices are modeled using thermal circuits. In thermal circuits, heat flow paths are expressed as circuits including thermal resistance and heat source (shown as current source), as illustrated in Figure 4.9. In addition, thermal capacitance is introduced in the modeling of transient thermal characteristics.

By giving the thermal resistance of transistors as a common value independent of the device layout, self-heating effects have been incorporated in widely used transistor models for IC design: SPICE model.[22] However, operation temperature cannot be accurately reproduced using a single common thermal resistance for a certain structure. Since interconnect wires act as additional heat dissipation paths, the operation temperatures of individual transistors depend on the layout of a circuit consisting of interconnects and transistors. Therefore, thermal resistance should be separated into device and interconnect components. As a result, the thermal resistances of devices and interconnects have been evaluated.[13,23] However, it is difficult to apply these models to large-scale circuits, because they require many geometrical parameters for all the transistors in a circuit. Thus, for the accurate evaluation of operation temperature in a FinFET-based circuit, it is indispensable to obtain thermal resistances of FinFETs without interconnects nor vias and those of interconnects and vias separately by simple methods.

In this section, a simple methodology for evaluating the maximum temperature of each FinFET connected by interconnect wires is described. The thermal resistances of a device are decomposed into four components, which correspond to heat

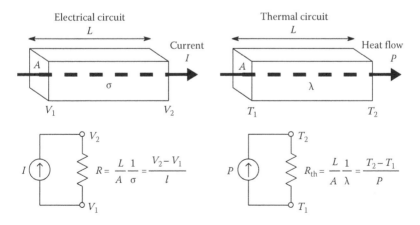

FIGURE 4.9 Correspondence between electrical and thermal circuits.

flow from the maximum temperature point to the source, drain, gate, and substrate. The four thermal resistances can be directly extracted from a set of temperature simulations of devices. The thermal resistances of interconnects and vias are obtained as analytical expressions.

4.5.1 THERMAL MODELING OF DEVICES

Figure 4.10 shows the equivalent thermal circuit of FinFETs. The equivalent thermal circuit has the five temperature nodes at the source, drain, gate, substrate, and hot spot, where the lattice temperature is the highest. By assuming that the heat source is located at the hot spot, all the heat flows inside a device are modeled by thermal resistances between the hot spot and the other four nodes: $R_{th,s}$, $R_{th,d}$, $R_{th,g}$, and $R_{th,sub}$.

The four thermal resistances are extracted from a set of device simulations under limited heat dissipation paths. For example, in the case of the extraction of $R_{th,s}$, operation temperature is calculated under the condition that only one heat dissipation path to the source exists. It should be noted that the thermal circuit can be utilized for various electron devices, because the model does not depend on its device geometry.

The extracted thermal resistances, $R_{th,s}$, $R_{th,d}$, $R_{th,g}$, and $R_{th,sub}$, for the 14 and 22 nm node bulk and SOI FinFETs[24] are listed in Table 4.1. All thermal resistances except for $R_{th,d}$ increase as device dimensions are downscaled, indicating that device downscaling results in greater self-heating effects. The thermal resistances of bulk FinFETs are smaller than those of SOI FinFETs owing to the larger heat dissipation

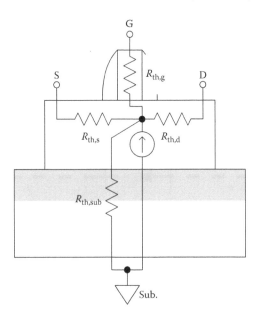

FIGURE 4.10 Equivalent thermal circuits of FinFETs.

TABLE 4.1

Thermal Resistances of Bulk and SOI FinFETs for 14 and 22 nm Nodes Extracted From the Device Simulations

	14 nm Node		22 nm Node	
	Bulk	**SOI**	**Bulk**	**SOI**
$R_{th,s}$ (K μW^{-1})	1.84	1.93	1.62	1.76
$R_{th,d}$ (K μW^{-1})	0.526	0.547	0.597	0.618
$R_{th,g}$ (K μW^{-1})	1.74	1.77	1.53	1.58
$R_{th,sub}$ (K μW^{-1})	1.59	1.72	1.24	1.45

Source: Takahashi, T. et al., *Jpn. J. Appl. Phys.*, 52, 064203, 2013.

through the p+ silicon slab region. Here, the STI thickness of bulk FinFETs and BOX thickness of SOI FinFETs is the same.

In Table 4.1, $R_{th,s}$ is significantly larger than $R_{th,d}$ in all devices, resulting from the asymmetric temperature distribution in the device (see Figure 4.7). Since the carriers are accelerated and lose their energy near the drain side of channel, the hot spot is in the drain region: the heat source of the equivalent thermal circuit is also in the drain region. Therefore, the physical distance between the heat source and the source node is longer than that between the heat source and the drain node. Since the physical distance corresponds to the magnitude of thermal resistance (Figure 4.9), $R_{th,s}$ is larger than $R_{th,d}$. Thus, the asymmetric thermal property of FinFETs originates from the asymmetric temperature distributions.

4.5.2 THERMAL MODELING OF INTERCONNECTS AND VIAS

Figure 4.11a shows the equivalent thermal circuit of interconnects. Two heat flow paths exist in interconnects: one along the interconnects and the other from interconnects to the substrate. These two heat flow paths are modeled by thermal resistance and conductance per unit interconnect length, r_{th} and g_{th}, respectively. The equivalent thermal circuits of interconnects are given as a distributed constant thermal circuit using r_{th} and g_{th}.

The r_{th} and g_{th} of interconnect wires are different from those of vias, because the geometrical structures of the wire and vias are different, as illustrated in Figure 4.11b. The r_{th} and g_{th} of interconnects and vias are given by

$$r_{th,int} = \frac{1}{\lambda_{int}A} = \frac{1}{\lambda_{int}H_{int}W_{int}}, \tag{4.3}$$

$$g_{th,int} = \lambda_{IL}S_{int}, \tag{4.4}$$

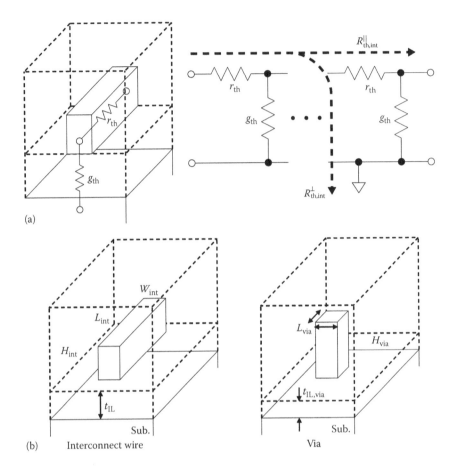

FIGURE 4.11 (a) Equivalent thermal circuit of interconnects. (b) Schematics of interconnect wires and vias.

$$r_{\mathrm{th,via}} = \frac{1}{\lambda_{\mathrm{int}} A} = \frac{1}{\lambda_{\mathrm{int}} L_{\mathrm{via}}^2}, \tag{4.5}$$

$$g_{\mathrm{th,via}} = \lambda_{\mathrm{IL}} S_{\mathrm{via}}, \tag{4.6}$$

where

A, λ_{int}, and λ_{IL} are the cross-sectional area of interconnects/vias, the thermal conductivity of interconnects, and the thermal conductivity of interlayer oxide, respectively

S_{int} and S_{via} are the shape factors per unit interconnect length of wires and vias, respectively

The shape factor S is defined as a coefficient that determines the heat flow between two surfaces: $Q_{12} = \lambda S \Delta T_{12}$, where λ and ΔT_{12} are the thermal conductivity of the material and the temperature difference between two surfaces, respectively. Therefore, S_{int} and S_{via} are determined between the substrate and wire/via surfaces. Since S is a geometrical parameter, S must be separately obtained for wires and vias. For interconnect wires, S_{int} has been obtained by solving Laplace's equation[25]:

$$S_{int} = 1.685 \times \left[\log_{10}\left(1 + \frac{t_{IL}}{W_{int}}\right) \right]^{-0.59} \left(\frac{t_{IL}}{H_{int}}\right)^{-0.078}. \tag{4.7}$$

In order to evaluate S_{via}, the via structure is approximated as a cylinder whose diameter is L_{via}. The assumption is reasonable, because actual vias do not have an accurate square pole shape owing to difficulties in fabricating nanoscale via holes. Since the analytical expression of S has been obtained for a cylinder connected to the substrate,[25] S_{via} is evaluated as the difference between the shape factors of two cylinders whose heights from the substrate are $t_{IL,via} + H_{via}$ and $t_{IL,via}$ (Figure 4.11b):

$$S_{via} = \frac{2\pi(t_{IL,via} + H_{via})}{\ln[4(t_{IL,via} + H_{via})/L_{via}]} - \frac{2\pi t_{IL,via}}{\ln[4t_{IL,via}/L_{via}]}. \tag{4.8}$$

As described earlier, the thermal models of interconnects and vias are separately given as distributed constant thermal circuits using r_{th} and g_{th}. Since r_{th} and g_{th} are parameters per unit interconnect/via length, it is difficult to estimate the impacts of interconnect/via parameters on operation temperature without thermal circuit simulations. Here, the thermal properties of interconnects and vias are analytically investigated by the solutions of a 1D equation of heat conduction under specific boundary conditions.

The size dependence of heat dissipation through each long interconnect wire is evaluated. The steady-state 1D equation of heat conduction for interconnects is given as

$$\lambda_{int} H_{int} W_{int} \frac{d^2 T(x)}{dx^2} - \lambda_{IL} S_{int}(T(x) - T_0) = 0. \tag{4.9}$$

Here, the x-axis is along interconnects, $T(x)$ is the interconnect temperature at x, and T_0 is the constant substrate temperature. Using $\Delta T(x) = T(x) - T_0$, Equation 4.9 is written as

$$\frac{d^2 \Delta T(x)}{dx^2} - m^2 \Delta T(x) = 0, \tag{4.10}$$

$$m \equiv \sqrt{\frac{\lambda_{IL} S_{int}}{\lambda_{int} H_{int} W_{int}}}. \tag{4.11}$$

A general solution of $\Delta T(x)$ obtained from Equations 4.10 and 4.11 is

$$\Delta T(x) = C_1 e^{mx} + C_2 e^{-mx}, \tag{4.12}$$

where C_1 and C_2 are constant coefficients. By extracting C_1 and C_2 under two boundary conditions, $\Delta T(x)$ can be obtained.

For each long interconnect wire, the boundary conditions are

$$-\lambda_{int} H_{int} W_{int} \left.\frac{d\Delta T(x)}{dx}\right|_{x=0} = P_{in}, \tag{4.13}$$

$$-\lambda_{int} H_{int} W_{int} \left.\frac{d\Delta T(x)}{dx}\right|_{x=\infty} = 0, \tag{4.14}$$

where P_{in} is the input power at the wire edge ($x = 0$). From Equations 4.12 through 4.14, the 1D temperature distribution along the interconnect wire, $\Delta T(x)$, is analytically obtained as

$$\Delta T(x) = \frac{P_{in}}{\lambda_{int} H_{int} W_{int} m} e^{-mx}. \tag{4.15}$$

The obtained $\Delta T(x)$ is plotted in Figure 4.12 for the 14 and 22 nm nodes at P_{in} of 50 μW. The wire edge temperature, $\Delta T(0)$, in the 14 nm node is much

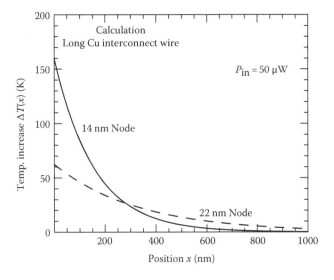

FIGURE 4.12 Temperature increase of a long interconnect wire, $\Delta T(x)$, as a function of distance from the heat source, x, for 14 and 22 nm node copper interconnects. (From Takahashi, T. et al., *Jpn. J. Appl. Phys.*, 52, 064203, 2013.)

higher than that in the 22 nm node, mainly owing to the smaller wire size and lower λ_{int} in the 14 nm node.[24] However, $\Delta T(x)$ in the 14 nm node more rapidly decreases and is lower than $\Delta T(x)$ in the 22 nm node when x is greater than 300 nm. The temperature decay along wires is characterized by the thermal healing length, $L_h = 1/m$.[23] L_h in the 14 nm node (156 nm) is approximately half of that in the 22 nm node (329 nm), resulting in the rapid temperature decay in the interconnect wires of the 14 nm node. These thermal properties in down-scaled interconnects are reflected on r_{th} and g_{th} of the thermal circuit: the wire size and λ_{int} dependence of r_{th} is stronger than that of g_{th} (Equations 4.3 and 4.4). Therefore, generally speaking, heat dissipation through interconnects in scaled technologies is worse.

4.5.3 Thermal Modeling of Circuits

The introduced thermal circuits and parameters are applied in a simple electrical circuit, as shown in Figure 4.13a. Two transistors are connected in parallel by the 300 nm-length interconnect wire. One of the transistors (Tr.1) is operated, whereas the other (Tr.2) is turned off. Therefore, the heat is generated only in Tr.1. The equivalent thermal circuit of the electrical circuit is shown in Figure 4.13b. The drain-to-channel thermal resistance of the OFF state transistor (Tr.2), $R_{th,dc}$, is defined as $(R_{th,s} + R_{th,d})/2$. Figure 4.14 shows the contour plot of lattice temperature for the circuit obtained by a device simulator.[24] Joule heat is generated in Tr.1 and flows along the interconnect wire and vias to Tr.2.

Figure 4.15 shows the relationship between the increase in the maximum lattice temperature of Tr.1 and input power of Tr.1 for the 14 nm node bulk FinFETs connected in parallel.[24] The temperature increases are calculated using a device simulator and thermal circuit model. The temperature increases are in good agreement between the two methods.

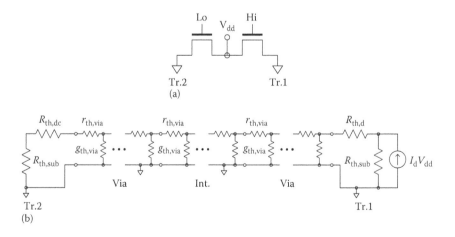

FIGURE 4.13 (a) Electrical circuit used for thermal modeling. (b) Equivalent thermal circuit of (a). The thermal circuits of interconnect (int) and vias are separately shown.

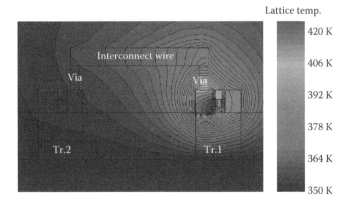

FIGURE 4.14 Contour plot of lattice temperature of Figure 4.13a, calculated by device simulator. (From Takahashi, T. et al., *Jpn. J. Appl. Phys.*, 52, 064203, 2013.)

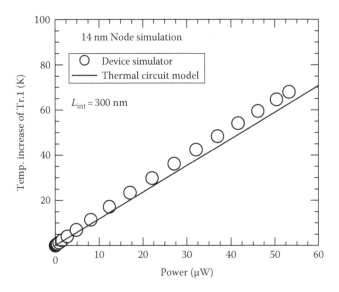

FIGURE 4.15 Increase in maximum lattice temperature of Tr.1 versus input power of Tr.1 for the electrical circuit shown in Figure 4.13a. The symbols and line indicate the results calculated using device simulator and thermal circuit model, respectively. (From Takahashi, T. et al., *Jpn. J. Appl. Phys.*, 52, 064203, 2013.)

4.5.4 THERMAL ADVANTAGE OF ADVANCED INTERCONNECT MATERIAL

Carbon-based materials, such as carbon nanotubes (CNTs) or graphene, have been intensively studied as future interconnect materials because of their superior electrical and thermal conductivities.[26,27] The thermal conductivity of individual multi-walled CNTs has been experimentally measured as approximately $3000\,W\,m^{-1}\,K^{-1}$,[28,29] which is much higher than the thermal conductivity of the interconnect material used in present LSIs (copper). On the other hand, it has been reported that the thermal conductivity of graphene nanoribbon degrades to 80 W m^{-1} K^{-1} because of the edge scattering of phonons.[30]

In order to evaluate the thermal advantages of high-thermal-conductivity interconnect material, the input thermal resistances of a finite-length interconnect wire are analytically evaluated. For the two heat dissipation paths of interconnect wires, two input thermal resistances, $R_{th,int}^{\parallel}$ and $R_{th,int}^{\perp}$, can be defined as shown in Figure 4.11a. When $R_{th,int}^{\parallel}$ and $R_{th,int}^{\perp}$ are evaluated, the temperature of the wire end is set to be T_0 in Equation 4.12. Therefore, one boundary condition for Equation 4.12 is

$$\Delta T(L_{int}) = 0. \tag{4.16}$$

The other condition is the same as that in Equation 4.13. From Equations 4.12, 4.13, and 4.16, $\Delta T(x)$ is obtained as

$$\Delta T(x) = \frac{P_{in}}{\lambda_{int} H_{int} W_{int} m} \left(-\frac{e^{mx}}{1 + e^{2mL_{int}}} + \frac{e^{-mx}}{1 + e^{-2mL_{int}}} \right). \tag{4.17}$$

$R_{th,int}^{\parallel}$ and $R_{th,int}^{\perp}$ are defined as the temperature increase at the wire edge ($x = 0$) divided by heat flow through the wire end ($x = L_{int}$) and substrate, namely,

$$R_{th,int}^{\parallel} \equiv \frac{\Delta T(0)}{P(L_{int})}, \tag{4.18}$$

$$R_{th,int}^{\perp} \equiv \frac{\Delta T(0)}{P_{in} - P(L_{int})}, \tag{4.19}$$

respectively. Here, the heat flow through the wire edge, $P(L_{int})$, is extracted from $\Delta T(x)$ as

$$P(L_{int}) = -\lambda_{int} H_{int} W_{int} \left. \frac{d\Delta T(x)}{dx} \right|_{x=L_{int}} = \frac{2}{e^{mL_{int}} + e^{-mL_{int}}} P_{in}. \tag{4.20}$$

From Equations 4.17 through 4.20, $R_{th,int}^{\parallel}$ and $R_{th,int}^{\perp}$ are obtained as

$$R_{th,int}^{\parallel} = \frac{1}{\lambda_{int} H_{int} W_{int} m} \times \frac{e^{mL_{int}} - e^{-mL_{int}}}{2}, \tag{4.21}$$

$$R_{th,int}^{\perp} = \frac{1}{\lambda_{int} H_{int} W_{int} m} \times \frac{e^{mL_{int}} - e^{-mL_{int}}}{e^{mL_{int}} + e^{-mL_{int}} - 2}, \tag{4.22}$$

respectively. The input thermal resistances of vias, $R_{th,via}^{\parallel}$ and $R_{th,via}^{\perp}$, are also obtained similarly.

Figure 4.16a and b shows the plots of $R_{th,int}^{\parallel}$ and $R_{th,int}^{\perp}$ as a function of inter-connect length for copper and CNT interconnects at the 14 nm node,[24] respec-tively. $R_{th,sub}$ of 14 nm node bulk FinFETs and $R_{th,via}^{\parallel}$ are also shown. In copper interconnects/vias, $R_{th,via}^{\parallel}$ is larger than $R_{th,sub}$, implying that the heat dissipation through the interconnect hardly affects the device temperature, because tempera-ture rapidly decays in the via. On the other hand, $R_{th,via}^{\parallel}$ and $R_{th,int}^{\parallel}$ of CNT intercon-nects/vias are much smaller than $R_{th,sub}$ owing to the higher thermal conductivity of CNT, resulting in better heat dissipation through interconnects and vias. Thus, for

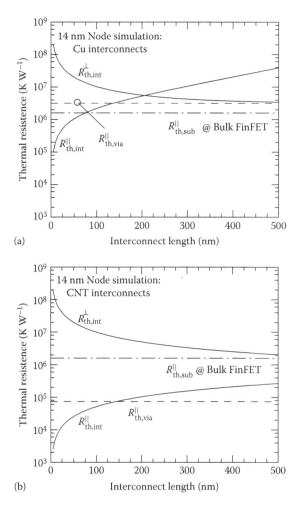

(a)

(b)

FIGURE 4.16 Input thermal resistance of a finite length interconnect wire, $R_{th,int}^{\parallel}$ and $R_{th,int}^{\perp}$, versus interconnect length for 14 nm node (a) copper and (b) CNT interconnects. (From Takahashi, T. et al., *Jpn. J. Appl. Phys.*, 52, 064203, 2013.) Input thermal resistance of vias, $R_{th,via}^{\parallel}$, and $R_{th,sub}^{\parallel}$, of 14 nm node bulk FinFETs are also shown for comparison.

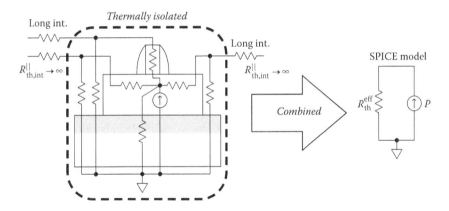

FIGURE 4.17   Equivalent thermal circuit of a thermally isolated system.

lowering operation temperature by utilizing heat dissipation through the interconnects and vias, $R^{\|}_{\mathrm{th,int}}$ and $R^{\|}_{\mathrm{th,via}}$ should be smaller than $R_{\mathrm{th,sub}}$.

Finally, the conventional SPICE models of self-heating effects are interpreted on the basis of the presented thermal circuit model. In the widely used transistor models, steady-state self-heating effects are represented using one thermal resistance, $R^{\mathrm{eff}}_{\mathrm{th}}$.[22] By utilizing an experimentally extracted $R^{\mathrm{eff}}_{\mathrm{th}}$, current degradation due to self-heating effect can be reproduced.[8] In the experimental $R^{\mathrm{eff}}_{\mathrm{th}}$ extractions, self-heating effects are usually evaluated under thermally isolated conditions, because interconnect wires are sufficiently longer than L_{h}. Figure 4.17 illustrates the equivalent thermal circuit under the thermally isolated condition. In this case, self-heating effects can be represented using a single $R^{\mathrm{eff}}_{\mathrm{th}}$ by combining thermal resistances of isolated device and interconnects/vias into $R^{\mathrm{eff}}_{\mathrm{th}}$. Thus, the conventional SPICE models are consistent with the present thermal circuit model. On the basis of the aforementioned discussion, it is concluded that the present thermal circuit model should be used when devices are thermally coupled, namely, when $R^{\|}_{\mathrm{th,int}}$ and $R^{\|}_{\mathrm{th,via}}$ are smaller than $R_{\mathrm{th,sub}}$.

REFERENCES

1. Auth, C., C. Allen, A. Blattner, D. Bergstrom, M. Brazier, M. Bost, M. Buehler et al. 2012. A 22 nm high performance and low-power CMOS technology featuring fully-depleted tri-gate transistors, self-aligned contacts and high density MIM capacitors. In *Symposium on VLSI Technology (VLSIT), 2012*, Honolulu, HI, pp. 131–132. doi:10.1109/VLSIT.2012.6242496.
2. Kolluri, S., K. Endo, E. Suzuki, and B. Kaustav. 2007. Modeling and analysis of self-heating in finfet devices for improved circuit and EOS/ESD performance. In *IEDM Technical Digest*, Washington, DC, pp. 177–180.
3. Takahashi, T., N. Beppu, and K. Chen. 2011. Thermal-aware device design of nanoscale bulk/SOI FinFETs: Suppression of operation temperature and its variability. In *IEEE International Electron Devices Meeting (IEDM), 2011*, Washington, DC, pp. 809–812. doi:10.1109/IEDM.2011.6131672.

4. Su, L. T., J. E. Chung, D. A. Antoniadis, K. E. Goodson, and M. I. Flik. 1994. Measurement and modeling of self-heating in SOI NMOSFET's. *IEEE Transactions on Electron Devices* 41(1): 69–75. doi:10.1109/16.259622.

5. Beppu, N., T. Takahashi, T. Ohashi, and K. Uchida. 2012. Impact of gate poly depletion on evaluation of channel temperature in silicon-on-insulator metal-oxide-semiconductor field-effect transistors with four-point gate resistance measurement method. *Japanese Journal of Applied Physics* 51(2): 02BC15. doi:10.1143/JJAP.51.02BC15.

6. Takahashi, T., T. Matsuki, T. Shinada, Y. Inoue, and K. Uchida. 2013. Comparison of self-heating effect (SHE) in short-channel bulk and ultra-thin BOX SOI MOSFETs: Impacts of doped well, ambient temperature, and SOI/BOX thicknesses on SHE. In *IEEE International Electron Devices Meeting, 2013*, Washington, DC, pp. 7.4.1–7.4.4. doi:10.1109/IEDM.2013.6724581.

7. Tenbroek, B. M., M. S. Lee, and W. Redman-White. 1996. Self-heating effects in SOI MOSFETs and their measurement by small signal conductance techniques. *IEEE Transactions on Electron Devices* 43(12): 2240–2248. doi:10.1109/16.544417.

8. Jin, W., W. Liu, S. K. Fung, P. C. Chan, and C. Hu. 2001. SOI thermal impedance extraction methodology and its significance for circuit simulation. *IEEE Transactions on Electron Devices* 48(4): 730–736. doi:10.1109/16.915707.

9. Wang, R., J. Zhuge, C. Liu, R. Huang, D.-W. Kim, D. Park, and Y. Wang. 2008. Experimental study on quasi-ballistic transport in silicon nanowire transistors and the impact of self-heating effects. In *IEEE International Electron Devices Meeting, 2008 (IEDM'08)*, San Francisco, CA, pp. 753–756. doi:10.1109/IEDM.2008.4796806.

10. Scholten, A. J., G. D. Smit, R. M. Pijper, L. F. Tiemeijer, H. P. Tuinhout, J.-L. van der Steen, A. Mercha, M. Braccioli, and D. B. Klaassen. 2009. Experimental assessment of self-heating in SOI FinFETs. In *IEDM Technical Digest*, Baltimore, MD, pp. 305–308.

11. Ota, K., M. Saitoh, C. Tanaka, Y. Nakabayashi, and T. Numata. 2011. Systematic understanding of self-heating effects in tri-gate nanowire MOSFETs considering device geometry and carrier transport. In *IEEE International Electron Devices Meeting, 2011 (IEDM'11)*, Washington, DC, pp. 513–516. doi:10.1109/IEDM.2011.6131597.

12. Pop, E., R. Dutton, and K. Goodson. 2003. Thermal analysis of ultra-thin body device scaling. In *IEDM Technical Digest*, Washington, DC, pp. 883–886.

13. Shrivastava, M., M. Agrawal, S. Mahajan, H. Gossner, T. Schulz, D. K. Sharma, and V. R. Rao. 2012. Physical insight toward heat transport and an improved electrothermal modeling framework for FinFET architectures. *IEEE Transactions on Electron Devices* 59(5): 1353–1363. doi:10.1109/TED.2012.2188296.

14. Hisamoto, D., T. Kaga, Y. Kawamoto, and E. Takeda. 1989. A fully depleted lean-channel transistor (DELTA)—A novel vertical ultra thin SOI MOSFET. In *International Electron Devices Meeting, Technical Digest, 1989 (IEDM'89)*, Washington, DC. doi:10.1109/iedm.1989.74182.

15. Okano, K., T. Izumida, H. Kawasaki, A. Kaneko, A. Yagishita, T. Kanemura, M. Kondo et al. 2005. Process integration technology and device characteristics of CMOS FinFET on bulk silicon substrate with sub-10 nm fin width and 20 nm gate length. In *IEEE International Electron Devices Meeting, Technical Digest, 2005*, Washington, DC, pp. 721–724. doi:10.1109/IEDM.2005.1609454.

16. Liu, W., K. Etessam-Yazdani, R. Hussin, and M. Asheghi. 2006. Modeling and data for thermal conductivity of ultrathin single-crystal SOI layers at high temperature. *IEEE Transactions on Electron Devices* 53(8): 1868–1876. doi:10.1109/TED.2006.877874.

17. Asheghi, M., K. Kurabayashi, R. Kasnavi, and K. E. Goodson. 2002. Thermal conduction in doped single-crystal silicon films. *Journal of Applied Physics* 91(8): 5079–5088. doi:10.1063/1.1458057.

18. Yamane, T., N. Nagai, S.-i. Katayama, and M. Todoki. 2002. Measurement of thermal conductivity of silicon dioxide thin films using a 3ω method. *Journal of Applied Physics* 91(12): 9772–9776. doi:10.1063/1.1481958.

19. Takahashi, T., N. Beppu, K. Chen, and S. O. Uchida. 2013. Self-heating effects and analog performance optimization of fin-type field-effect transistors. *Japanese Journal of Applied Physics* 52: 04CC03. doi:10.7567/JJAP.52.04CC03.

20. Mason, S. 1954. Power gain in feedback amplifier. *Transactions of the IRE Professional Group on Circuit Theory* 1(2): 20–25. doi:10.1109/TCT.1954.1083579.

21. Gupta, M. S. 1992. Power gain in feedback amplifiers, a classic revisited. *IEEE Transactions on Microwave Theory and Techniques* 40(5): 864–879. doi:10.1109/22.137392.

22. BSIMSOI ver.4.5.0. http://www-device.eecs.berkeley.edu/bsim/?page=BSIMSOI, accessed August 18, 2015.

23. Goodson, K. E. and M. I. Flik. 1992. Effect of microscale thermal conduction on the packing limit of silicon-on-insulator electronic devices. *IEEE Transactions on Components, Hybrids, and Manufacturing Technology* 15(5): 715–722. doi:10.1109/33.180035.

24. Takahashi, T., S. Oda, and K. Uchida. 2013. Methodology for evaluating operation temperatures of fin-type field-effect transistors connected by interconnect wires. *Japanese Journal of Applied Physics* 52: 064203. doi:10.1143/JJAP.52.064203.

25. Sunderland, J. E. and K. R. Johnson. 1964. Shape factors for heat conduction through bodies with isothermal or convective boundary conditions. *ASHRAE Transactions* 70: 237–241.

26. Nihei, M., A. Kawabata, S. Sato, T. Nozue, T. Hyakushima, M. Norimatsu, T. Murakami, and D. Kondo. 2008. Carbon nanotube via interconnects with large current carrying capacity. In *Ninth International Conference on Solid-State and Integrated-Circuit Technology, 2008 (ICSICT'08)*, Beijing, China, pp. 541–543. doi:10.1109/ICSICT.2008.4734598.

27. Li, H., C. Xu, N. Srivastava, and K. Banerjee. 2009. Carbon nanomaterials for next-generation interconnects and passives: Physics, status, and prospects. *IEEE Transactions on Electron Devices* 56(9): 1799–1821. doi:10.1109/TED.2009.2026524.

28. Kim, P., L. Shi, A. Majumdar, and P. L. McEuen. 2001. Thermal transport measurements of individual multiwalled nanotubes. *Physical Review Letters* 87: 215502. doi:10.1103/PhysRevLett.87.215502.

29. Horibe, M., M. Nihei, D. Kondo, and A. K. Awano. 2004. Influence of growth mode of carbon nanotubes on physical properties for multiwalled carbon nanotube films grown by catalytic chemical vapor deposition. *Japanese Journal of Applied Physics* 43(10): 7337–7341. doi:10.1143/JJAP.43.7337.

30. Liao, A. D., J. Z. Wu, X. Wang, H. Dai, and E. Pop. 2011. Thermally limited current carrying ability of graphene nanoribbons. *Physical Review Letters* 106: 256801. doi:10.1103/PhysRevLett.106.256801.

5 Spintronics-Based Nonvolatile Computing Systems

Tetsuo Endoh

CONTENTS

5.1 Introduction 105
5.2 Nonvolatile Logic LSIs.. 106
    5.2.1 Nonvolatile Microprocessor.. 106
    5.2.2 Nonvolatile Recognition Processor.. 110
5.3 MTJ-Based Memory LSIs 113
    5.3.1 STT-MRAM with Fast Write Cycle 113
    5.3.2 Benchmark of STT-MRAMs' Power Consumption 117
5.4 Summary 120
References... 121

5.1 INTRODUCTION

Current computing systems based on volatile logic circuits and working memories suffer from two problems that limit the performance improvements that can be realized by semiconductor device miniaturization as shown in Figure 5.1a. One is the large power consumption in cache memories and the main memory. The other is the speed gap between the last level (LL) cache by static random access memory (SRAM) and the main memory by dynamic random access memory (DRAM) and that between the main memory and the storage memory by solid-state drive (SSD) or hard disk drive (HDD). As the length of the channel decreases, the SRAM memory cell realizes a large subthreshold leakage. This trend, together with the requirement to increase the caches' capacity in order to boost the computer's performance, drastically increases the caches' power consumption, especially in the LL cache having the largest capacity. This power increase is a bottleneck in the goal to realize increased computer performance using device scaling. The DRAMs in the main memory rely on a single-ended read scheme, with their cycle time being limited to several tens of nanoseconds, which is much slower than that of the LL cache, whose cycle time is as fast as a few nanoseconds. Storage memories are also much slower than the main memories. These differences in speed are also a serious bottleneck to realizing improvements in the computer performance.

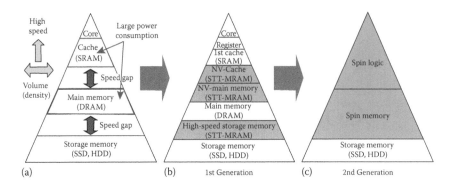

FIGURE 5.1 Restructuring of computer hierarchy: (a) conventional memory hierarchy, (b) NV first-generation computer, and (c) NV second-generation computer.

To overcome this situation, the concept of a nonvolatile (NV) computing system has been proposed [1]. A two-step transition to a complete NV computer is assumed. In the near future, the SRAM in the LL cache will be replaced by a spin-transfer-torque magnetoresistive random access memory (STT-MRAM). In addition, the first-generation model will see the two speed gaps being filled with STT-MRAMs. Although the computer is made such that it is partially NV, this generation can resolve the main bottlenecks in computers that are currently used. This is because when used as the LL cache, the operation power and the standby power are shown to be much lower than the conventional SRAM (see Section 5.3.2). Moreover, when the speed gaps are filled in, the access time of STT-MRAM is shown to be faster than that of DRAM (see Section 5.3.1). Afterward, as the second generation, all of the logic circuits and memories are made to be NV using spintronics technologies. This full NV computer can further reduce its power because of the logic-in-memory architecture. In the architecture, memory and logic functions are combined by placing magnetic tunnel junction (MTJ) memory devices on top of the CMOS logic circuits. This combination makes the MTJs perform some of the logic functions, leading to a reduction in the device count. The global wirings that run between the memory and the logic in the conventional architecture are also eliminated. Both the approaches are effective in reducing the chip size, performance, and power.

5.2 NONVOLATILE LOGIC LSIs

5.2.1 NONVOLATILE MICROPROCESSOR

Several techniques are applied to conventional microprocessors to reduce their power. One of the most effective methods for reducing the static power is the power-gating technique, in which the power supply voltage is shut down to a circuit block when it is not working [2]. However, the application of the conventional power-gating method to volatile computers has the drawback of large latencies at power on and power off [3]. The logic circuits include many memories that store the data of the internal states that are needed when operating the logic functions. Therefore, before the power supply is shut down, the data in the memories must be saved into external low-power memories that generally exist far away from the logic parts, leading to high latency. After power on, the saved data must be loaded from the external memory to the internal memories of the logic states. This also results in high latency. Therefore, there is a trade-off between the power and the performance (latencies in power on and power off) with the application of the conventional power-gating method to volatile logic large-scale integration circuits (LSIs). Figure 5.2 shows the state machine of conventional volatile microprocessors (MPUs). C0 represents the operating mode, while C1 and C2 represent the clock-gating modes. Although they can reduce some of the dynamic power of the clock, the large static power cannot be reduced. C5 is the deep sleep mode of the power-gating method and can significantly reduce the static power. However, the wake-up time from this mode (C5) to the operating mode (C0) is long, taking several tens of microseconds.

This trade-off can be avoided if the data of the internal states in logic LSIs are stored in local NV memories, because an abrupt shutdown of the power supply voltage is possible without relying on external low-power memories. The logic circuits' operation can also start immediately after power on, because the internal state data that are required to resume logic operations are loaded from the local NV memories soon after power-on. Figure 5.3 shows the block diagram of the 32 bit NV MPU in which the data of the program counter, the internal states, register file, data memories, and pipeline registers are stored in the local MTJ-based NV memories [4]. The total circuit is power gated by p-channel field-effect transistor (PFET) headers. The basic circuit element of the NV memories is an NV latch that is shown in Figure 5.4. By adding power-on initializers, the latch can load data stably from the MTJs to the CMOS circuit even with a fast (10 ns) power on, as shown in the simulated waveforms in Figure 5.4. Figure 5.5 shows a microphotograph of the NV MPU chip that was fabricated using 90 nm CMOS and 100 nm MTJ processes. The chip's features are summarized in the table in Figure 5.5. The standby power is reduced to 1.2 mW with a power supply voltage of 1.0 V. This power is solely consumed by the PFET headers with a higher threshold voltage and the gate signal controller. Figure 5.6 shows the NV function check results of the 32 bit NV MPU. The contents of 32 bit pipeline registers before and after power off are checked, and we verified that the 32 bit pattern "00100320" resumed successfully after power off. Figure 5.7 shows the waveform of the internal power supply voltage that was power-gated by the PFET headers. It further shows that the exit and entry delays (latencies) are about 3 μs. These delays are more than one order of magnitude faster than the conventional volatile MPU with external memory backup.

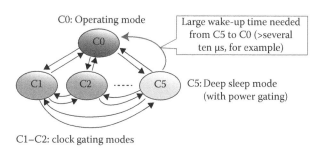

FIGURE 5.2 State machine of the conventional volatile microprocessor.

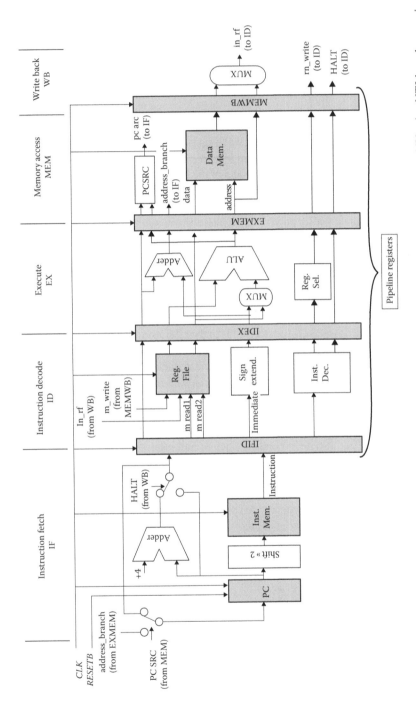

FIGURE 5.3 Block diagram of the NV microprocessor (MPU). The memories in the gray-colored blocks are made NV using NTJ-based memories.

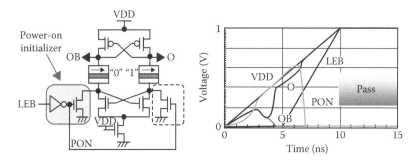

FIGURE 5.4 An NV latch circuit implemented in the NV MPU.

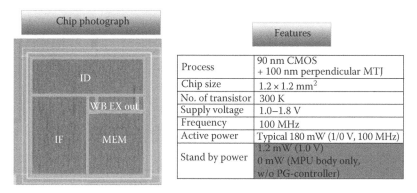

FIGURE 5.5 Microphotograph of the 32-bit NV MPU chip fabricated using 90 nm CMOS and 100 nm MTJ processes. Its features are summarized in the table.

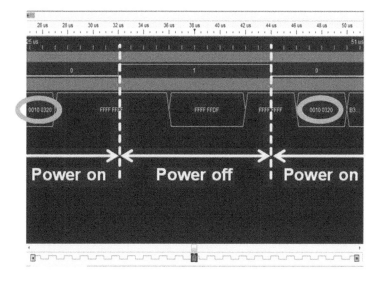

FIGURE 5.6 NV function check of the 32-bit NV MPU. The contents of the 32-bit pipeline registers before and after power off are checked.

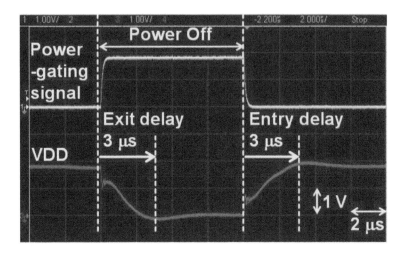

FIGURE 5.7 Measurement results of the exit and entry latencies for the 32-bit NV MPU.

5.2.2 NONVOLATILE RECOGNITION PROCESSOR

The recognition function of intelligent systems has attracted significant interest. In these situations, the best vector matching has been widely utilized as the fundamental method of recognition when searching for the most similar candidate from a large number of multidimensional feature vectors obtained from previous studies. Moreover, real-time operation is anticipated for time-critical applications such as automotive car control, gesture recognition, and object tracking, as shown in Figure 5.8. Owing to the high computational cost arising from the data processing of all these feature vectors, it is not feasible to realize real-time operation by relying only on the software approach. Therefore, high-speed recognition processors have been developed using parallel processing computing. In these recognition processors, large embedded memories are usually required to store feature vectors, as shown in Figure 5.9. Because they are volatile, the large power required to retain the data is a disadvantage, especially for mobile intelligent systems.

To reduce the power consumption, NV memories are required for storing the feature vectors. Furthermore, fine-grained power-gating schemes need to be properly applied

(a)   (b)   (c)

FIGURE 5.8 Several time-critical applications that recognize searches for the most similar feature vector from a large amount of previous studies: (a) automotive car control, (b) gesture recognition, and (c) object tracking.

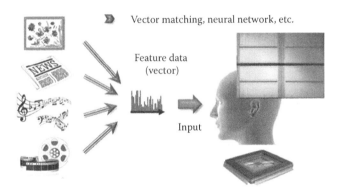

Vector matching, neural network, etc.

Feature data
(vector)

Input

FIGURE 5.9   Large embedded memories are required in the hardware intelligent systems.

to the cell arrays in order to minimize the operating power. The architecture required to search for the feature vector that is most similar to the input vector is shown in Figure 5.10 [5]. This architecture is suitable when applying fine-grained power-gating schemes to the NV memories for storing the feature vectors. The search process consists of two steps. In the first step, as shown in Figure 5.10a, the input vector consisting of 128 8-bit components is compared with the centroid vectors in the clusters. A cluster refers to a group of feature vectors that are similar to each other. A vector α is considered more similar than γ to a vector β if the distance between α and β is smaller than that between α and γ in the 128-dimensional space. In this first step, we determine the cluster to which the input vector is to belong. In the next step, as shown in Figure 5.10b, the distances between the input vector and the feature vectors that belong to the cluster determined in the first step are measured and compared to each other to find the vector that is most similar to an output vector. In the first step, the power supply voltage to the memory cells that store all of the feature vectors is shut off to significantly reduce the power consumption required for the search. To do this, in the second step, we again shut off the power supply voltage to the memory cells that store feature vectors in other clusters. The 4T2MTJ STT-MRAM cell is used as an NV memory cell that can be applied with a fine-grained power-gating technique [6].

Figure 5.11 shows the physical layout of the memory array in a recognition processor that is used to store feature vectors and centroid vectors when the 8 bit fine-grained power-gating scheme is applied. Each feature vector consists of 128 8-bit cells. There are two memory planes: the right one is for the four centroid vectors and the left one is for the feature vectors that are grouped into four clusters. The array is designed using a 4T2MTJ NV memory cell to store the feature vectors, and the centroid vectors, which are suitable for low-power searches of the most similar vector. The 4T2MTJ memory cell is illustrated in Figure 5.12. Figure 5.13 shows the power-reduction effect of the NV recognition processor with the cell array in Figure 5.11, and it was compared with that of the conventional architecture using SRAM. The power can be reduced by 85% using the NV memory with the power-gating scheme when compared with the reduction that is possible using the conventional SRAM with a four-cluster design. By increasing the number of clusters to 16, the reduction rate can be increased to 97%.

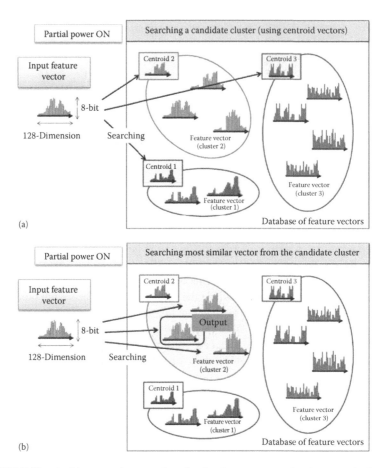

(a)

(b)

FIGURE 5.10 Architecture that searches for the feature vector that is most similar to an input vector: (a) searching a candidate cluster (using centroid vectors) and (b) searching most similar vector from the candidate cluster.

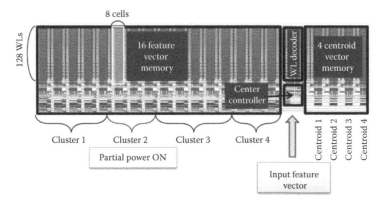

FIGURE 5.11 **(See color insert.)** Memory array architecture for storing feature vectors and centroid vectors with a fine-grained power-gating technique applied.

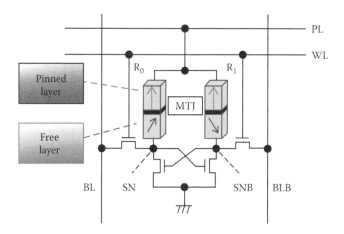

FIGURE 5.12 The structure of the 4T2MTJ cell.

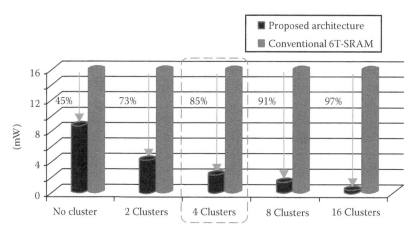

FIGURE 5.13 Power comparison between the NV recognition processor and the conventional one based on SRAM.

5.3 MTJ-BASED MEMORY LSIs

5.3.1 STT-MRAM with Fast Write Cycle

The STT-MTJ is regarded as the only candidate for realizing an NV working memory because of its superior switching endurance features [7]. In order to make the main memory NV, a small footprint STT-MRAM cell, which is referred to as the 1T1MTJ cell, has been developed [8]. However, it is difficult for the single-ended read cell to realize a NV cache memory, because the access time is not sufficiently fast. Differential pair-type STT-MRAM cells have been developed for high-performance applications [9–11]. However, the quick switching of the STT-MTJ requires a relatively large current, and the probability of switching within a short period of time decreases as it is driven by a moderate current [12,13]. Therefore, it is difficult to realize a high-speed and high-density STT-MRAM, because large MOSFETs need to be hybridized using MTJs whose channel widths are sufficiently large to enable the flow of large currents that can switch the MTJs quickly.

In this situation, a new write method for differential pair-type STT-MRAM cells has been proposed, which can realize high-speed and high-density memory [14]. The general configuration of differential pair-type STT-MRAM cells is shown in Figure 5.14, in which a CMOS latch and MTJs are combined. The data are written from or read to a pair of bit lines (BLs) (BL and /BL) through a pair of switching transistors. Data are supplied from a pair of bit lines (BLs) through the switching transistors gated by the word line (WL). The conventional method used to write to this type of cell is illustrated in Figure 5.15. The write data are supplied from a pair of BLs, and they are rapidly written to a CMOS latch. MTJ switching is slower than CMOS latch switching; therefore, the switching transistors need to be open for a period that is longer than the CMOS latch time until the MTJ switching is complete, because the data used to switch the MTJs also come from the BLs. With this method, the write-cycle time to the STT-MRAM is limited by the MTJ switching time, which is very large unless the switching transistors are sufficiently large. Figure 5.16 explains a new method that can be employed to write to the differential pair-type STT-MRAM cell. This is called the background write method. The concept is based on the scheme whereby external data is written through a pair of BLs to a CMOS latch in a fast write cycle. The MTJs are slowly switched using the updated data in the CMOS latch

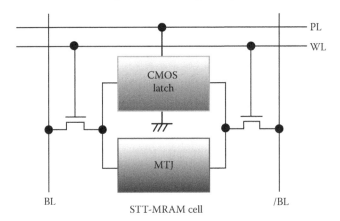

FIGURE 5.14 Structure of a differential pair-type STT-MRAM cell.

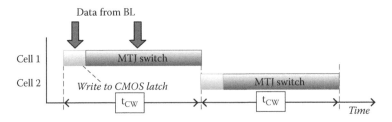

FIGURE 5.15 Conventional write method for a differential pair-type STT-MRAM cell.

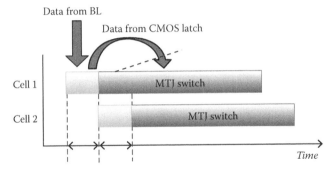

FIGURE 5.16 Background write method for a differential pair-type STT-MRAM cell.

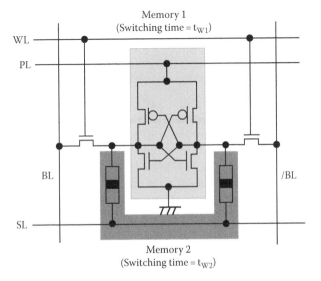

FIGURE 5.17 6T2MTJ cell that achieve MTJs' switch using the data in the CMOS latch.

internally even after the switching transistors are closed. Because this MTJ switching can be performed within a memory cell without any additional help processes running in the background, it does not affect subsequent write cycles to other cells.

Figure 5.17 shows the structure of a memory cell to which the background write method was applied for a particular design. In the structure, a pair of MTJs is connected in series between the storage nodes in the CMOS latch, and the connected node is controlled by a source line (SL). It should be noted that the states of the two MTJs could be reversed by letting a current flow through the serially connected MTJs according to the state of the CMOS latch if the pinned layers of the two MTJs are connected to the floating SL. Figure 5.18 illustrates an actual control to two cells that are accessed serially in fast write cycles. When WL_0 level is increased, PL_0 is also set high in order to supply power to the cell. SL remains low for a short time (less than 1 ns) to enable data to be loaded from the MTJs to the CMOS latch (the SL floats from the controller afterwards). After the data from a pair of BLs are written to the CMOS latch,

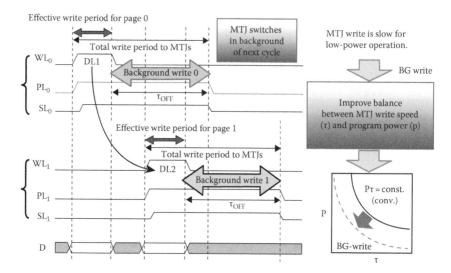

FIGURE 5.18 Background write method for a differential pair-type STT-MRAM cell.

WL_0 is lowered to terminate the external write to the cell. However, PL_0 remains high to switch the MTJs using the data in the CMOS latch. When the switching is complete, PL_0 is lowered in order to eliminate the static current flowing in the cell. When WL_0 level is decreased, the next write operation can be applied to any cell in the same cell array, as shown for WL_1, PL_1, and SL_1 in Figure 5.18. This second write operation does not affect the MTJ switching operation in the first cell that was accessed.

Figure 5.19 shows a PL/SL driver corresponding to the memory cells. Several cells along a WL are grouped as a grain in which PLs and SLs are connected

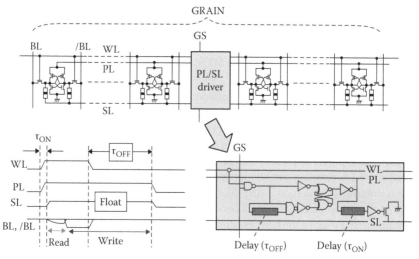

FIGURE 5.19 Grain and PL/SL driver for the background write scheme.

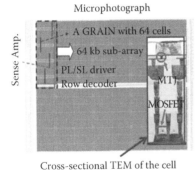

Features of 1 Mb BG write STT-MRAM	
Process	90 nm CMOS + 100 nm MTJ
Cell size	$1.13 \times 3.06 = 3.46\ \mu m^2$
Macro size (Mb)	$5.14\ mm^2$
Cell efficiency	70.5%
Supply voltage	1.3 V
Organization	64K work × 16 bit
Grain size	64 bit
Access mode	Page/random
Function	Asynchronous SRAM
Random read cycle	1.5 ns
Random write cycle	2.1 ns

FIGURE 5.20 Chip microphotograph and the features of 1 MB STT-MRAM with background write scheme.

together to be controlled by the PL/SL driver, which is placed in the middle of the grain. The driver includes two timers. One is to measure τ_{ON}, which is necessary to load data at power on. The other is to measure, τ_{OFF}, which is necessary for switching the MTJs. We designed and fabricated a 1 Mb STT-MRAM with the background write scheme using 90 nm CMOS and 100 nm MTJ processes. Figure 5.20 shows the microphotograph of the chip and its features, and shows that we achieved a 1.5 ns random read cycle and a 2.1 ns random write cycle. Using this background write scheme, we successfully achieved a fast-cycle STT-MRAM that can be applied to L3 and L2 caches.

5.3.2 BENCHMARK OF STT-MRAMS' POWER CONSUMPTION

Although a large static leak power is consumed in the state-of-the-art SRAM cell array, STT-MRAMs can be used to eliminate the static leak power in cell arrays when the power-gating technique is applied to the cell array. However, compared to the SRAM cell, a large power is consumed when switching the STT-MRAM cell. Therefore, it is important for the replacement of the SRAM by the power-gated STT-MRAM to contribute to the development of lower-power computing systems (see Figure 5.21).

	SRAM	Power-gated STT-MRAM
Static current is standby	Large	Zero
Load/save current from/to MTJ	Zero	Large
Total current	?	?

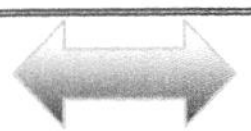

Which is smaller?

FIGURE 5.21 It is not trivial whether or not the power of STT-MRAM is lower than that of SRAM.

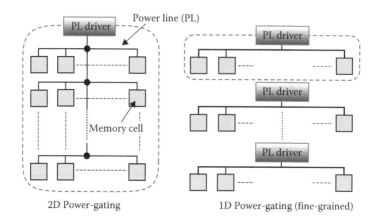

2D Power-gating 1D Power-gating (fine-grained)

FIGURE 5.22 Two-dimensional and one-dimensional power-gating techniques applied to NV memory.

Benchmark studies that compared the power between several differential pair-type STT-MRAMs and SRAM have been performed [15]. Before applying the benchmark, we reviewed the general concept of power-gating techniques that can be applied to memory cells. Figure 5.22 illustrates two kinds of power-gating architectures that are applied to memory cell arrays. In the two-dimensional (2D) power-gating method, the power supply voltage of a whole cell array is controlled. In the one-dimensional (1D) power-gating method, the power supply voltages of cells along the WL or BL are controlled. Therefore, the 2D power-gating method is coarse grained and the 1D power-gating method is fine grained. In general, the 1D power-gating method provides a lower operation power, because fewer cells are powered, including the accessed ones. Although the area penalty pertaining to the PL drivers is large compared with that of the 2D power gating, the 1D power gating will be an optimum option in the trade-off between the performance and the area penalty when the grain size is properly designed. There are two kinds of 1D power-gating schemes, as shown in Figure 5.23. One is the column (BL) 1D power-gating scheme, in which the cells in a grain are along a BL. The other is the row (WL) 1D power-gating scheme, in which the cells in a grain are along a WL. The row (WL) 1D power-gating approach is more suitable for multibit access NV memories, because several cells corresponding to bits that are accessed simultaneously can be assigned to the cells in a grain, which will minimize the operation power.

Figure 5.24 shows the three differential pair-type STT-MRAM cells and the 6TSRAM cell whose write currents were simulated for comparison purposes: (a) 4T2MTJ cell [11], (b) 6T2MTJ-1 [9], (c) 6T2MTJ-2 [10], and (d) the conventional 6TSRAM; 32 MB (256 MB) memories are virtually designed using the cells. The three NV memories using cells (a)–(c) are applied with a 32 bit fine-grained row 1D power-gating scheme; 64 bytes (512 bits) cells are accessed simultaneously in all memories. The memories are therefore designed considering the L3 cache application. Figure 5.25 shows the simulation results for which we compared the averaged

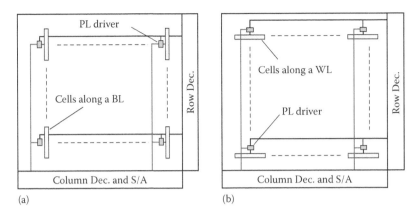

FIGURE 5.23 Two kinds of 1D power gating, that is, (a) the column (BL) 1D power gating and (b) the row (WL) 1D power gating.

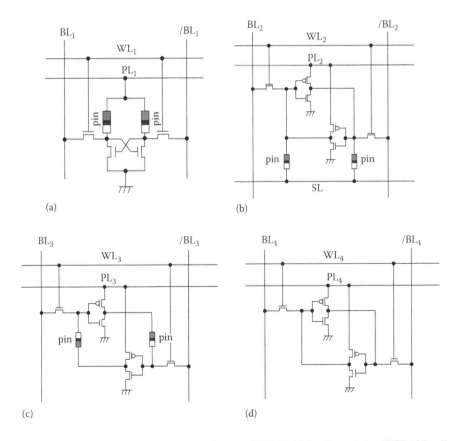

FIGURE 5.24 The three differential pair-type STT-MRAM cells and the 6TSRAM cell whose powers are compared: (a) 4T2MTJ (Tohoku Univ.), (b) 6T2MTJ-1 (Tokyo Tech), (c) 6T2MTJ-2 (Toshiba), and (d) 6TSRAM.

Memory cell		NMOS/MTJ Hybride type	NMOS/MTJ Add-on type	CMOS/MTJ Hybrid type	LOP	LSTP
		4T2MTJ	6T2MTJ-1	6T2MTJ-2	6TSRAM	
V_{dd}		0.9 V	1.2 V	1.9 V	0.9 V	
Total current/32 MB Including only write cycle (most severe condition for STT-MRAMs)@15 ns		70 mA	121 mA	164 mA	1.46 A	138 mA
Components	Write current/64 bytes	63 mA	108 mA	143 mA	0.18 mA	
	(Write current/cell)	123 μA	210 μA	279 μA	0.36 μA	
	Array control current/64 bytes	7 mA	13 mA	21 mA	4 mA	
	Static current/32 MB	0	0	0	1.46 A	134 mA

FIGURE 5.25 Comparison of operating current in the write cycle for the four memories obtained using the cells in Figure 5.24.

operation currents in a 15-ns write cycle using the 90 nm CMOS and 100 nm MTJ process technologies. Two different 6TSRAMs were simulated. One is a low operating-power (LOP) SRAM that is used in laptop computers, etc., while the other is a low standby-power SRAM that is used in mobile phones, etc. The power supply voltages (V_{dd}) for the three STT-MRAMs are the minimum values that can switch the MTJs in the corresponding cells. The total currents were broken down into three components: a write current to 64 bytes cells (write current per cell), an array control current per 64 bytes, and a static current per 32 MB. The array control current refers to the current that is consumed during the dynamic operation of cell-array control signals such as WL, BL, SL, and PL.

The results indicate that all of the STT-MRAMs that were obtained by applying the fine-grained power-gating scheme consume lower operation currents than the LOP SRAM. They also show that the operation currents of the 4T2MTJ STT-MRAM and the 6T2MTJ-1 STT-MRAM are even lower than that of the LSTP SRAM. We confirmed that the write currents to a cell in the STT-MRAMs are about 350–800 times larger than that of the SRAM. However, the static leak current in the 32 MB cell array in the 6TSRAM is so large that the total current becomes larger than that of the STT-MRAMs. It should be noted that of the four memories, the smallest 4T2MTJ STT-MRAM consumes the smallest write current. The 4T2MTJ STT-MRAM, whose performance is comparable with that of the LOP SRAM, consumes about half of the current of the LSTP SRAM in 90 nm technology.

5.4 SUMMARY

Figure 5.26 summarizes several MTJ-based NV logic and memory LSIs that were developed by the Endoh group at Tohoku University. Some of them are explained in this textbook. The LSIs that are not explained in this textbook are the NV latch [16], the PFET 1T1MTJ STT-MRAM [17], and the 4T2MTJ STT-MRAM [6]. The STT-MTJ

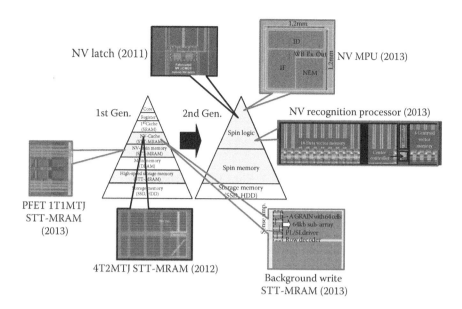

FIGURE 5.26   The NV logic and memory LSIs that were developed by the Endoh group at Tohoku University.

has superior features that have the potential to make the total computer system NV in an attempt to increase the performance of computers by device scaling. However, there remain several challenges that need to be addressed in order to make this technology feasible for future low-power computing systems. All of these challenges will be overcome by combining the progresses made in material, device, circuit, and system design, and we have presented examples of the design of such circuits in this chapter.

REFERENCES

1. T. Endoh, T. Ohsawa, H. Koike, T. Hanyu, and H. Ohno, Restructuring of memory hierarchy in computing system with spintronics-based technologies, *2012 Symposium on VLSI Technology Digest of Technical Papers*, Honolulu, HI, pp. 89–90, June 2012.
2. S. Mutoh, T. Douseki, Y. Matsuya, T. Aoki, S. Shigematsu, and J. Yamada, 1-V power supply high-speed digital circuit technology with multithreshold-voltage CMOS, *IEEE J. Solid-State Circuits*, 30(8), 847–854, August 1995.
3. Intel® Atom™ Processor Z5xx Series, Datasheet, Section 2, June 2010.
4. H. Koike, T. Ohsawa, N. Sakimura, R. Nebashi, Y. Tsuji, A. Morioka, S. Miura et al., A power-gated MPU with 3-microsecond entry-exit delay using MTJ-based nonvolatile flip-flop, *IEEE Asian Solid-State Circuits Conference (A-SSCC)*, Singapore, Singapore, pp. 317–320, November 2013.
5. Y. Ma, T. Shibata, and T. Endoh, An MTJ-based nonvolatile associative memory architecture with intelligent power-saving scheme for high-speed low-power recognition applications, *IEEE International Symposium on Circuits and Systems (ISCAS)*, Beijing, China, pp. 1248–1251, May 2013.

6. T. Ohsawa, H. Koike, S. Miura, H. Honjo, K. Kinoshita, S. Ikeda, T. Hanyu, H. Ohno, and T. Endoh, A 1Mb nonvolatile embedded memory using 4T2MTJ cell with 32b fine-grained power gating scheme, *IEEE J. Solid-State Circuits*, 48(6), 1511–1520, June 2013.

7. Y. Fujisaki, Review of emerging new solid-state non-volatile memories, *Jpn. J. Appl. Phys.*, 52, 040001, 2013.

8. N. D. Rizzo, D. Houssameddine, J. Janesky, R. Whig, F. B. Mancoff, M. L. Schneider, M. DeHerrera et al., A fully functional 64 Mb DDR3 ST-MRAM built on 90 nm CMOS technology, *IEEE Trans. Mag.*, 49(7), 4441–4446, July 2013.

9. S. Yamamoto and S. Sugahara, Nonvolatile static random access memory using magnetic tunnel junctions with current-induced magnetization switching architecture, *Jpn. J. Appl. Phys.*, 48, 043001, 2009.

10. K. Abe, K. Nomura, S. Ikegawa, T. Kishi, H. Yoda, and S. Fujita, Hierarchical nonvolatile memory with perpendicular magnetic tunnel junctions for normally-off computing, *2010 International Conference on Solid State Devices and Materials (SSDM)*, Tokyo, Japan, pp. 1144–1145, September 2010.

11. T. Ohsawa, F. Iga, S. Ikeda, T. Hanyu, H. Ohno, and T. Endoh, High-density and low-power nonvolatile static random access memory using spin-transfer-torque magnetic tunnel junction, *Jpn. J. Appl. Phys.*, 51, 02BD01, 2012.

12. T. Aoki, Y. Ando, D. Watanabe, M. Oogane, and T. Miyazaki, Spin transfer switching in the nanosecond regime for CoFeB/MgO/CoFeB ferromagnetic tunnel junctions, *J. Appl. Phys.*, 103, 103911, 2008.

13. A. Driskill-Smith, D. Apalkov, V. Nikitin, X. Tang, S. Watts, D. Lottis, K. Moon et al., Latest advances and roadmap for in-plane and perpendicular STT-RAM, *2011 Third International Memory Workshop* (*IMW*), Monterey, CA, May 2011.

14. T. Ohsawa, S. Miura, K. Kinoshita, H. Honjo, S. Ikeda, T. Hanyu, H. Ohno, and T. Endoh, A 1.5nsec/2.1nsec random read/write cycle 1Mb STT-RAM using 6T2MTJ cell with background write for nonvolatile e-memories, *2013 Symposium on VLSI Circuits Digest of Technical Papers*, Kyoto, Japan, pp. C110–C111, June 2013.

15. T. Ohsawa, S. Ikeda, T. Hanyu, H. Ohno, and T. Endoh, Power reduction by power gating in differential pair type spin-transfer-torque magnetic random access memories for low-power nonvolatile cache memories, *Jpn. J. Appl. Phys.*, 53, 04ED04, 2014.

16. T. Endoh, S. Togashi, F. Iga, Y. Yoshida, T. Ohsawa, H. Koike, S. Fukami et al., A 600MHz MTJ-based nonvolatile latch making use of incubation time in MTJ switching, *2011 International Electron Devices Meeting* (*IEDM*), Washington, DC, pp. 75–78, December 2011.

17. H. Koike, T. Ohsawa, S. Miura, H. Honjo, S. Ikeda, T. Hanyu, H. Ohno, and T. Endoh, Wide operation margin capability of 1kbit STT-MRAM array chip with 1-PMOS and 1-bottom-pin-MTJ type cell, *2013 International Conference on Solid State Devices and Materials* (*SSDM*), Fukuoka, Japan, pp. 1094–1095, September 2013.

6 NEMS Devices

Yoshishige Tsuchiya and Hiroshi Mizuta

CONTENTS

6.1 Introduction 123
6.2 NEM Switches 124
    6.2.1 Introduction 124
    6.2.2 Pull-In and Pull-Out 125
    6.2.3 Suspended-Gate Field Effect Transistors 126
    6.2.4 NEM Relays 127
    6.2.5 Energy-Reversible NEM Switches 130
    6.2.6 Summary and Future Prospects 131
6.3 NEM Memories 131
    6.3.1 Introduction 131
    6.3.2 From MEMS to NEMS 132
    6.3.3 Self-Buckling Floating Gate NEM Memory 133
    6.3.4 Suspended-Gate NEM Memories 137
    6.3.5 NEMory Cell 140
    6.3.6 Energy-Reversible Nonvolatile NEM Memories 140
    6.3.7 Summary and Future Prospects 142
6.4 NEM Resonator Sensors 144
    6.4.1 Introduction 144
    6.4.2 Mass Sensitivity Simulation 145
    6.4.3 Integrated NEM Resonator Sensors 148
    6.4.4 Summary and Future Prospects 150
6.5 Conclusions 150
References 151

6.1 INTRODUCTION

Technological development of silicon CMOS device fabrication technology has much affected the development of micro-electro-mechanical systems (MEMS) technology. Recent advancement of nanoscale silicon device technology has also been bringing the MEMS technology into nanoscale: nano-electro-mechanical systems (NEMS). The scaling trend of the characteristic length of the electromechanical resonators is plotted in Figure 6.1 together with that of the CMOS physical gate length [1]. The characteristic length of the NEMS will merge with that of the silicon CMOS in the next decade, indicating hybrid NEMS-CMOS devices will be more realistic and

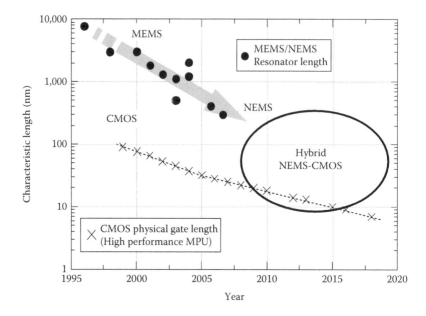

FIGURE 6.1 Miniaturization trend of semiconductor-based electromechanical resonators superposed on the CMOS downscaling trend from International Technology Roadmap for Semiconductors (ITRS). (From International Technology Roadmap for Semiconductors, http://www.itrs.net/, accessed August 28, 2015.)

become a key technological element in the "More-than-Moore" era where Si nanodevices become more powerful and efficient with additional functionality.

Considering the NEMS as a promising candidate of advanced More-than-Moore devices, we identify three main applications, nano-electro-mechanical (NEM) switches, NEM memories, and NEM sensors. Among those NEM devices, high-speed and energy-efficient operations are expected for switching and memory devices and high sensitivity and good temperature stability for sensing devices. The following sections review the past and recent developments of these three NEM devices in order. As our interest is primarily possible integration of NEMS with Si CMOS devices, in this chapter, we have selected the devices whose fabrication process is well compatible with the conventional silicon device fabrication technology. Therefore, our review is mainly on the devices using electrostatic actuation and we have excluded other NEM devices employing electromagnetic, thermal, or piezoelectric transductions.

6.2 NEM SWITCHES

6.2.1 Introduction

While scaling down of silicon CMOS devices and very large scale Integration (VLSI) has attained a big success for decades, the emerging challenge is to reduce the overall power consumption of the integrated circuits. In particular for logic applications, the off-state leakage current in the very short channel largely contributes to the overall power consumption. An ultimate ideal switch features (1) zero impedance and infinite current in the "on" state, (2) infinite impedance and zero current in the "off" state, and (3) the ultrafast transition between two states with "zero" switching time. However, in the current MOSFET technology where the thermionic energy barrier is modulated electronically in switching, the subthreshold swing is theoretically limited to 60 mV/decade due to the high-energy tail of the carrier distribution function. One of the candidates for a steep-slope switch with the subthreshold swing of less than 60 mV/decade, Tunnel FET (TFET) [3], is detailed in Chapter 7. NEM switch is another candidate for the steep-slope switch, and their switching characteristics are much closer to the ideal switch. After a brief description of fundamentals of electromechanical switching, several types of NEM switches are reviewed in the following sections.

6.2.2 PULL-IN AND PULL-OUT

In NEM switches, the phenomena called pull-in and pull-out due to electrostatic actuation are used in switching operation. Figure 6.2a and b shows a parallel plate capacitor model to explain the pull-in and pull-out phenomena in electromechanical devices. A suspended movable electrode plate is separated from a fixed electrode by the initial air gap of g_0. When the voltage V is applied between the two electrodes,

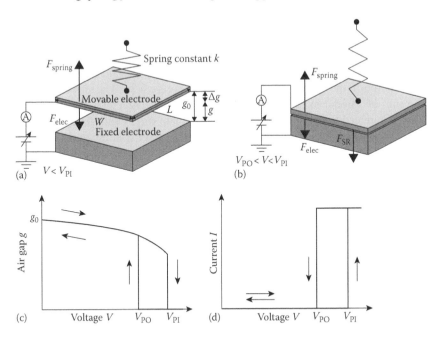

FIGURE 6.2 A schematic diagram of the air-gap parallel plate capacitor model at the voltage of the movable gate $V < V_{PI}$ before pull-in (a) and at $V_{PO} < V < V_{PI}$ after pull-in (b). The air gap distance g (c) and the current flowing between the two electrodes (d) are schematically drawn as a function of V in pull-in and pull-out. The hysteresis in (c) and (d) are due to short-range forces acting after pull-in.

the suspended electrode is displaced to a distance of Δg. In this state, the electrostatic force between the plates, F_{elec} is

$$F_{elec} = \frac{1}{2}\frac{\varepsilon_0 WLV^2}{g},$$ (6.1)

where
   W is the width
   L is the length of the suspended plate
   ε_0 is the vacuum dielectric constant
   $g = g_0 - \Delta g$ is the actual air gap under the voltage V

Assuming the spring constant of the suspended plate to be k, the mechanical restoring force F_{spring} acting on the suspended plate is

$$F_{spring} = k(\Delta g) = k(g_0 - g).$$ (6.2)

The critical displacement beyond which $F_{elec} > F_{spring}$ is $\Delta g = g_0/3$. Therefore, the pull-in voltage V_{PI} is defined as

$$V_{PI} = \sqrt{\frac{8kg_0^3}{27\varepsilon_0 WL}}.$$ (6.3)

Once the suspended plate is pulled in, the additional short range force F_{SR} should be considered in the force balance as shown in Figure 6.2b. The short range force is the sum of van der Waals force and Casimir force. Therefore, even though the voltage is reduced to less than V_{PI}, the movable plate stays on until the mechanical restoring force becomes larger than the sum of F_{elec} and F_{SR} at the pull-out voltage, V_{PO}. Therefore, V_{PO} is lower than V_{PI}, resulting in a hysteric behavior for the air gap g as a function of the voltage V shown in Figure 6.2c. The difference between V_{PI} and V_{PO} indicates how strong the short range force between the plates is in this system. If we monitor the current flowing between the two parallel plates, very abrupt change of the current is expected, as shown in Figure 6.2d, according to the pull-in and pull-out. In NEM switches, this abrupt switching is used to achieve a very low subthreshold swing close to zero.

6.2.3 SUSPENDED-GATE FIELD EFFECT TRANSISTORS

A movable object is applied for the gate of a field-effect-transistor (FET), which is called the suspended-gate (SG-) FET reported by Abele et al. [4]. A schematic of the cross section and scanning electron micrograph of the device are shown in Figure 6.3a and b. The current of the FET channel is modulated due to the change of the position of the biased movable gate. While the idea of the suspended-gate field effect transistor (SG-FET) was already mooted in 1965 by Nathanson and Wickstrom [5], Abelé et al. [4] have fabricated the first time by using MEMS/NEMS fabrication technology.

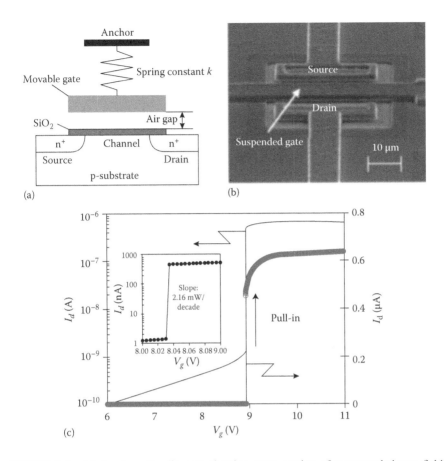

FIGURE 6.3 (a) A schematic diagram showing cross-section of a suspended-gate field effect transistor (SGFET). (Redrawn from Abelé, N. et al., Suspended-gate MOSFET: Bringing new MEMS functionality into solid-state MOS transistor, *IEDM 2005 Technical Digest* Washington, DC, 2005, p. 479.) (b) Scanning electron micrograph of the SGFET and (c) I_d–V_g characteristics of the SGFET show the subthreshold slope of ~2 mV/decade at pull-in. (Reprinted with permission from Abelé, N., Fritschi, R., Boucart, K., Casset, F., Ancey, P., and Ionescu, A.M., Suspended-gate MOSFET: Bringing new MEMS functionality into solid-state MOS transistor, *IEDM 2005 Technical Digest*, Washington, DC, p. 479, © 2005 IEEE.)

Drain current (I_d)–gate voltage (V_g) characteristics of the SG-FET are shown in Figure 6.3c. An abrupt change of I_d is clearly observed at V_g ~ 9 V with the very steep slope with the subthreshold swing, SS of ~2 mV/decade.

6.2.4   NEM RELAYS

Electromechanical relay devices simply consist of an actuation mechanism and a mechanical contact. While two terminal devices are possible as shown in Figure 6.2a and b, three or more terminal relays incorporating "gate" electrodes are practically more common. A pioneering work in a micromechanical switch fabricated by bulk micromachining technology was reported in 1978 [6], but it was only in late 1980s that improved MEMS switches attracted much attention, thanks to the rapid development of the surface micromachining technology. While the MEMS switches are considered to be used in RF MEMS to modulate radio frequency (RF) or microwave transmission, an idea to develop NEM relays for logic application has been mooted after the progress of nanofabrication technology as well as the demand on low energy consumption in logic circuits.

Akarvardar et al. proposed logic gates using NEM relays and investigated their operation theoretically [7]. Figure 6.4a is a schematic of the three-terminal NEM relay. The cantilever is actuated by the gate electrode underneath and the current flows when the bump at the top of the cantilever comes in contact with the drain electrode. A complementary operation is achievable if the two NEM relays are connected as shown in the circuit diagram in Figure 6.4b. One of two cantilevers connected to the ground is called n-relay and another connected to the voltage source V_{DD} is p-relay. DC transfer characteristics of the complementary NEM inverter are calculated as shown in Figure 6.4c. Their calculation suggests that NEM relays are not

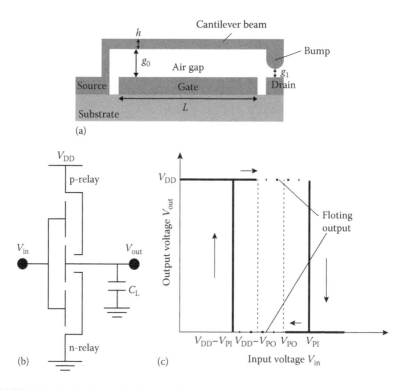

FIGURE 6.4 (a) A schematic diagram of a three-terminal NEM relay. (b) A circuit diagram of a pair of the NEM relays for complementary operation. (c) DC transfer characteristics of the complementary NEM relay. The output is floated in between the pull-out of the p-relay and the pull-in of the n-relay, and vice versa. (Redrawn from Akarvardar, K. et al., Design considerations for complementary nanoelectromechanical logic gates, *IEDM 2007 Technical Digest*, Washington, DC, 2007, p. 299.)

faster than MOSFETs but show good area scalability and excellent dynamic power features with the switching energy of sub-0.1 fJ at $V_{DD} = 1.5$ V, which are competitive compared with MOSFETs. In addition to abrupt DC transfer characteristics, hysteresis and floating output points are seen due to the difference of the pull-in and pull-out voltages. This peculiar behavior suggests the switch could be used as a memory device as well. This point will be revisited in the later section on NEM memories.

So far there are a number of experimental attempts to realize electrostatically actuated MEM/NEM switches; these are summarized in a review by Pott et al. [8]. Jang et al. [9] made TiN-based cantilever NEM switches and observed abrupt switching with SS of less than 3 mV/decade. Furthermore, 40% reduction of operation voltage was demonstrated by introducing insulating liquid medium in packaging [10]. On the other hand, Kam et al. [11] prototyped a three-terminal MEMS relay where poly-SiGe was used for structural material of the doubly-clamped beam, and TiO_2-coated W is used for contacting electrodes. The contact resistance of ~10 kΩ at 20°C–200°C, and endurance of more than 10^9 on/off cycles were demonstrated at $V_{DD} = 5$ V. Czaplewski et al. [12] also developed an NEM switch in a CMOS compatible manner and reported the switching time of 400 ns at the operating voltage of 5 V. Jeon et al. [13] used their technology to develop a dual-ended "see-saw" relay device to realize complementary switching operation that is required for digital logic operation. Individual devices achieved abrupt switching with the swing of less than 0.1 mV/decade and those characteristics are very symmetric about $V_{DD}/2$, offering large operating voltage margin and small crowbar current in the inverter characteristics.

One of the most important aspects we have to consider in logic application is the device scalability. King Liu et al. [14] have investigated the scaling rule of the MEM/NEM relays and summarized the relevant device parameters as shown in Table 6.1, where the constant-field scaling methodology is used with the scaling factor of $\kappa > 1$. A remarkable feature is that the power consumption is reduced by $1/\kappa^3$ in scaling. To ensure the operation of NEM relays, F_{spring} must be larger than F_{SR}. This condition determines

TABLE 6.1
Summary of Scaling Factors for NEM Relay Switches in Constant Field Scaling

NEMS Relay Parameter	Constant Field Scaling
Spring constant	$1/\kappa$
Actuation area	$1/\kappa^2$
As-fabricated gap	$1/\kappa$
Mass	$1/\kappa^3$
Switching energy	$1/\kappa^3$
Pull-in voltage	$1/\kappa$
Pull-in delay	$1/\kappa$
Device density	κ^2
Power density	1

Source: Data taken from King Liu, T.-J. et al., Prospects for MEM logic switch technology, *IEDM 2010 Technical Digest*, San Francisco, CA, 2010, p. 424.

the lower limit of energy required for switching. The surface adhesion energy of an ultrasmall contact area (less than 50 nm × 50 nm) is estimated by considering atomic bonding at the uneven contact region. Therefore, the minimum energy for a relay is expected to be down to 1 aJ, which is more than 10 times lower than that of ~100 aJ for CMOS [14]. On the other hand, we must consider switching speed in order to make a fair comparison between NEM and CMOS devices. With respect to the results in Reference 11 and their analytical theory, scaled NEM relays will offer better energy efficiency for applications that require the performance of up to 100 MHz.

6.2.5 ENERGY-REVERSIBLE NEM SWITCHES

In line with the trend to develop low power logic devices, Akarvardar et al. [15] proposed a novel idea of energy-reversible (ER) complementary NEMS logic gates. A schematic of the device concept is shown in Figure 6.5. A movable elastic beam is placed between two side electrodes: one is biased at V_{DD} and another is grounded. The top of the cantilever is electrically isolated from the rest of the elastic beam and can work to connect either of side electrodes with the output electrode. In the on state when the input electrode is grounded, the elastic beam is attracted to the left-hand side electrode so that the top of the movable will connect the output electrodes at the left-hand side, resulting in the output of V_{DD}. When the V_{DD} is applied to the input, the repulsive force between the beam

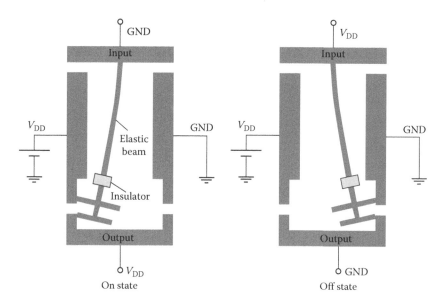

FIGURE 6.5 Schematics of an energy-reversible NEM switch. The left-hand side one is in the on state where the top of the cantilever bridges the gap between the left-hand side and output electrodes. The top of the cantilever is released when the repulsive force between the beam and left-hand side electrode overcomes the adhesion force due to the applied voltage to the input, and then the beam travels to bridge the right-hand side electrode with the output, shown in the right-hand side figure. (Redrawn from Akarvardar, K. et al., Energy-reversible complementary NEM logic gates, *DRC 2008 Technical Digest*, Santa Barbara, CA, 2008, pp. 69–70.)

and the left-hand side gate and attractive one between the beam and right-hand side gate make the beam move to the right-hand side, and as a result, the output is grounded. An advantage of the ER NEM switch is that the beam can travel from one side to the other by using elastic energy stored in the beam, which is basically reversible except for energy dissipation due to damping. While the conventional NEM switches require relatively large electrostatic force for beam actuation, ER NEM switches do not; so the V_{DD} can be significantly reduced. Analysis in Reference 15 suggests that the ER NEM inverter can achieve the delay time of 10.8 ns with $V_{DD} = 0.59$ V instead of $V_{DD} = 2.24$ V for a conventional NEM relay.

6.2.6 SUMMARY AND FUTURE PROSPECTS

So far, various types of NEM switches have been proposed and also prototyped and have shown their advantages of low standby power operation in the individual device level. Another big advantage of the NEM switches is that their fabrication processes are well compatible with silicon VLSI fabrication technology. In fact NEM switches are listed in International Technology Roadmap for Semiconductors as an emerging alternative device for logic applications [2]. The most recent trend is to model NEM switches by using finite element analysis (FEA) and then to perform circuit-level simulation where the models of NEM switches can be considered as an element of the circuit. Spencer et al. [16] demonstrated a model of integrated MEM relay circuits by modeling the devices prototyped in References 11 and 13, and suggested that scaled NEM relays have potential to reduce the energy consumption of various VLSI blocks.

Another interesting approach is to consider an NEM switch as an alternative of sleep transistors that are used for IC power management. Enachescu et al. [17] have employed a model of the SG-FET developed by Tsamados et al. [18] and designed a 3D stacked integrated circuit. According to their circuit simulation, the static power consumption of the SGFET-integrated IC can be reduced more than two orders of magnitude less than that of the equivalent system with conventional MOSFETs. These results suggest that the advantage of the NEM switches on ultralow standby power is maintained even in the circuit level. In further scaling of NEM switches, effects of van der Waals and Casimir forces become more dominant, and as is pointed out in Reference 14, the magnitude of those short range forces become equivalent with the atomic binding energy. Atomic-scale consideration is to be more important for future extremely scaled NEM switches.

6.3 NEM MEMORIES

6.3.1 INTRODUCTION

Our common information and communication technology (ICT) tools normally require two types of solid-state semiconductor memory devices: random access memories (RAMs) that deal with information at high speed and nonvolatile memories (NVMs) that store information with longer retention time. Currently, different technologies are employed to meet the requirements: dynamic and static random access memories (DRAM and SRAM) for RAMs, and flash memory for NVMs. This is due to the difficulty of realizing high-speed programming and erasing operations and sufficient retention time at the same time. For example, in floating gate flash memories, electrons tunnel through the thin oxide between the floating gate and the FET channel underneath in programming and erasing operations. The repetition of the electron tunneling deteriorates the quality of the oxide, then eventually the floating gate becomes unable to store sufficient amount of charges due to serious electron leakage through the oxide. Therefore, the thickness of the tunneling oxide cannot be scaled less than about 7 nm, resulting in the lowering of programming and erasing speed to less than ~1 µs.

Most of recent research on emerging memory devices such as ferroelectric RAM (FeRAM), magnetic RAM (MRAM), phase change RAM (PCRAM), and resistive RAM (ReRAM) aim to develop a high-speed and nonvolatile "universal" memory, which could meet both the requirements. NEM memories, which incorporate NEM in their operation, have also been developed within the context of realizing high-speed operation and nonvolatility at the same time. The biggest advantage of employing NEMS is again process compatibility with conventional silicon device technology. While any of the emerging memories rely on a specific property of novel materials to secure their nonvolatile nature, NEM memories are able to show clear bistability without introducing any exotic materials.

6.3.2 FROM MEMS TO NEMS

An idea to use mechanical bistability of a mechanically movable object for nonvolatile memory was proposed by Halg [19]. The author fabricated self-buckling bridges integrated with lateral electrodes. For the beam with the dimensions of 20 µm in length (L), 3 µm in width (W), and 30 nm in thickness (t), mechanical bistability was demonstrated with the applied voltage of 30 V. The resonance frequency of the beam with $L = 30$ µm $\times$ $W = 3$ µm $\times$ $t = 30$ nm is estimated ~7 MHz.

The characteristic speed of the electromechanical object is roughly estimated as the resonant frequency:

$$f_{res} = 2\pi\sqrt{\frac{k}{m}}, \tag{6.4}$$

where
  k is the stiffness constant
  m is the mass of the mechanical object

When the size of the resonator becomes smaller, the mass decreases so that the characteristic frequency becomes higher. Drastic improvement of MEMS fabrication technology in recent decades has brought about smaller beams with higher resonant frequencies. The resonance of a Si MEMS beam reported in 1996 [20] was 77 MHz for the beam with $L = 7$ µm $\times$ $W = 100$ nm $\times$ $t = 50$ nm. After 7 years, the resonance of 1.1 GHz was reported for a SiC beam with $L = 1.1$ µm $\times$ $W = 50$ nm $\times$ $t = 50$ nm [21], indicating even the operational speed of mechanical objects becomes comparable with that of electronic devices.

6.3.3 SELF-BUCKLING FLOATING GATE NEM MEMORY

Tsuchiya et al. [22] proposed a novel memory device featuring a self-buckling floating gate. A schematic of the device structure in the off and on states is shown in Figure 6.6. The self-buckling floating gate is placed in the cavity between the control gate and FET channel underneath. The operation principle is as follows. First, by applying a relatively larger voltage between the control gate and the substrate, charges are injected to the silicon nanodots embedded in the self-buckling beam either through the silicon supports or by pull-in of the control gate with the upward-bent floating gate shown in Figure 6.6a. Here we assume the beam is positively charged, and the voltage level in programming

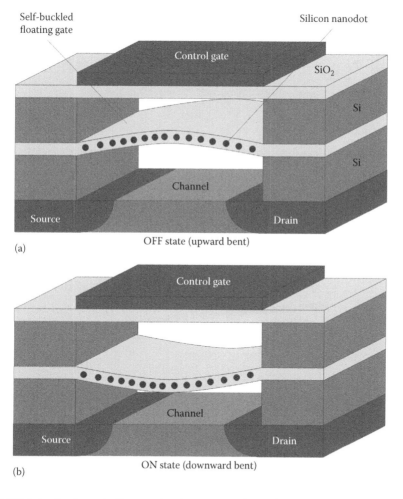

FIGURE 6.6 A schematic diagram showing cross-sections of the on (a) and off (b) states of the self-buckling floating gate NEM memory proposed by Tsuchiya et al. [22]. The charged buckled beam placed between the control gate and FET channel is flip-flopped depending on the control gate voltage, resulting in the hysteresis in the I_d–V_g characteristics of the FET underneath. (From Tsuchiya, Y. et al., *J. Appl. Phys.*, 100, 094306, 2006.)

and erasing operation is lower than that in charging. When positive voltage is applied to the control gate, repulsive force that acts between the self-buckling beam and control gate flips the upward-bent beam into downward bent (Figure 6.6b). When the control gate is then negatively biased, attractive force lets the beam back to the upward-bent state. Whether the beam is upward bent or downward bent is readable by the FET whose threshold voltage is shifted due to the position of the charged beam.

The device is modeled and analyzed by using multiphysics FEM [23–25]. Figure 6.7a shows the model structure used in the simulation. The dimensions of the self-buckling beam at the flat position are 1 μm in length (L_b), 500 nm in width (W_b), and 50 nm in thickness (t_b), while the thickness of the upper air gap t_{uc} is 30 nm and that of lower one t_{lc} is 50 nm. SiO$_2$ is assumed as a material for the buckling beam and chromium for the

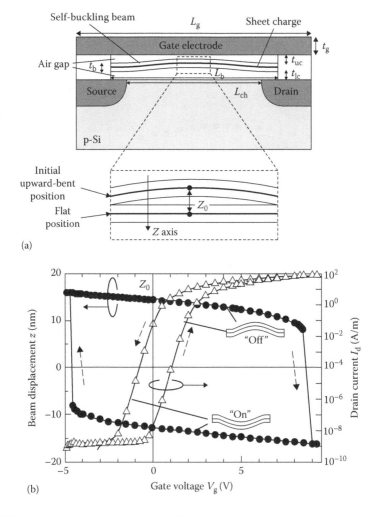

FIGURE 6.7 (a) A schematic diagram of the model used in simulation of the electrical characteristics of the self-buckling floating gate NEM memory. (b) The beam displacement z and drain current I_d as a function of the gate voltage V_g.

gate electrode. The fixed charges in the movable floating gate are modeled by the sheet of charges with the area density of 8.5×10^{-8} C/cm². In order to model the bistable self-buckling floating gate, inner stress is induced in the flat beam, corresponding to create the energy potential barrier between two energy minima. Loading along z direction releases the inner stress, resulting in an upward-bent beam. Figure 6.7a illustrates the definition of the initial beam displacement z_0 which is the distance between the centers of self-buckling beam and the flat beam as shown in the blow-up in Figure 6.7a. The beam displacement as a function of the gate voltage is plotted in Figure 6.7b. Initially at $V_g = 0$, the beam is in the upward-bent "off" state with $z_0 \sim 15$ nm. With increasing V_g, the displacement decreases and at $V_g \sim 8.6$ V, sudden change of the displacement takes place corresponding to the flip of the buckling beam. The beam stays in the downward-bent "on" state even though V_g is reduced back to zero. Another flip is seen at $V_g \sim -4.7$ V so that the beam is back to the upward-bent state. The corresponding drain current I_d of the FET channel is also plotted in Figure 6.7b. Here the length of FET channel L_{ch} is 900 nm, the source–drain voltage $V_{SD} = 1$ V, and the doping concentrations of the channel and electrodes are 10^{16} and 2×10^{17} cm^{-3}, respectively. Clear V_{th} shift is seen to be able to distinguish the "on" and "off" states at $V_g = 0$.

To estimate the programming and erasing speed, transient characteristics are also calculated by solving the equation of motion considering inertia, damping, and mechanical energy terms. Figure 6.8 shows the time evolution of the beam displacement. In this calculation, we start from the downward-bent beam and the damping constant β of 1×10^{10} s is employed. While staying in the same state at $V_g = -4.5$ V, the beam is flipped after applying $V_g = -4.7$ V into the upward-bent state. It is reasonable to see the switching speed becomes faster when increasing the magnitude of the gate voltage. The switching speed within this voltage range is 40–55 ns, which is significantly shorter than that of conventional floating gate flash memories [24].

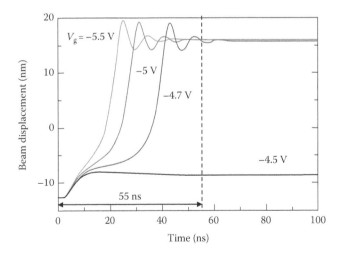

FIGURE 6.8 Transient behavior of the beam displacement in applying gate voltages. The beam can flip by applying voltage negatively larger than −4.7 V. The overshoot observed corresponds to bouncing of the beam. The estimated switching time is ~55 ns.

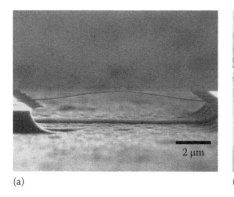
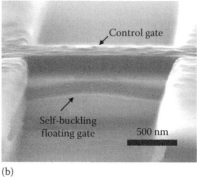

(a)   (b)

FIGURE 6.9 Scanning electron micrographs of (a) self-buckled SiO_2 beam fabricated by releasing thermal strain induced in thermal oxidation and (b) silicon-nanodot-embedded self-buckling beam placed in the cavity under the control gate.

A typical fabrication process of a self-buckling SiO_2 beam is as follows [22]. First a thermal oxide is grown on a Si substrate followed by patterning of a narrow beam part and anchors on the both sides of the beam by lithography. The pattern is transferred onto the thermal oxide by anisotropic dry etching. Subsequent isotropic Si dry etching undercuts the silicon underneath to suspend the beam. During this isotropic etching, the stress induced in thermally grown SiO_2 is released so that the beam buckles naturally. Figure 6.9a shows an example of the self-buckling SiO_2 beam of length of 10 μm. By integrating the process, we have succeeded in fabricating a memory cell of the self-buckling floating gate NEMS memory as shown in Figure 6.9b [26]. In fabrication, we grow or deposit all the required layers first, followed by patterning and etching. As media to store charges in a self-buckling beam, silicon nanodots are employed and deposited on top of the thermal oxide by using the pulsed-gas plasma process developed in Reference 27. The following CVD SiO_2 deposition embeds the Si nanodots in the SiO_2 beam. A sacrificial poly-Si layer deposition is followed by a chromium layer that is for the control gate. After lithographic patterning, the two-step dry etching process is carried out twice to release the control gate first and then the floating gate. As the lower half of the SiO_2 floating gate is grown thermally, the beam is released with buckling after the second two-step etching process as is clearly seen in Figure 6.9b.

Scalability of memory devices is of paramount importance toward high-density integration. According to our simulation [23,25] the switching voltage V_S is expressed as

$$V_S \propto L^{-4}TZ_0^3 d_{gap}\sigma_{fg}^{-1} \tag{6.5}$$

,

where
    L is the length
    T is the thickness
    Z_0 is the initial displacement of the beam
    d_{gap} is the distance between the gate and the substrate
    σ_{fg} is the stored charge density of the floating gate.

The equation suggests that the reduction of L is compensated if Z_0 and T are reduced proportionally in miniaturization to keep V_s unchanged. However, the reduction of Z_0 leads to the smaller threshold voltage shift that degrades the on/off ratio of the drain current in reading operation. As a result, the further scaling below $L \sim 100$ nm is limited in this memory device due to the trade-off between the reduction of the switching voltage and the degradation of the drain current on-off ratio.

6.3.4 SUSPENDED-GATE NEM MEMORIES

While the floating gate is movable in the self-buckling memory, a suspended-gate MOSFET memory proposed by Abelé et al. [28] features a movable control gate. The device structure is basically the same as that of the SG-MOSFET shown in Figure 6.3a and b, but the operational mode is different. In memory operation, the movable suspended gate is used to inject or eject charge carriers into the gate dielectric. I_d-V_g characteristics of this SG MOSFET memory are shown in Figure 6.10. When the positive gate voltage V_g is applied, an abrupt increase of the drain current I_d is observed due to pull-in at the voltage V_{PI}. When V_g is decreased, the suspended gate is pulled out but I_d is kept at a higher current level. When the gate is negatively biased, an abrupt drop of the I_d is observed, corresponding to another pull-in taking place. An idea is to use the large hysteresis causing a large threshold voltage shift for a nonvolatile memory. Mechanisms of device operation are explained by considering interplays between defects in the oxide and at the interface and carriers injected or

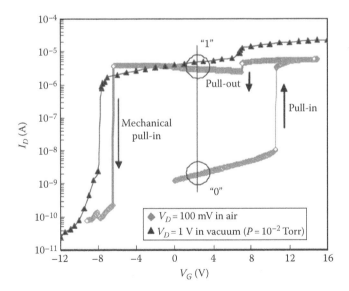

FIGURE 6.10 I_d-V_g characteristics of suspended-gate MOSFET memory reported by Abelé et al. (Reprinted with permission from Abelé, N., Villaret, A., Gangadharaiah, A., Gabioud, C., Ancey, P., and Ionescu, A.M., 1T MEMS memory based on suspended gate MOSFET, *IEDM 2006 Technical Digest*, San Francisco, CA, p. 479, © 2006 IEEE.)

ejected at the pull-in of the suspended gate. A reasonable retention time and good endurance up to 10^5 cycles are demonstrated.

Garcia-Ramirez et al. proposed a novel suspended-gate NEMS memory device featuring the silicon nanodots as carrier storage in the insulating layer and named the device as suspended-gate silicon nanodot memory (SGSNM) [29]. Figure 6.11a shows a schematic diagram of the device structure in data retention and in programming and erasing (P/E). While both P/E and reading are through the tunneling oxide under the floating gate in conventional floating gate flash memories, P/E are carried out by the suspended gate from the top of the floating gate in SGSNM but reading is by the FET from the bottom, allowing us to optimize the structures for those operations separately. Pull-in and pull-out characteristics of the suspended gate are simulated by using FE analysis. Figure 6.11b shows the displacement of the suspended gate as a function of V_g for an Al beam with $L = 1$ μm $\times W = 300$ nm $\times t = 30$ nm and

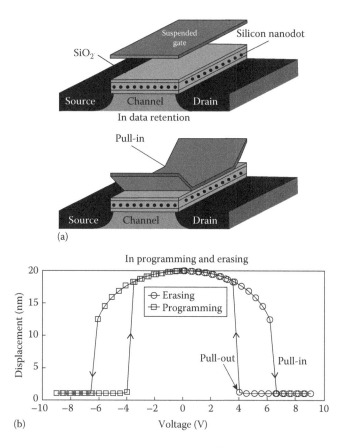

FIGURE 6.11 (a) Schematics of the suspended-gate silicon nanodot memory proposed by Garcia-Ramirez et al. [29]. The upper diagram represents the state in data retention and the lower one in programming and erasing. (b) Displacement of the suspended gate is plotted as a function of the gate voltage. The negative voltage is applied in programming while the positive in erasing.

with the air gap of 20 nm. Considering the short range forces in simulation, the pull-in and pull-out characteristics are seen in both positive and negative voltage ranges. In programming operation, negative V_g is applied so that the electrons are injected after the pull-in and stay in the silicon nanodots even after pull-out and at $V_g = 0$. In erasing operation, positive V_g is applied so that the electrons are ejected after pull-in, then the system returns to the initial state after pull-out.

In order to model a full memory cell of the SGSNM, we develop a hybrid simulation model where movable gates and electron tunneling through the oxide can be taken into account [30]. Figure 6.12a is an equivalent circuit model of the SGSNM where the suspended gate is implemented by a parallel connection of a

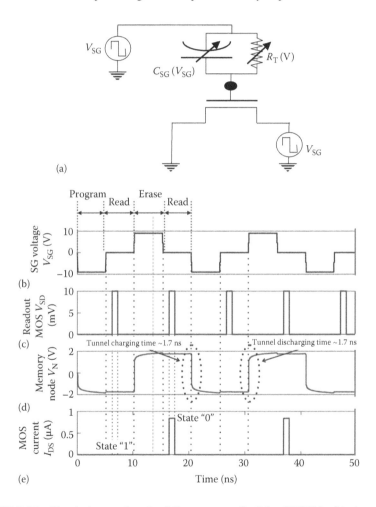

(a)

(b)

(c)

(d)

(e)

FIGURE 6.12 Simulation results of a full memory cell of the SGSNM with the equivalent circuit model in (a). The following time charts show (b) the voltage applied to the suspended-gate V_{SG}, (c) the source–drain voltage applied to the readout MOSFET V_{SD}, (d) the voltage at the memory node V_N, and (e) the source–drain current of the readout MOSFET I_{DS}, respectively.

variable capacitance C_{SG} that represents capacitance change due to gate actuation, and a variable resistance R_T that represents tunneling current when the suspended gate is pulled in. C_{SG} is modeled by an FEM simulation as shown in Figure 6.11b, while R_T is calculated separately by solving 1D Schrödinger equation and Poisson's equation self-consistently. The modeling results are implemented in a hardware language to be integrated with circuit simulation. The rest of Figure 6.12 is dedicated to chart diagram of the parameters of the SGSNM model [31]. Voltage pulses shown in Figure 6.12b are applied to the suspended-gate V_{SG}. In between P/E operations, the V_{SG} is kept zero, while the drain voltage V_{SD} is pulsed to read the current of the FET (Figure 6.12c). The memory node voltage V_N changes abruptly between the low and high voltage states just after P/E pulses (Figure 6.12d). These two states "1" and "0" are successfully distinguished by the readout FET (Figure 6.12e). Looking at the transient behavior of V_N in switching, the time needed for charging or discharging via tunneling is estimated to be 1.7 ns. Considering the mechanical pull-in time of ~0.8 ns estimated separately by FEM transient analysis, P/E time of the SGSNM is estimated to be ~2.5 ns, which indicates the P/E speed is 100 times higher than that of conventional floating gate flash memories.

6.3.5 NEMory Cell

An interesting work is to design an NEM nonvolatile memory (NEMory) cell [32] that does not include FETs but is able to be fabricated with silicon-compatible processes. Figure 6.13a shows a schematic of a NEMory cell that consists of a metal NEM cantilever structure in cavity and a dielectric oxide/nitride/oxide (ONO) stack. The mechanically movable cantilever made of metal is located in a cavity formed between the metal layer and ONO stack. The ONO stack is used to store a fixed number of charges to adjust the position of the hysteresis curve in order to achieve bistability at $V_{g1} = 0$ as shown in Figure 6.13b. Operational principle of the NEMory is as follows. In initialization process, the cantilever is pulled down by applying large negative bias to the metal electrode below the ONO stack. In programming, negative voltage is applied to the cantilever to make V_{g1} less than the V_{PO}. In reading operation, a positive voltage is applied to the top electrode, so the cantilever pulled up to the top electrode only if it was not pulled down to the ONO stack. According to the simulation, the cell with a 75 nm long cantilever is expected to show the programming and erasing speed of less than 1 ns under the biasing condition of sub 1 V [33].

6.3.6 Energy-Reversible Nonvolatile NEM Memories

Another interesting concept is a novel nonvolatile NEM memory based on energy-reversible NEM switches proposed by Akarvardar and Philip Wong [34]. Figure 6.14a shows a schematic of top view of the device structure, which consists of a movable undoped cantilever and p+-doped and n+-doped side gates. The operation procedure is as follows. Initially $V_{XY} = 0$, corresponding $V_n = V_p = E_G/2q$, and the cantilever close to n+ side are assumed. The initial values of V_n and V_p are determined by the work function difference between heavily doped and undoped silicon materials. With increasing V_{XY}, V_p increases while V_n decreases. When V_{XY} exceeds the positive

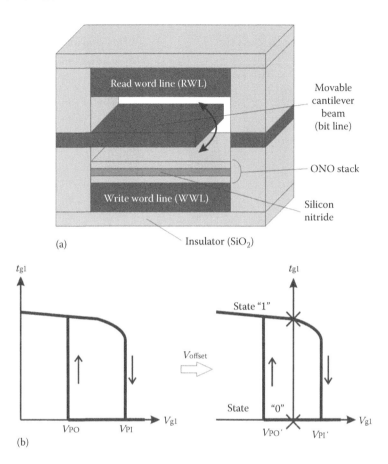

FIGURE 6.13 (a) A schematic diagram of a NEMory cell proposed by Choi et al. [32]. (b) The gap thickness between the movable cantilever beam and the top of the oxide–nitride–oxide (ONO) stack is plotted as a function of the voltage in between. The charge stored in the ONO stack offsets the hysteresis curves to realize the bistability at $V_{g1} = 0$. (Redrawn from Choi, W.Y. et al., Compact nanoelectro-mechanical nonvolatile memory (NEMory) for 3D integration, *IEDM 2007 Technical Digest*, Washington, DC, 2007, pp. 603–606.)

switching voltage $V^+ = E_G/2q - V_{PO}$, corresponding to $V_n < V_{PO}$, the cantilever is released from the n^+ side and moves to the p^+ side. When $V_{XY} > V^+$, the condition of $V_p > V_{PO}$ is already satisfied so that the cantilever can arrive to the p^+ side. Even after reducing V_{XY} to zero, the cantilever stays in the p^+ side, "1" state. Similar but biasing opposite direction down to $V_{XY} < V^- = V_{PO} - E_G/2q$ is required to switch the beam back to "0" state. Although reading operation is destructive, the scheme is well designed with respect to low power operation and is suitable for crossbar array architecture. Their calculation suggests that the cell is estimated to operate at 0.18 V, dissipate 10 aJ switching energy, and achieve less than 10 ns switching delay.

Motivated by this work, Boodhoo et al. have been developing a silicon energy-reversible switch by using SOI-compatible technology [35]. A schematic and an SEM image of the prototyped devices are shown in Figure 6.14b. A round shape

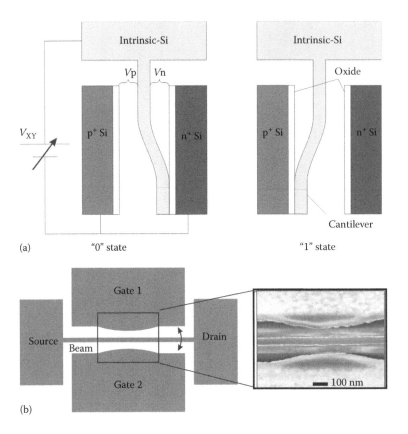

FIGURE 6.14   (a) Schematics of an energy-reversible nonvolatile NEM memory. Two configurations of the cantilever depicted corresponds to "0" and "1" states, respectively. (Redrawn from Akarvardar, K. and Philip Wong, H.-S., *IEEE Electron Dev. Lett.*, 30(6), 626, 2009.) (b) A schematic of a double-clamped beam energy-reversible nonvolatile memory reported by Boodhoo et al. [35]. An SEM image of the prototype of the suspended beam with the round shape gates is shown as a blow-up.

designed for the side gates is "stiction controller," which could control the magnitude of the short range force acting between the side gate and the beam. Up to now, the pull-in voltage observed in prototypes is found to be consistent with that estimated in simulation [35].

6.3.7   SUMMARY AND FUTURE PROSPECTS

Various attempts of using NEMS for nonvolatile memory applications are reviewed. NEM memories are advantageous to achieve high-speed P/E operations toward sub-nano seconds and serious nonvolatility at the same time. The operational speed becomes higher when the characteristic dimensions of the mechanically movable structures become smaller. In addition, recent works suggest that low power operation is achievable with NEMS. On the other hand, the biggest concern is its scalability. Under current technologies, NEM memories are not competitive with the flash memories with respect to the measure of cell per unit area. Therefore, specific applications that require low power, serious nonvolatile nature, or operation in a harsh environment would be preferable.

Recently, crossbar architecture has attracted much attention in overcoming the physical spatial limitation of current FET-based cell architectures such as 1T or 1T1C. A NEMS version of crossbar architecture was first proposed by Rueckes et al. [36] in line with their works on carbon nanotube NEM memories. Here we revisit the crossbar architecture assuming to fabricate with silicon-based NEM technology. A schematic of a silicon nanowire crossbar NEM memory is illustrated in Figure 6.15a. A set of laterally aligned nanowires is placed on the substrate. Then another set of laterally aligned nanowire assembly is crossly placed on top and suspended. Figure 6.15b shows schematics of "on" and "off" states at the cross point where bistability arises from the interplay of the elastic energy of the suspended nanowire and van der Waals energy between the nanowires. When the voltage is applied, the suspended nanowire is pulled in so that the distance between the nanowires at the cross point is very close. At this "on" state, the van der Waals force and elastic force are balanced. By applying the voltage with the same sign and magnitude to the both nanowires at the same time, the suspended one is pulled out due to the repulsive force and return to the initial "off" state. The memory state at the cross point is able to be read by measuring the junction resistance. There are a number of technological issues still to be overcome, but we believe this architecture is worth considering the aim of further NEM memory cell integration.

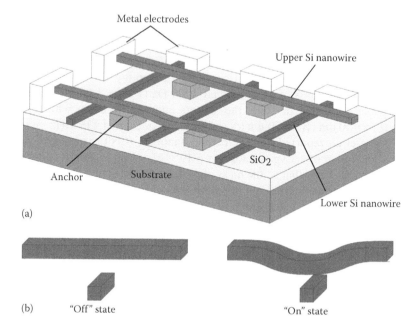

FIGURE 6.15   (a) A schematic diagram of a cross-point NEM memory by using silicon nanowire arrays. (b) Two configurations correspond to on and off states, respectively.

Another thing we want to point out here is a trend on nonvolatile logic applications. As was already seen even in this chapter, a number of NEM devices have shown dual natures: as a switch and as a memory. An idea of nonvolatile logic devices is based on the requirement on reconfigurable circuits and also related to neuron computing that requires variable threshold voltage. We believe NEM memories would have an opportunity for this type of application.

6.4   NEM RESONATOR SENSORS

6.4.1   INTRODUCTION

The other important application of NEMS is sensing. As was already introduced in the earlier sections, NEM structures have very small active masses so that their resonant frequency is very high up to the order of several GHz. In addition, the quality factor Q of the NEM resonator is as high as 10^3–10^5, which is considerably higher than that of electrical circuit resonators.

The use of electromechanical resonance for mass sensing has already been quite common. Due to the attachment or adsorption of molecules onto the resonator, the shift of the resonant frequencies is observed, and the frequency shift is directly related to the inertial mass of the molecules. Historically, the acoustic vibration modes of crystals, thin film resonators, and microscale cantilevers have been employed for the resonator devices. In particular, quartz crystal microbalance (QCM), which uses acoustic vibration of quartz with Q of up to 10^6, is widely used for mass sensing, and biomolecular sensors based on QCM technology are already in production.

An interesting research trend in this field is to investigate an ultimate sensitivity of the NEMS-based resonators. After the development of doubly-clamped SiC beam resonators [21], Yang et al. applied the resonator for mass sensing and demonstrated the best mass resolution of ~7 zg (zepto-gram = 10^{-21} g), which is equivalent to the mass of an individual 4000 Da (1 Da = 1.66×10^{-24} g) [37]. In this paper published in 2006, the authors employed a beam with a size of 2.3 µm (L) × 150 nm (W) × 70 nm (t) and with a frequency of 133 MHz. Six years later, Chaste et al. reported that the carbon nanotube (CNT) resonators showed the mass resolution of 1.7 yg (yoctogram = 10^{-24} g), which is almost equal to the atomic mass unit 1 Da, that is, the mass of proton [38]. Note that the CNT resonator has the dimensions of 150 nm in length and 1.7 nm in diameter. All of these results clearly suggest that NEM resonators are very promising toward single molecule or single atom level mass sensing.

In parallel with the research aiming at the ultimate sensitivity, there has been another trend to develop the technology to integrate NEMS with Si CMOS-based technology. The pioneering paper by Nathanson and Wickstrom [5] already reported the integration of a resonant suspended beam with FET underneath. However, since then, though both on the Si platform, MEMS and CMOS technologies have been developed separately because of the large difference of their feature sizes. Recently, as is shown in Figure 6.1, the feature size of the mechanical objects becomes closer and closer to that of CMOS. Grogg et al. made an attempt to integrate a MEMS resonator with a MOSFET intensively in the device layer in 2007 [39]. The works in 2008 by Durand et al. were based on Silicon-On-Nothing technology where SOI for CMOS devices is used for fabricating integrated NEM resonators [40]. Nanometer-scale in-plane resonant suspended-gate MOSFETs proposed by Durand et al. [41] is one of the most remarkable examples of NEMS-CMOS integration. An in-plane FET is placed just next to a suspended beam so that the structure is to be the lateral version of the resonant suspended-gate FET. The change of the capacitance between the suspended beam and FET channel is directly amplified through the transconductance of the RSGFET. Significant improvement of the signal/noise ratio has been demonstrated in comparison with conventional capacitance measurements. Considering the integration even in the research phase is quite natural and beneficial in terms of using one the most advanced technology resources available and also of reducing the time for commercialization. As the main focus of this book is on silicon nanoelectronics, in the following sections, we will introduce past and recent attempts to develop NEM resonator sensors in the context of NEMS-CMOS integration.

6.4.2 MASS SENSITIVITY SIMULATION

Modeling plays a vital role in developing novel integrated systems in order to estimate overall performance of the systems prior to actual fabrication. However, there are two issues. Firstly, simulation technologies have also been developed separately between CMOS and MEMS devices; therefore, it is not so easy to simulate a hybrid system where both elements are integrated. Secondly, there was no modeling study to simulate the performance for sensing in device simulations. Hassani et al. built a model of in-plane RSGFET device structure and simulated the performance of the device for chemical and biological sensing applications [42]. A schematic of the model of the device is shown in Figure 6.16a. The dimensions of the beam used in this study are 1 mm (L) × 50 nm (W) × 100 nm (t), and the gap between the beam and the channel is 50 nm. Figure 6.16b shows schematics of events which would happen at the surface of the silicon nanowire sensor head. Normally first the silicon nanowires are coated by linker molecules that are able to make a link between the surface of Si and a target molecule. This process is called functionalization. In modeling this functionalization, a layer of a material with a constant density is added. Next in sensing, the target molecules are adsorbed on top of linker molecules, but the absorption is not uniform as is indicated in Figure 6.16b. Therefore, to model the sensing events, the density of the coated material is changed with a fixed thickness. As shown in Figure 6.17a, three different coating configurations are considered for possible inhomogeneous coating. The resonant frequency is calculated using model analysis. While the top only (TO) and top and bottom (TB) configurations show a decrease in the frequency after the coating, the resonant frequency of the all around (AA) configuration increases with increase in the coating layer thickness. This result suggests that the change of the mechanical stiffness is more dominant than the increase of the mass in the AA configurations. Figure 6.17b summarizes the simulation

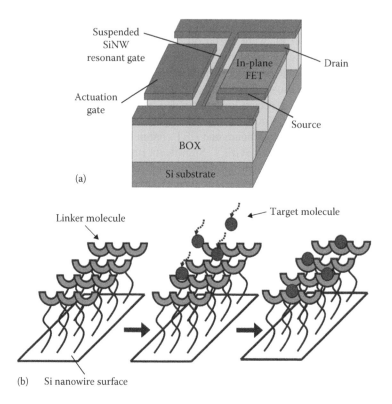

(a)

(b) Si nanowire surface

FIGURE 6.16 (a) A schematic diagram of the in-plane resonant suspended-gate MOSFETs where the NEM resonator is capacitively coupled with an in-plane FET that can directly amplify the signal of electromechanical oscillation of the beam. (b) Schematic models of surface functionalization and subsequent sensing of molecules used in this study. Functionalization is modeled by increasing thickness with a fixed material density, while sensing is by increasing density with a fixed thickness.

results of sensing condition for each configuration. In general, the added mass δM is expressed in a linearized manner as

$$\delta M \approx \frac{\partial M_{\text{eff}}}{\partial \omega_0} \delta \omega_0 = R^{-1} \delta \omega_0 = S \delta \omega_0, \tag{6.6}$$

where
    ω_0 is the resonance frequency
    M_{eff} is the effective mass of the beam
    R is the mass responsivity

$$R = \frac{\partial \omega_0}{\partial M_{\text{eff}}}, \tag{6.7}$$

which is the inverse of the mass sensitivity S. From Figure 6.17b, S of the NEMS resonator is estimated to be ~0.05 zg/Hz. This value is about nine orders smaller

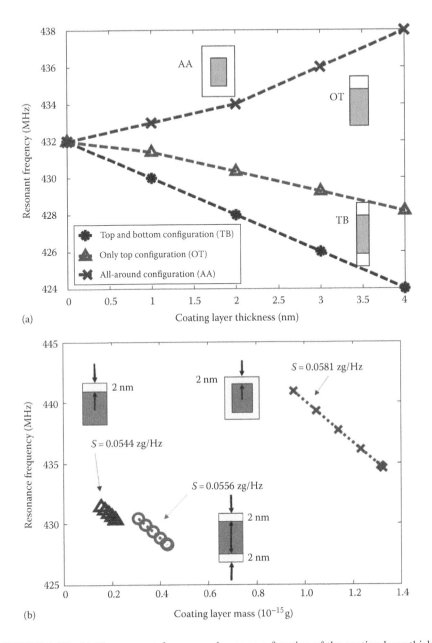

FIGURE 6.17 (a) The resonant frequency change as a function of the coating layer thickness in functionalization modeling. Various coating configurations shown schematically in the figure are modeled. The result indicates the functionalization on the sidewall of the resonator significantly affects the change of stiffness of the resonator. (b) The resonant frequency change in sensing modeling as a function of the mass of the coating layer. The inverse of the mass sensitivity S is estimated to be ~0.05 zg/Hz.

than that of the QCM commercial mass sensors mentioned before. Here the minimum measurable frequency shift $\delta\omega_0$ is estimated from the bandwidth at the half maximum, which is expressed as

$$\delta\omega_0 = \frac{\omega_0}{Q}. \tag{6.8}$$

Assuming the $\omega_0 = 2\pi \times 440$ MHz and $Q \sim 10,000$, the minimum measurable mass is estimated to be ~14 ag (attogram = 10^{-18} g) corresponding to the mass of typical protein molecules or nanoparticles. This can be improved with further miniaturization of the beam as well as with wider bandwidth of the measurement system. In addition, hybrid circuit analysis has been performed for this NEMS-MOSFET integrated resonator sensor by using commercial software designed for the hybrid simulation The results are quite consistent with the one obtained by 3D FEM modal analysis.

6.4.3 INTEGRATED NEM RESONATOR SENSORS

In this section, some of recent demonstrations of NEMS resonators toward NEMS-CMOS integrated systems are introduced. Ollier et al. reported scaled single-crystal Si NEMS resonators integrated monolithically with CMOS at the front end level for the first time [43]. Figure 6.18a shows an SEM image of the fabricated devices

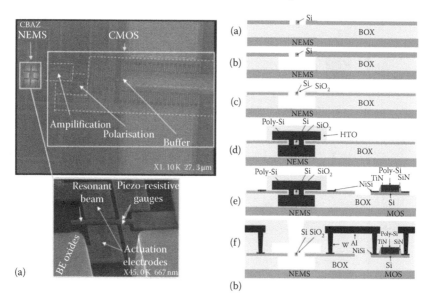

FIGURE 6.18 (a) An SEM image of a NEM resonator integrated with CMOS-based circuitry on the same SOI wafer. The NEM resonator part is shown in a blow-up in detail. (b) A fabrication process of the NEMS resonator in a manner compatible with SOI-CMOS fabrication technology. (Reprinted with permission from Ollier, E., Dupré, C., Arndt, G., Arcamone, J., Vizioz, C., Duraffourg, L., Sage, E. et al., Ultra-scaled high-frequency single-crystal Si NEMS resonators and their front-end co-integration with CMOS for high sensitivity applications, *IEEE MEMS 2012*, Paris, France, p. 1368, © 2012 IEEE.)

integrated on FDSOI. A low noise amplifier, biasing stage, and buffer for signal processing are built in the CMOS part, while the NEMS resonator is just next to the circuitry on the same wafer. A cross-beam structure with side actuation electrodes is employed for the resonator as shown in a blow-up in Figure 6.18a. The bar across the resonant beam is used for piezoresistive gauges for transduction of the mechanical motion of the beam. The process flow of the device fabrication is schematically illustrated in Figure 6.18b. After patterning the NEMS resonators, the beam is released by undercutting the BOX oxide underneath and then passivated by thermal oxide. Then the air gap is filled with poly-Si followed by deposition of oxide to protect the NEMS resonator from the following CMOS fabrication and metallization processes. After metallization is over, a hole is made to expose the surface of the poly-Si and then the filler is etched out to release the resonator again. This process allows us to make a NEMS beam of which the surface is covered by high-quality thermal oxide. For the resonant beams with the dimensions of 1.2 mm (L) × 100 nm (W) × 40 nm (t), a clear resonance peak with the frequency of 104 MHz and the Q-value of 935 has been observed.

Bartsch et al. proposed junctionless resonant-body silicon nanowire FET and fabricated on an FD-SOI-CMOS platform [44]. An SEM image of the device structure is in the inset of Figure 6.19. In operation, the suspended silicon nanowire works as a NEM resonator and an electromechanical transducer at the same time. The depletion charges in the suspended Si nanowire junctionless transistor by the double side gates are modulated by the mechanical oscillation of itself. The subsequent study suggests that this transduction mechanism is advantageous in terms of signal-to-noise ratio compared to the second-order piezoresistive modulation [45]. A resonator with

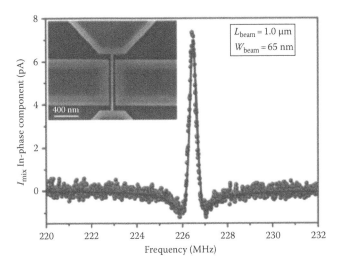

FIGURE 6.19 A resonance peak obtained from a junctionless resonant-body silicon nanowire FET proposed by Bartsch et al. [44]. The inset shows a plan view of the resonator fabricated on FD-SOI-CMOS platform. (Reprinted with permission from Bartsch, S.T., Dupre, C., Ollier, E., and Ionescu, A.M., Resonant-body silicon nanowire field effect transistor without junctions, *IEDM 2012 Technical Digest*, San Francisco, CA, p. 351, © 2012 IEEE.)

the dimensions of 1 μm (*L*) × 65 nm (*W*) × 45 nm (*t*) shows the resonant frequency of 226 MHz as shown in Figure 6.19. The authors have also suggested that by using their junctionless silicon nanowire resonators, δ*M* of 180 kDa is possible at 300 K, which is better when compared to the simulation study.

A suspended junctionless transistor is also fabricated in an integration-friendly manner by Arab Hassani et al. The device is entirely heavily doped throughout the source, drain, channel, and double side gates. They have also succeeded in observing the resonance [46].

6.4.4 SUMMARY AND FUTURE PROSPECTS

So far a number of attempts have been reviewed to realize ultrahigh resolution mass detectors by employing silicon NEM resonators. While researches on exploring the fundamental limit of mass detection are still going on, investigation of Si NEM resonators and their integration technology with Si CMOS platform is more active in recent years as is shown. Actually a number of co-integrated devices have been already successful at the device level. The next crucial step toward commercialization could be to develop a tool to design large integrated circuits including NEMS components. Therefore, device modeling and circuit-level design technology for NEMS-CMOS integrated systems will become more important as well as there need to be continuous efforts in developing device level design and fabrication technology. Petrescu et al. have developed circuit models for NEMS resonators and conducted circuit simulation of NEMS-CMOS integrated system [47] toward "electric nose" applications.

On the other hand, when we consider real biosensing application, to establish the technology to bridge the Si technology to bionanotechnology is of paramount importance. In particular, when we consider using the devices for sensing specific molecules, we have to functionalize the surface of the NEM beam. This functionalization process has not been well established to guarantee a level of reproducibility. There are some early attempts on suspended nanowire devices [48,49], but we believe, more detailed investigation on what actually happens at the surface and interface in surface functionalization process is very important.

In relation to the point we mentioned, on NEM devices for further commercialization, packaging technology is another key thing to address. While this is a common issue for overall NEM devices, it is particularly important for NEM resonators as their performance improves when they are operated under vacuum, rather than when exposed to atmospheric pressure. In parallel with developing the front end of device fabrication, improvement of back end process such as wiring and packaging should be taken into account in the future commercialization of a NEMS-embedded silicon chip.

6.5 CONCLUSIONS

Recent progress of silicon-based NEM devices has been reviewed. Three major applications have been focused: NEM switches, NEM memories, and NEM resonator sensors. NEM switches have been identified to have great advantage for low power applications. In addition to its intrinsic nature as a steep-slope switch, a new concept of energy-reversible switching is interesting for further reduction of power consumption. Switches for IC power management are one of target applications of NEM switches. NEM memories are advantageous for high-speed programming and erasing operations without losing their serious nonvolatility. In addition, low power operation is expected with extremely scaled NEM memories. To overcome their scaling issue, different circuit architecture such as crossbar structure is worth considering. Nonvolatile logic application is one of interesting fields for NEM memories to consider. NEM resonator sensors are a promising device to detect ultrasmall amount of mass. An active integration between NEMS and CMOS technologies is well into progress from both fabrication and modeling points of view. The integrated systems are very promising for future smart sensor system applications.

REFERENCES

1. H. Mizuta, M. A. G. Ramirez, Y. Tsuchiya, T. Nagami, S. Sawai, S. Oda, and M. Okamoto, Multi-scale simulation of hybrid silicon nano-electromechanical (NEM) information devices, *J. Automat. Mobile Robot. Intell. Syst.* 3, 58 (2009).
2. International Technology Roadmap for Semiconductors, http://www.itrs.net/, accessed August 28, 2015.
3. A. M. Ionescu and H. Riel, Tunnel field-effect transistors as energy-efficient electronic switches, *Nature* 479, 329 (2011).
4. N. Abelé, R. Fritschi, K. Boucart, F. Casset, P. Ancey, and A. M. Ionescu, Suspended-gate MOSFET: Bringing new MEMS functionality into solid-state MOS transistor, *IEDM 2005 Technical Digest*, Washington, DC, p. 479 (2005).
5. H. C. Nathanson and R. A. Wickstrom, A resonant gate silicon surface transistor with high Q bandpass properties, *Appl. Phys. Lett.* 7, 84 (1965).
6. K. E. Petersen, Dynamic micromechanics on silicon: Techniques and devices, *IEEE Trans. Electron Dev.* ED-25(10), 1241 (1978).
7. K. Akarvardar, D. Elata, R. Parsa, G. C. Wan, K. Yoo, J. Provine, P. Peumans, R. T. Howe, and H.-S. P. Wong, Design considerations for complementary nanoelectromechanical logic gates, *IEDM 2007 Technical Digest*, Washington, DC, p. 299 (2007).
8. V. Pott, H. Kam, R. Nathanael, J. Jeon, E. Alon, and T.-J. King Liu, Mechanical computing redux: Relays for integrated circuit applications, *Proc. IEEE* 98(12), 2076–2094 (2010).
9. W. W. Jang, J. O. Lee, J.-B. Yoon, M.-S. Kim, J.-M. Lee, S.-M. Kim, K.-H. Cho, D.-W. Kim, D. Park, and W.-S. Lee, Fabrication and characterization of a nanoelectromechanical switch with 15-nm-thick suspension air gap, *Appl. Phys. Lett.* 92, 103110 (2008).
10. J.-O. Lee, M.-W. Kim, S.-D. Ko, H.-O. Kang, W.-H. Bae, M.-H. Kang, K.-N. Kim, D.-E. Yoo, and J.-B. Yoon, 3-Terminal nanoelectromechanical switching device in insulating liquid media for low voltage operation and reliability improvement, *IEDM 2009 Technical Digest*, Baltimore, MD, p. 227 (2009).
11. H. Kam, V. Pott, R. Nathanael, J. Jeon, E. Alon, and T.-J. King Liu, Design and reliability of a micro-relay technology for zero-standby-power digital logic applications, *IEDM Technical Digest*, Baltimore, MD, p. 809 (2009).
12. D. A. Czaplewski, G. A. Patrizi, G. M. Kraus, J. R. Wendt, C. D. Nordquist, S. L. Wolfley, M. S. Baker, and M. P. de Boer, A nanomechanical switch for integration with CMOS logic, *J. Micromech. Microeng.* 19, 085003 (2009).
13. J. Jeon, V. Pott, H. Kam, R. Nathanael, E. Alon, and T.-J. King Liu, Perfectly complementary relay design for digital logic applications, *IEEE Electron Dev. Lett.* 31(4), 371 (2010).

14. T.-J. King Liu, J. Jeon, R. Nathanael, H. Kam, V. Pott, and E. Alon, Prospects for MEM logic switch technology, *IEDM 2010 Technical Digest*, San Francisco, CA, p. 424 (2010).

15. K. Akarvardar, D. Elata, R. T. Howe, and H.-S. Philip Wong, Energy-reversible complementary NEM logic gates, *DRC 2008 Technical Digest*, Santa Barbara, CA, pp. 69–70 (2008).

16. M. Spencer, F. Chen, C. C. Wang, R. Nathanael, H. Fariborzi, A. Gupta, H. Kam et al., Demonstration of integrated micro-electro-mechanical relay circuits for VLSI applications, *IEEE J. Solid-State Circuits*, 46(1), 308 (2011).

17. M. Enachescu, G. Voicu, and S. D. Cotofana, Advanced NEMS-based power management for 3D stacked integrated circuits, *2010 International Conference on Energy Aware Computing (ICEAC)*, Cairo, Egypt, December, 16–18, 2010.

18. D. Tsamados, Y. S. Chauhan, C. Eggimann, K. Akarvardar, H.-S. Philip Wong, and A. M. Ionescu, Finite element analysis and analytical simulations of suspended Gate-FET for ultra-low power inverters, *Solid-State Electron.* 52, 1374–1381 (2008).

19. B. Halg, On a micro-electro-mechanical nonvolatile memory cell, *IEEE Trans. Electron Dev.* 37(10), 2230 (1990).

20. A. N. Cleland and M. L. Roukes, Fabrication of high frequency nanometer scale mechanical resonators from bulk Si crystals, *Appl. Phys. Lett.* 69, 2653 (1996).

21. X. M. H. Huang, C. A. Zorman, M. Mehregany, and M. L. Roukes, Nanodevice motion at microwave frequencies, *Nature* 421, 496–496 (2003).

22. Y. Tsuchiya, K. Takai, N. Momo, T. Nagami, H. Mizuta, S. Oda, S. Yamaguchi, and T. Shimada, Nanoelectromechanical nonvolatile memory device incorporating nano-crystalline Si dots, *J. Appl. Phys.* 100, 094306 (2006).

23. T. Nagami, H. Mizuta, N. Momo, Y. Tsuchiya, S. Saito, T. Arai, T. Shimada, and S. Oda, Three-dimensional numerical analysis of switching properties of high-speed and non-volatile nanoelectromechanical memory, *IEEE Trans. Electron Dev.* 54, 1132 (2007).

24. T. Nagami, Y. Tsuchiya, S. Saito, T. Arai, T. Shimada, H. Mizuta, and S. Oda, Electro-mechanical simulation of switching characteristics for nanoelectromechanical memory, *Jpn. J. Appl. Phys.* 48, 114502 (2009).

25. T. Nagami, Y. Tsuchiya, K. Uchida, H. Mizuta, and S. Oda, Scaling analysis of nano-electromechanical memory devices, *Jpn. J. Appl. Phys.* 49, 044304 (2010).

26. N. Momo, T. Nagami, S. Matsuda, Y. Tsuchiya, S. Saito, T. Arai, Y. Kimura, T. Shimada, H. Mizuta, and S. Oda, Fabrication and characterization of nanoscale suspended floating gates for NEMS memory, *IEEE 2006 Silicon Nanoelectronics Workshop*, Honolulu, HI, June 11–12, 2006.

27. T. Ifuku, M. Otobe, A. Itoh and S. Oda, Fabrication of nanocrystalline silicon with small spread of particle size by pulsed gas plasma, *Jpn. J. Appl. Phys.* 36, 4031 (1997).

28. N. Abelé, A. Villaret, A. Gangadharaiah, C. Gabioud, P. Ancey, and A. M. Ionescu, 1T MEMS memory based on suspended gate MOSFET, *IEDM 2006 Technical Digest*, San Francisco, CA, p. 479 (2006).

29. M. A. Garcia-Ramirez, H. Yoshimura, Y. Tsuchiya, and H. Mizuta, Suspended gate silicon nanodot memory, *ESSDERC/CIRC Fringe (ESS-Fringe)*, Edinburgh, Scotland, September 2008.

30. M. A. Garcia-Ramirez, Y. Tsuchiya, and H. Mizuta, Hybrid circuit analysis of a suspended-gate silicon nanodot memory (SGSNM) cell, *Microelectron. Eng.* 87, 1284–1286 (2010).

31. M. A. Garcia-Ramirez, Y. Tsuchiya, and H. Mizuta, Hybrid numerical analysis of a high-speed non-volatile suspended gate silicon nanodot memory, *J. Comput. Electron.* 10(1), 248–257 (2011).

32. W. Y. Choi, H. Kam, D. Lee, J. Lai, and T.-J. King Liu, Compact nanoelectro-mechanical non-volatile memory (NEMory) for 3D integration, *IEDM 2007 Technical Digest*, Washington, DC, pp. 603–606 (2007).

33. W. Y. Choi, T. Osabe, and T.-J. King Liu, Nano-electro-mechanical nonvolatile memory (NEMory) cell design and scaling, *IEEE Trans. Electron Dev.* 55(12), 3482 (2008).
34. K. Akarvardar and H.-S. Philip Wong, Ultralow voltage crossbar nonvolatile memory based on energy-reversible NEM switches, *IEEE Electron Dev. Lett.* 30(6), 626 (2009).
35. L. Boodhoo, Y. P. Lin, H. M. H. Chong, Y. Tsuchiya, T. Hasegawa, and H. Mizuta, Energy reversible Si-based NEMS switch for nonvolatile logic systems, *Proceedings of the IEEE NEMS 2013*, Suzhou, China, pp. 558–561 (2013).
36. T. Rueckes, K. Kim, E. Joselevich, G. Y. Tseng, C.-L. Cheung, and C. M. Lieber, Carbon nanotube-based nonvolatile random access memory for molecular computing, *Science* 289, 94 (2000).
37. Y. T. Yang, C. Callegari, X. L. Feng, K. L. Ekinci, and M. L. Roukes, Zeptogram-scale nanomechanical mass sensing, *Nano Lett.* 6, 583 (2006).
38. J. Chaste, A. Eichler, J.Moser, G. Ceballos, R. Rurali, and A. Bachtold, A nanomechanical mass sensor with yoctogram resolution, *Nat. Nanotechnol.* 7, 301 (2012).
39. D. Grogg, D. Tsamados, N. D. Badila, and A. M. Ionescu, Integration of MOSFET transistor in MEMS resonators for improved output detection, *14th International Conference on Solid-State Sensors, Actuators and Microsystems* (Transducers 2007), Lyon, France, p. 1709 (2007).
40. C. Durand, F. Casset, B. Legrand, M. Faucher, P. Renaux, D. Mercier, D. Renaud, D. Dutartre, E. Ollier, P. Ancey, and L. Buchaillot, Characterization of In-IC integrable in-plane nanometer scale resonators fabricated by a silicon-on-nothing advanced CMOS technology, *IEEE MEMS 2008*, Tucson, AZ, p. 1016 (2008).
41. C. Durand, F. Casset, P. Renaux, N. Abelé, B. Legrand, D. Renaud, E. Ollier, P. Ancey, A. M. Ionescu, and L. Buchaillot, In-plane silicon-on-nothing nanometer-scale resonant suspended gate MOSFET for In-IC integration perspectives, *IEEE Electron Dev. Lett.* 29(5), 494 (2008).
42. F. A. Hassani, C. Cobianu, S. Armini, V. Petrescu, P. Merken, D. Tsamados, A. M. Ionescu, Y. Tsuchiya, and H. Mizuta, Numerical analysis of zeptogram/Hz-level mass responsivity for in-plane resonant nano-electro-mechanical sensors, *Microelectron. Eng.* 88(9), 2879–2884 (2011).
43. E. Ollier, C. Dupré, G. Arndt, J. Arcamone, C. Vizioz, L. Duraffourg, E. Sage et al., Ultra-scaled high-frequency single-crystal Si NEMS resonators and their front-end co-integration with CMOS for high sensitivity applications, *IEEE MEMS 2012*, Paris, France, p. 1368 (2012).
44. S. T. Bartsch, C. Dupre, E. Ollier, and A. M. Ionescu, Resonant-body silicon nanowire field effect transistor without junctions, *IEDM 2012 Technical Digest*, San Francisco, CA, p. 351 (2012).
45. S. T. Bartsch, M. Arp, and A. M. Ionescu, Junctionless silicon nanowire resonator, *IEEE J. Electron Dev. Soc.*, 2(2), 8–15 (2014).
46. F. Arab Hassani, Y. Tsuchiya, H. Mizuta, C. Dupré, E. Ollier, S. T. Bartsch, and A. M. Ionescu, Dual-gate junction-less FET-detection for in-plane nano-electro-mechanical resonators, *ICICDT 2013*, Pavia, Italy, p. 115 (2013).
47. V. Petrescu, J. Pettine, D. M. Karabacak, M. Vandecasteele, M. C. Calama, and C. Van Hoof, Power-efficient readout circuit for miniaturized electronic nose, *ISSCC 2012 Technical Digest*, San Francisco, CA, p. 318 (2012).
48. M. A. Ghiass, S. Armini, M. Carli, A. M. Caro, V. Cherman, J. Ogi, S. Oda, Z. Moktadir, Y. Tsuchiya, and H. Mizuta, Temperature insensitive conductance detection with surface-functionalised silicon nanowire sensors, *Microelectron. Eng.* 88, 1753–1756 (2011).
49. S. Armini, V. Cherman, A. Volodin, S. Lenci, S. F. Pieri, D. Wouters, J. Moonens et al., Selective surface functionalization of Si and poly-SiGe resonators for a monolithic integration of bio- and gas sensors with CMOS, *MRS Spring Meeting Symposium B*, San Francisco, CA, April 2012.

7 Tunnel FETs for More Energy-Efficient Computing

Adrian M. Ionescu

CONTENTS

7.1 Introduction: Steep Slope Devices, Voltage Scaling, and Energy Efficiency....155
7.2 Tunnel FETs: Principle and State-of-the-Art.. 161
    7.2.1 Silicon- and Germanium-Based Tunnel FETs................................. 162
    7.2.2 III–V-Based Tunnel FETs 165
7.3 Toward a Density-of-States Electronic Switch Exploiting 2D–2D Tunneling...168
7.4 Tunnel FETs for Low-Power Analog and Sensing Applications 173
7.5 Conclusions... 174
Acknowledgment 176
References..176

7.1 INTRODUCTION: STEEP SLOPE DEVICES, VOLTAGE SCALING, AND ENERGY EFFICIENCY

In the history of the silicon CMOS technology, its dominance, supported by continuous enhancement of the delivered performance by aggressive scaling with associated low cost, always prevailed over other technology candidates. However, as today we are approaching the physical limits of MOSFETs dimensions scaling, other challenges have appeared and tend to become important drivers of the digital computing technology. Among these, one is the *energy efficiency* of digital computing and corresponding metrics to characterize it from device to system level.

The scaling down of the supply voltage, V_{dd}, and of the threshold voltage, V_{th}, of digital CMOS have significantly slowed down when entering the sub-100 nm technology nodes (e.g., V_{dd} was 1 V for 45 nm node and is 0.8 V for 22 nm node). These limitations are critically related to one of the physically nonscalable parameters of MOSFETs: the subthreshold swing, S (see depiction in Figure 7.1 of key static figures of merit of a digital switch). Attempts to scale down further the threshold voltage, V_{th}, would generate an exponential increase in the off-current, I_{off}, of at least 10× for every 60 mV of V_{th} reduction. More concretely, a V_{th} decrease toward 0.1 V, compared to the current value in advanced CMOS, would unacceptably increase the value of I_{off} by more than three orders of magnitude, which would make intolerable the value of the standby power

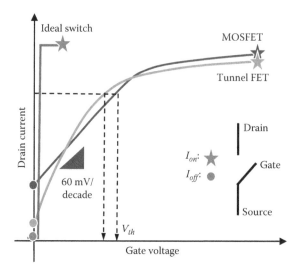

FIGURE 7.1 Qualitative depiction of the drain current versus gate voltage characteristic of the MOSFET, tunnel FET, and ideal three-terminal switch. A steeper transition between *off* and *on* state offers a lower V_{th} (extracted here at constant current) and lower I_{off} for the tunnel FET compared to the MOSFET.

consumption. As the dynamic power component is roughly proportional to V_{dd}^2, the power consumption of electronic devices increases when the complexity of electronics chips increases without a corresponding scaling of V_{dd}. Moreover, as in advanced technology nodes, the standby power ($\sim I_{off} \times V_{dd}$) is comparable or even dominant over the dynamic power; the power issue becomes even more critical. Therefore, future prospects and solutions for addressing the power issue should address both the static and the dynamic power components of digital computing and, essentially, find out practical ways to scale down the voltage supply. Figure 7.1 suggests that a steep slope switch could offer a solution to the V_{th} voltage scaling and low standby power issues, but the question remains about the effect on the overall energy efficiency.

When studying the balance between the static and dynamic energy in digital CMOS, one finds there is a minimum of the total energy per switched bit of information as a function of the voltage supply, resulting from the following relation:

$$E_{tot} = E_{leak} + E_{dyn} = \frac{I_{off} V_{dd}}{\alpha f} + C_L V_{dd}^2 \qquad (7.1)$$

where E_{tot}, E_{leak}, and E_{dyn} are the total, leakage, and dynamic energy, respectively.

The coordinates of this minimum (V_{dd_min}, E_{tot_min}) measure the energy efficiency capability of a given digital technology, Figure 7.2. Therefore, lowering both coordinates of this minimum is one of the challenges of the future beyond CMOS devices. One way to reduce both V_{dd_min} and E_{tot_min} is to lower the inverse subthreshold slope or the subthreshold swing, $S = [dI_d/dV_{gs}]^{-1}$, of the electronic

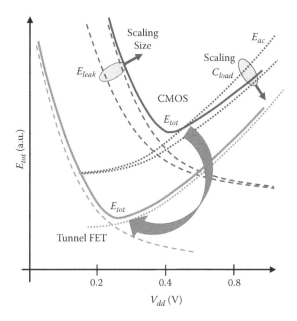

FIGURE 7.2 (See color insert.) Total switching energy, E_{tot} (in arbitrary units, a.u.), versus supply voltage, V_{dd}, for CMOS and tunnel FET switches. The scaling of CMOS size increases the E_{leak} while the reduction of the load capacitance, C_L, decreases E_{dyn}. Overall, E_{tot_min} and V_{dd_min} are significantly lower for tunnel FETs, which opens new available design space for energy-efficient design.

switch [1]. Various device innovations have been proposed to reduce the subthreshold swing below its fundamental limits in a MOSFET via the reduction of m and n factors in its corresponding analytical equations:

$$S = \underbrace{\frac{dV_{gs}}{d\Psi_S}}_{m} \underbrace{\frac{d\Psi_S}{d(\log_{10} I_d)}}_{n} \tag{7.2}$$

$$S_{MOSFET} \cong \left(1 + \frac{C_d}{C_{ox}}\right) \ln 10 \frac{kT}{q} \rightarrow \frac{kT}{q} \ln 10 \cong 60 \text{ mV/decade}\Big|_{T=300\text{ K}} \tag{7.3}$$

where

Ψ_S is the surface potential
V_{gs} is the gate voltage
kT/q is the thermal voltage
C_d and C_{ox} are the depletion and the oxide capacitance, respectively

Equation 7.2 is general, applying to any three-terminal devices having a gated channel, while Equation 7.3 points out the swing expression and its thermal limit, in a MOSFET.

Another way to understand why S is a crucial energy efficient parameter is the definition of the average subthreshold swing S_{avg} (Equation 7.1):

$$S_{avg} = \frac{V_{th} - V_{Goff}}{\log(I_{th}/I_{off})}\bigg|_{subthreshold\ logic} \cong \frac{V_{dd}}{\log(I_{on}/I_{off})}. \qquad (7.4)$$

Expression 7.4 suggests that an effective way of reducing the voltage supply, V_{dd}, without performance loss (same I_{on}) is to increase the turn-on steepness, which means decreasing S_{avg}. Therefore, devices with a steep subthreshold are expected to enable V_{dd} scaling and future energy-efficient digital integrated circuits.

A *subthermal subthreshold swing* (i.e., with a value below 60 mV/decade at $T = 300$ K) by reducing the n factor requires the modification of the carrier injection mechanism; impact ionization (IMOS switch [2]) and quantum mechanical band-to-band tunneling (BTBT) mechanisms in tunnel FETs (tunnel field-effect transistors) have been proposed as solutions for very steep transitions between off and on states. In case of the IMOS devices, to our best knowledge, the degradation of the V_{th} due to hot carrier effects (related reliability issues) was not conveniently solved. Therefore, in this category of devices, the tunnel FETs remain the most promising steep slope solid state device without major reliability issues reported to date. Another alternative to lower S is to reduce below 1 the body factor, m. In theory, this can be achieved by using the recently proposed negative capacitance effect [3,4] or by using electromechanical gates [5,6] with movable electrodes, where instability points between electrical and mechanical forces are used to define superabrupt transitions between *off* and *on* states. However, despite experimentally demonstrated values of swing lower than 2 mV/decade, with little dependence on temperature, electromechanical devices have their own limitations such as hysteretic characteristics, limitation of voltage scaling, reliability, and need of controlled ambient for robust operation (packaging). The negative capacitance effect reports are still in infancy, and there is still debate about their practical use in hysteretic or nonhysteretic operation.

More recently, another class of solid-state devices offering a steep slope based on the metal-insulator-transition (MIT) in some special classes of functional oxides like vanadium dioxide (VO_2) have been demonstrated [7,8]. In [8], we have demonstrated subthermal switching over three orders of magnitude of current in a two-terminal VO_2 switch. However, the main limitation of MIT switches is currently related to their high *off* state current and to the hysteretic characteristics related to the electrothermal actuation (formation of conductive filaments).

For the purpose of a quick qualitative comparison, Figure 7.3 depicts the expected transfer characteristics, I_d–V_{gs}, of some of the known steep slope switches and the projected qualitative promise in terms of I_{on} and I_{off}, compared to CMOS. The tunnel FET is defined as the solid-state device that has some of the lowest I_{off} (only a NEM relay can offer a better solution because of the in-series air-gap) and, certainly, the highest potential for voltage scaling in a technology compatible with the existing CMOS technology platforms [1,9] and exploiting the main CMOS technology boosters [10]. The main challenges of tunnel FETs remain their rather low I_{on}, for which various approaches for enhancements will be discussed in this chapter. On the other

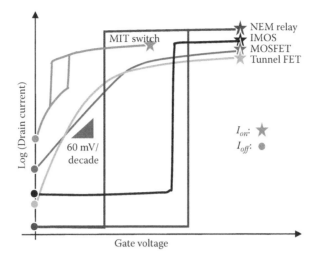

FIGURE 7.3 **(See color insert.)** Qualitative representation of the transfer characteristics, log(drain current) versus gate voltage, for different classes of steep slope switches exploiting a three-terminal configuration: NEM relay, IMOS, tunnel FET, and MIT switch, as compared with MOSFET and its 60 mV/decade limit of S. Some of these devices can offer very steep transitions between the *off* and the *on* states but the voltages at which these transitions take place are not scalable. Some other devices are hysteretic and need special design of their circuit applications. The figures also depict the projected comparative performance in terms of I_{on} and I_{off} for all these devices.

hand, some authors [11] have provided theoretical evidence that there could be a trade-off between the steepness of S (dictated by the barrier thickness modulation) over more than four decades of current and the achievable conductance or I_{on} level; therefore, the tunnel FET optimization and performance limits are more complex to understand.

From a physical point of view, the tunneling rate, T, expression has been proposed by many authors [1] as an analytical vehicle to identify the main parameters that should be optimized in order to enhance to the I_{on}. In the same time, the exponential dependence of T_{WKB} (in the Wentzel–Kramer–Brillouin expression) on the various parameters and on the gate voltage suggests the abrupt nature of the tunnel FET switch and the possible engineering directions for optimizing this device.

$$T_{WKB} \approx \exp\left(-\frac{4\lambda\sqrt{2m^*}E_g^{1.5}}{3\hbar(\Delta\Phi + E_g)}\right) \qquad (7.5)$$

As the tunneling current is proportional to T_{WKB}, the I_{on} can be optimized via bandgap (E_g), electrostatic control (λ), and effective tunneling mass (m^*) engineering. Interesting analytical derivations and discussions for achieving a steep slope have been reported in [12] where it was noted that an exponential increase of the tunneling current is maintained, since *barrier thinning with gate voltage remains dominant*, and, additionally, a *band-pass filtering* behavior of the energy can be also uniquely observed in this type of switch. These two contributions of the current change with the gate voltage can be particularly distinguished in the analytical expression of tunnel FET transconductance:

$$g_m = \frac{\partial I_d}{\partial V_g} = \frac{2e^2}{h} \left(\frac{\partial T_{WKB}}{\partial E_v^{ch}} F\left(E_v^{ch}\right) + T_{WKB} \frac{\partial F\left(E_v^{ch}\right)}{\partial E_v^{ch}} \right) \tag{7.6}$$

where $F(E_v^{ch})$ is energy integral of $f_s(E)\, f_d(E)$ between E_c^s and E_v^{sch} [12]. It is worth noting that drain current, I_d; the transconductance, g_m; and the subthreshold swing, S, are related via the following expression:

$$S = \frac{\ln(10) I_d}{g_m}. \tag{7.7}$$

Recent reports on tunnel FET show significant progress made in terms of increased I_{on}/I_{off}, low voltage (below 0.5 V) and low power DC operation, and improved energy efficiency (swing improvement over many decades). However, there is still a lack of experimental demonstration of dynamic operation at frequencies approaching the GHz, aggressive scaling of the device dimensions (technology demonstrators with tunnel FETs aligned with the current sub-22 nm nodes), operational reliability (specific reliability issues), variability (importance of process variations and needed parameter control for tunnel FETs as compared to CMOS), and CMOS technological/architectural compatibility.

For a more quantitative estimation of the progress made in the quest for the new subthermal switch with performance beyond or complementing the one of nanometer-scaled CMOS, the following performance factors must be prioritized [5]:

- *Steepness of the switch*: Reducing the limit of 60 mV/decade to below 10 mV/decade over at least four decades of current to enable sub-0.2 V voltage supply scaling.
- *On/off current ratio* higher than 10^5.
- *Off current lower than 0.1 nA/μm at $V_g = 0$ V and $V_d = V_{dd}$*: Note also that the gate current should remain negligible compared to drain and source currents over all range of gate and drain voltages.
- *Current density*: The target is to achieve complementary n- and p-type tunnel FETs with well-balanced current density in the range 0.1–1 mA/μm.
- *Operation speed* of the order of hundreds of MHz to 1 GHz.
- *Energy efficiency metrics* such as: S_{avg} (E_{total_min}, V_{dd_min}) or similar metrics that combine static and dynamic energy or address power components and delay at both device and circuit level. Here, possible targets are: $S_{avg} < 30$ mV/decade over more than four decades at room temperature, a reduction of E_{total_min} by 10× compared to CMOS, and V_{dd} significantly below 0.5 V (0.25 V is frequently cited in roadmaps [13]).
- *Variability* comparable with CMOS for similar gate length (technology node).
- *Manufacturability* on technology platforms compatible with silicon CMOS.

7.2 TUNNEL FETs: PRINCIPLE AND STATE-OF-THE-ART

Tunnel FET is a solid-state semiconductor device designed as a gated p–i–n (or p–n) diode operated in reverse bias and exploiting the quantum-mechanical BTBT at one of the junctions (Figure 7.4). The device configuration benefits from a very low leakage current (I_{off}) and, at the same time, the carrier BTBT injection mechanism at the source potentially allows very steep subthreshold slope values [14–16].

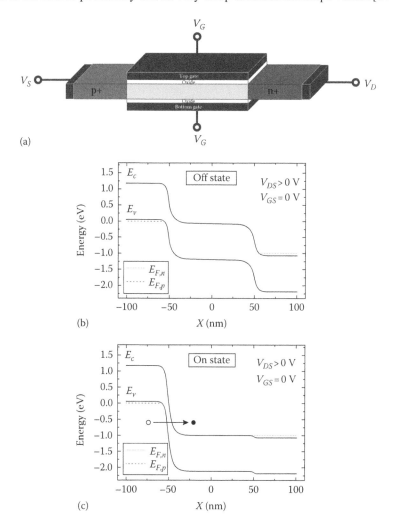

FIGURE 7.4 (a) Tunnel FET embodied in a double-gate (DG) thin-film configuration: if not particularly optimized, the device is ambipolar and tunneling can take place both at source and drain junctions, depending on the applied biases, (b) simulated band diagram in the *off* state of DG tunnel FET with a channel length of 100 nm and with tunneling at the source side (n-type device with $V_{DS} > 0$ and $V_{GS} = 0$), and (c) simulated band diagram in the *on* state of DG tunnel FET with a channel length of 100 nm and with tunneling at the source side (n-type device with $V_{DS} > 0$ and $V_{GS} > 0$).

Figure 7.4a depicts a homojunction silicon tunnel FET with double-gate configuration; Figure 7.4b and c shows the band diagrams in the *off* and *on* states, highlighting the gate control on the energy bands and the ability to narrowing down the tunneling barrier.

It is worth noting that the scaling of tunnel FET threshold and supply voltages can be more aggressive than that of a MOSFET with same channel dimension. Moreover, tunnel FET electrical characteristics have much less dependence on the channel length than the ones of MOSFET. Moreover, the tunnel FET architecture can be implemented in many material systems, with various engineering of the tunneling junction energy bands, in both homojunction and heterojunction embodiments, all exploiting well-known CMOS technology boosters: high-k dielectric, ultrathin body, abrupt junctions, multigate electrostatic control, local strain, etc., to achieve improved performance.

7.2.1 SILICON- AND GERMANIUM-BASED TUNNEL FETS

One of the attractive features of tunnel FETs is that they can exploit in an *additive* way the main technology boosters available on advanced CMOS platforms in order to improve their performance. In Figure 7.5, we start from a nonoptimized silicon-based tunnel FET and we show the gradual improvements obtained in terms of the steepness of S and I_{on} when additive technology boosters are used [17].

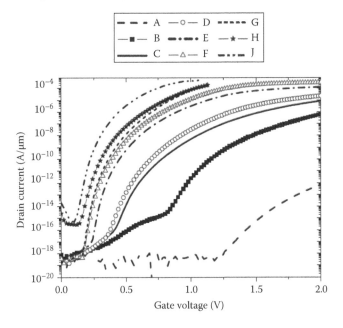

FIGURE 7.5   The effect of additive boosters applied to a silicon tunnel FET. (A) Based device with a single gate, silicon dioxide gate dielectric and nonoptimized parameters. (B) Like A with high-k dielectric. (C) Like B with narrower junction. (D) Like C with thinner body. (E) Like D with higher source doping. (F) Like E with double gate. (G) Like F with oxide only over intrinsic region. (H) Like G with shorter length. (J) Like H with bandgap = 0.8 eV at the tunnel junction. The drain bias used in simulations is V_{DS} = 1 V.

The starting point of our additive booster reported in Figure 7.5 is a silicon tunnel FET with an asymmetrically doped n–i–p structure. The gate dielectric thickness is 3 nm, and the drain and intrinsic region doping levels are 5×10^{18} and 10^{15} atoms/cm^3, respectively. The parameters that are individually optimized are: high-k gate dielectric, more abrupt doping profile at the tunnel junction, thinner device body, higher source doping, double gate, gate oxide aligned to the intrinsic region, and finally, shorter intrinsic region (and gate) length. Curve H in Figure 7.5 shows the optimized characteristics for an all-silicon tunnel FET. The last booster, curve J in Figure 7.5, consists of a smaller bandgap applied at the tunnel junction, which can represent an improvement resulting from a lateral strain profile within the device [18] or a heterostructure in which a lower bandgap material is used for the tunnel FET source [19,20]. As boosters are applied, the device threshold voltage is clearly shifted to lower gate voltages, and accompanying this is a significant increase in on-current and a decrease in swing. While the nonoptimized tunnel FET in curve A has a point swing of 69.5 mV/decade, the optimized device in curve H has a point swing of 23 mV/decade and an average swing of 59.7 mV/decade. The device with a reduced bandgap of 0.8 eV at the tunnel junction in curve J has further improved swings of 18 mV/decade (point) and 40 mV/decade (average). Off-state leakage is higher for the devices of curves H and J due to the shorter intrinsic region and gate length, but is still in the fA/μm range. This gives a static power of the order of 10^3–10^4 times lower than the one of today's advanced CMOS nodes.

Si nanowires (NWs) fabricated with bottom-up [21] and top-down approach [22] showing subthreshold swings below 60 mV/decade have been demonstrated. However, all-Si tunnel FETs show small I_{on} compared to MOSFETs due to the relatively large bandgap of silicon (Si). Strained Si, germanium (Ge) as well as SiGe compounds are of special interest, since their integration in standard CMOS technology is feasible and they can offer a smaller bandgap. Strained Si tunnel FETs have been studied in [23,24], showing improved performance. Epitaxially grown Ge on insulator (GOI) structures [25] and strained Ge channels [20] appeared to have superior on-currents of up to 300 μA/μm; however, these devices suffer from a substantial I_{off} increase. Heterostructure tunnel junctions with small bandgap material in the source and larger bandgap material in channel and drain regions offer a trade-off solution between high I_{on} and low I_{off}. Planar Ge-source Si-channel TFETs [26] showed improved I_{on}/I_{off} ratio up to 3×10^6 for low-voltage drain voltage (0.5 V), and step-like NW structures on pseudomorphically grown SiGe on Si substrates revealed I_{on}/I_{off} ratios exceeding 10^7 [27]. Even if these results have been encouraging, none of the experimental works have comprehensively addressed the limits of performance that tunnel device embodiments can ultimately offer but rather some technological paths for a certain quantitative improvement of the device characteristics.

Among the critical technological parameters for steep transfer characteristics, the tunneling junction plays a crucial role for tunnel FETs. Steep tunnel junctions (with steepness smaller than 5 nm/decade) can be achieved by in situ doping and novel techniques for activation of implanted dopants [28]. Another method to achieve a high doping concentration and a steep junction at the source is the utilization of silicides in combination with dopant segregation [29,30]. Further improvement can be achieved by using counter-doped pockets between source and channel [31], leading to increased I_{on} and steeper Leonelli et al. [32] from IMEC have achieved high I_{on} for p-TFET using silicidation and dopant segregation, but the value of S at room temperature was around 100 mV/decade.

Vandenberghe et al. [33] proposed the value of the drain current I_{60} as a figure of merit for sub-60 mV/decade devices, defined as the highest current where the input characteristics exhibit a transition from sub-60 mV/decade to over-60 mV/decade behavior. For sub-60 mV/decade devices to be competitive with MOSFETs, the authors suggested that I_{60} has to be in the 1–10 μA/μm range. In the field of Si- and Ge-based tunnel FETs, promising results using strained Si NW structure have been reported in [34,35]. In these works, the devices showed $I_{on} = 60$ μA/μm for the p-TFET, with gate-all-around (GAA) architecture, and $I_{on} = 9$ μA/μm for the n-TFET, with tri-gate architecture. A minimum S of 30 mV/decade for n-type tunnel FET and 62 mV/decade for p-type tunnel FET were achieved at 300 K for same devices (Figure 7.6a). These complementary devices allowed the demonstration of full functional funnel FET inverters with gain in the voltage transfer characteristics (VTC) superior to the one of CMOS at voltages below 0.5 V (Figure 7.6b). From the same group of Juelich, Richter et al. [36] proposed and compared two experimental TFET structures both based on arrays of strained SiGe on SOI NWs. Of particular interest

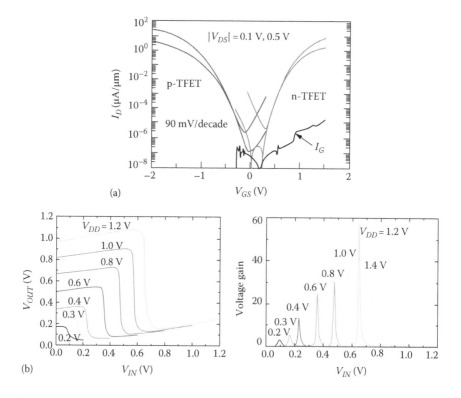

FIGURE 7.6 (a) Characteristics of strained Si NW array complementary tunnel FETs, providing a minimum S of 30 mV/decade for the n-FET, and of 90 mV/decade for the p-TFETs at 300 K. (b) VTC with voltage gain for NW tunnel FET inverters, functioning down to 0.2 V.

is a tri-gated NW TFET structure, which combines the concept of a heterostructure tunnel junction with the idea of an enlarged tunnel junction area. The heterostructure tunnel junction consisted of an in situ boron-doped $Si_{0.5}Ge_{0.5}$ source and an intrinsic Si channel. The tunnel direction was vertical and the tri-gate structure allowed for tunneling on the sidewalls and the front plane; even though the experimental results are not setting records, the concept of the device is scalable, being a step forward toward optimized tunnel FET on CMOS platforms.

Recently, CEA-LETI reported NW tunnel FETs [37] in a CMOS-compatible process flow featuring compressively strained $Si_{1-x}Ge_x$ ($x = 0, 0.2, 0.25$) NWs, $Si_{0.7}Ge_{0.3}$ source and drain and high-k/Metal gate. As expected, their NW architecture improved the electrostatic control, while low bandgap channel (SiGe) offered increased BTBT rates. Even if the fabricated devices exhibit the highest I_{on} ever reported (up to 760 μA/μm at $V_{DS} = -0.9$ V and $V_{GS} = -2$ V) for an average S of less than 80 mV/decade, the unusually high level of I_{on}, the quasiconstant swing versus gate voltage, and the reported temperature dependence of S suggest that these devices are not pure tunnel FETs and a trap-assisted tunneling and/or a parasitic MOSFET can better explain the reported high performance.

7.2.2 III–V-BASED TUNNEL FETS

Tunnel FETs can take advantage of the low bandgap of III–V materials to achieve significantly higher values of the I_{on} current than in all-silicon; this observation triggered many experimental and simulation efforts. High I_{on} values have been reported by University of Lund product: GaSb–InAsSb heterostructure NW tunnel FET (Figure 7.7) [38]; in their work, the *broken band alignment* allowed for BTBT without a barrier, leading a maximum drive current of 310 μA/μm at $V_{DS} = 0.5$ V and transconductance up to $g_m = 250$ mS/mm at $V_{DS} = 300$ mV. Similar levels of I_{on} have been reported by Zhou et al. from Notre Dame, in a vertical InAs/GaSb tunnel FET device with a broken band alignment: 180 μA/μm for $V_{GS} = V_{DS} = 0.5$ V and $I_{on}/I_{off} = 3 \times 10^3$ [39]. However, in both cases, the I_{on}/I_{off} ratio was poor (<10^4), essentially because of the increased I_{off} current, as expected in low bandgap materials. Mohata et al. [40] demonstrated tunnel FETs in the same material system, but exploiting a molecular beam epitaxy (MBE) mesastructure. They achieved similar values of I_{on} (~135 μA/μm) but with improved I_{on}/I_{off} ratio of 2.7×10^4 at $V_{DS} = 0.5$ V and $V_{GS} = 1.5$ V.

In combination with the increased interest in III–V tunnel FETs, various additive performance boosters, especially concerning improved electrostatic control and device geometry optimization, have been explored by different authors. The NW approach has been particularly preferred by many authors because of its optimum electrostatic control and wrapped-gate architectures, required for tunnel FET optimization. According to Schenk et al. [41], depending on geometry and doping, tunnel FET devices can feature a mix of *in*-junction and *off*-junction tunneling components: whereas the former cannot yield a subthermal slope, the latter can eventually produce a point slope of 25 mV/decade. The same authors suggested limits of I_{on} current in tunnel FETs by studying based on TCAD simulations Si, InAs, and Si–InAs device embodiments; they found that the maximum I_{on} in a Si–InAs hetero-TFET is about three orders of magnitude less than world-record CMOS.

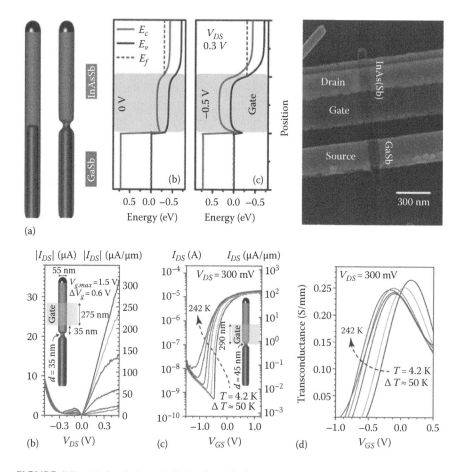

FIGURE 7.7 (a) Depiction of GaSb–InAsSb heterostructure NW tunnel FET with the broken band alignment: energy diagrams and top view of fabricated devices, (b), (c), and (d) experimental current and transconductance characteristics. (Reprinted from Dey, A.W. et al., *IEEE Electron Dev. Lett.*, 34, 211, 2013.)

Another smartly optimized device has been proposed by Notre Dame University [41]: their structure was a vertical n-channel tunnel FETs with tunneling normal to the gate based on an n+ InGaAs/p+ InP heterojunction and exhibiting simultaneously a high I_{on}/I_{off} ratio of 6×10^5, a minimum subthreshold swing S of 93 mV/decade, and an $I_{on} = 20$ μA/μm at $V_{DS} = 0.5$ V. The authors used as technology boosters a ultra-thin equivalent oxide thickness (EOT) of ~1.3 nm and a process based on plasma-enhanced chemical vapor deposition SiN mesa passivation. However, their structure would be difficult to scale and/or produce complementary devices for logic ICs.

In terms of steep swing, Intel [42,43] reported a heterojunction tunnel FET designed with TaSiO$_x$ gate dielectric, TiN/Pd metal gate, and In$_{0.7}$Ga$_{0.3}$As pocket (see Figure 7.8a) capable of providing a subthermal swing $S < 60$ mV/decade at supply voltage $V_{dd} = 0.3$ V and achieve $I_{on} = 1$ μA/μm and $I_{off} = 200$ pA/μm. In their findings, they interestingly showed a comparison with one of the best existing InGaAs FET

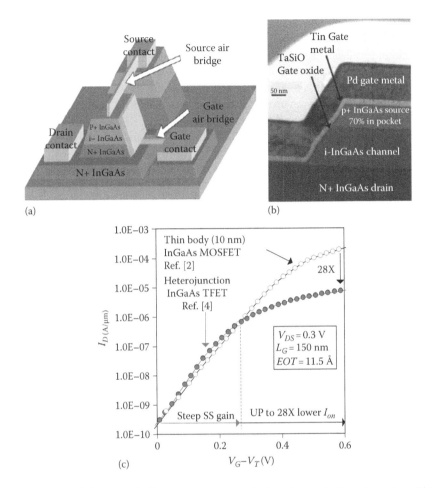

(a)

(b)

(c)

FIGURE 7.8 (a) A schematic diagram of the III–V single gate tunnel FET with a 5 μm thick body. The gate is formed on the vertical sidewall of the device. To reduce parasitic leakages, the gate and source pads are isolated via mesa etch. (b) Cross-section of the fabricated devices, (c) I_D versus gate overdrive ($V_G - V_T$) comparison of the best experimental heterojunction InGaAs TFET and thin body (T_{BODY} = 10 nm) $In_{0.7}Ga_{0.3}As$ MOSFET at matched I_{off} (200 pA), L_G, EOT, and V_{DS} = 0.3 V. Due to steeper SS, the InGaAs TFET shows gain over the InGaAs MOSFET at low overdrive, but reduced I_{on} at $V_G - V_T$ > 0.27 V. (Reprinted after Dewey, G. et al., Fabrication, characterization, and physics of III–V heterojunction tunneling Field Effect Transistors (H-TFET) for steep sub-threshold swing, *IEEE International Electron Devices Meeting (IEDM)*, 2011, pp. 33.6.1–33.6.4; Dewey, G. et al., III–V field effect transistors for future ultra-low power applications, *Symposium on VLSI Technology*, 2012, pp. 45–46.)

having a 10 nm thin body, pointing out clear performance advantages of tunnel FETs at voltages below 0.3 V (Figure 7.8b).

Compared to such III–V tunnel FET approaches, IBM Zürich reported GAA NW InAs/Si heterostructure p-type devices [44] (Figure 7.8) with I_{on} of 6 μA/μm ($|V_{GS}|$ = $|V_{DS}|$ = 1 V) and a room-temperature subthreshold swing, S, of ~160 mV/decade over at least three orders of magnitude in current, with one of the highest I_{on}/I_{off} of about 10^6.

With further optimization (reduction of EOT, control of interface states density, D_{it}, diameter of NWs below 20 nm, and reduction of contact resistances) such structure is scalable and probably one of the closest to future ultralarge scale industrial integration.

Finally, the challenge for tunnel FETs is to *simultaneously* demonstrate the high I_{on} and low I_{off} in combination with a very small swing S over a large current range (more than four decades of current), and so far, no tunnel FET has truly reached this goal to date. It is clear that all-Si homojunction tunnel FETs are not likely to reach the required I_{on} performance nor all-III–V heterostructure devices, in terms of I_{off} and I_{on}/I_{off}. Therefore, the solution is expected from heterostructures with bandgap engineering, embodied as NWs with reduced diameter dimensions (below 20 nm), which undisputedly provide the ultimate electrostatic control and scaling potential. In addition, the integration of III–V TFET devices on Si platform is of outmost importance and creates a credible path for industrial exploitation. Recently, in the European E²-SWITCH project [44,45], IBM has demonstrated that templated selective epitaxy (TSE) is a promising novel approach for the selective epitaxial integration of nanoscale III–V materials on Si (Figure 7.9), offering a high flexibility in the selection of materials, a direct solution to diameter aggressive scaling, and the possibility of combining several different material structures on the same Si platform.

7.3 TOWARD A DENSITY-OF-STATES ELECTRONIC SWITCH EXPLOITING 2D–2D TUNNELING

Despite the great progress and promise of the tunnel FETs reported previously, one can see that we are still far from a drastic reduction of the operating voltage (such as 0.1 V or less) and really abrupt subthreshold swings (below 10 mV/decade at room temperature). One stringent question arising is: for such tunnel FET devices, is there anychance to realistically achieve a truly deep subthermal electronic switch that can fulfill ultra-abrupt off-to-on switching characteristics and aggressive voltage scaling? The answer is of more fundamental nature. As discussed in the previous sections, several factors have to be taken into account in the BTBT rate, which determines the drain current dependence on V_{GS}, and a closed-form analytic formula is difficult to obtain without making some approximations. Indeed, a generally valid expression for the tunneling current dependence on the junction parameters is the following [46]:

$$I_t \approx \int_{E_C^n}^{E_V^p} T_t \left| f_V^p(E) - f_C^n(E) \right| D_V^p(E) D_C^n(E) dE \tag{7.8}$$

where

E_V^p and E_C^n are the maximum valence and minimum conduction bands in the source and the channel, respectively

f_V^p and f_C^n are the Fermi distributions in the valence and conduction band, for the source and the channel, respectively

D_V^p and D_C^n are the density of states (DOS) in the valence and conduction bands for the source and the channel, respectively.

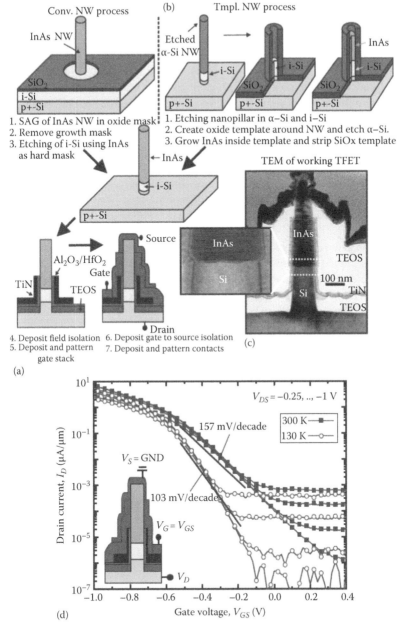

FIGURE 7.9 Illustration of two different integration methods: conventional NW growth (a), where InAs NWs are grown by selective area epitaxy in oxide openings and subsequently used as a hard mask for dry-etching into the Si. In the template-assisted method (b), InAs is grown selectively within an oxide tube. After NW growth, the two processes are similarly continued. (c) Cross-sectional transmission electron microscopy (TEM) of a finished (tmpl.NW) device is shown including a magnification of the InAs/Si interface in the inset. (d) Resulting tunnel FET experimental characteristics.

The term $f_V^p - f_C^n$ is also called the Fermi window, which is controlled by the applied bias. Note that the first parameter, T_t, is the transparency of the tunnel barrier, on which the majority of the previous studies have focused concerning its control and optimization through λ, m, and E_g. We observe that the drain current variation in tunnel FETs is mainly controlled by a joint effect of three factors: (1) the *transparency of the tunneling barrier*, (2) the *Fermi window*, and (3) *the DOS*. In the same way, the subthreshold slope will depend on the same three factors, which jointly determine the subthermal switching slopes in TFETs ($S < 60$ mV/decade) due to their energy dependence. The *energy filtering* was considered the best way to realize of tunnel FETs, as reported in the previous sections.

The DOS term (3) in Equation 7.8 was not exploited by any of the tunneling devices discussed in the previous sections. Interestingly, this equation captures a diversity of device optimization that can be summarized in what we call here DOS switch, as observed by Agarwal and Yablonovitch [47]. First, one can observe that, surprisingly, in a typical 3d bulk TFET, the nature of the turn on is actually quadratic in the gate voltage and it is possible to significantly improve this if the dimensionality is reduced. These authors explored the nature of the band overlap for the various dimensionalities and demonstrated that in a 2D–2D pn junction, the switching process is closer to an ideal step function. Confining each side of the pn junction is also advantageous, because it significantly increases the on-state conductivity at low voltages. A detailed discussion on the concept of backward diode as key step in understanding and optimizing a tunnel FET exploiting the effect of dimensionality and the sharpness of the band edges is offered in [48].

Lattanzio et al. have proposed for the first time a tunnel FET device exploiting the effect of dimensionality in the so-called electron-hole bilayer tunnel FET (EHBTFET) [49], illustrated in Figure 7.10a through c. The same authors have studied in depth this new DOS switch structure in [50–52] by making use of BTBT between two *electrostatically induced* electron and hole inversion layers. The need of a highly abrupt tunneling junction, a technological difficulty for high performance in tunnel FETs, is elegantly eliminated. The 2D nature of both electron and hole inversion layers are extremely favorable for obtaining a deep subthermal tunnel FET as the step-like DOS of 2D carrier gases allows a very steep switching slope. Another key advantage of this device concept is that the same structure can be operated as both n- and p-type switches, so the EHBTFET naturally has a complementary nature and it is easy to design and implement all elementary CMOS logic blocks using this device concept. Figure 10d reports the very abrupt simulated characteristics and the I_{on} and I_{off} performance for an EHBTFET in ultrathin Ge layer.

It is worth noting that using a silicon embodiment of EHBTFET suffers from the relatively large and indirect gap, which cripples the BTBT rate, and hence the *on* current. Furthermore, in the specific case of EHBTFET, since both electron and hole levels are quantized; the potential required to align the subbands reach very high values, which is clearly undesirable for a low supply voltage device. In germanium, on the other hand, both direct and indirect BTBTs can be observed, which allows much higher tunneling currents. However, such embodiment can also suffer from the indirect (phonon-assisted) BTBT that occurs at lower voltages than direct BTBT due to lower bandgap and higher effective masses.

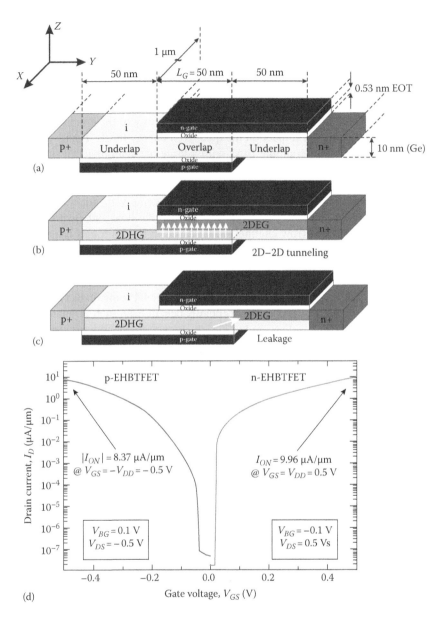

FIGURE 7.10 EHBTFET device structure after Lattanzio et al. [49,52]. The device proposed has a channel thickness $t_{CH} = 10$ nm, n+/p+ contact regions are here extended to 200 nm to account for the open boundaries in the contacts. (a) Indication of overlap and underlap regions (b) 2D–2D BTBT between quantized electron and hole gases. (c) Possible leakage path in the *off* state caused by the electron WF penetration into the underlap region. (d) Simulated device transfer characteristics of an **EHBTFET** embodied in germanium showing switching characteristics for both n- and p-type devices with a swing S smaller than 10 mV/decade at room temperature and operation below 0.5 V.

Since the BTBT rate for an indirect transition is much lower compared to a direct one, it could effectively act as a "leakage" mechanism that degrades device figures of merit. A detailed discussion of the importance of the leakage mechanisms in EHBTFET and ways to fix them was proposed by Alper et al. [53]. In the light of these arguments, the incorporation of III–V materials such as InAs, which have lower direct bandgap, is critical to obtain high I_{on} currents and significantly optimized EHBTFETs in the future [54]. Although the very low effective mass in InAs might indeed be an issue, the high nonparabolicity of the conduction bands in InAs will alleviate the impact on the quantized energy levels. Finally, it seems almost indispensable to combine InAs with a large bandgap material in order to keep I_{off} current levels under control. Another important aspect that is related to the material selection is the device dimensions, in particular the channel thickness. Using ultranarrow channels (below 10 nm) seems very appealing, at first view, since it would increase the coupling between the hole and electron layers. However, using a too narrow channel would result in too high bias requirements to align the electron and hole quantized energy levels as correctly pointed out first in [55], and supported by the extensive study given in [56].

Finally, one can observe that the discovery of electronic-grade 2D crystals has added a new material solutions to the list of conventional semiconductors used for transistors and the tunnel FETs can conceptually benefit from such 2D materials and device structures. Jena has discussed in [57] the tunneling phenomenon in structure with reduced dimensions and observed two major trends for tunnel FET optimization (see Figure 7.11): (1) *lateral tunnel FETs* that show evolution toward thinner structures where the tunneling current exploits better electrostatic control but quantum confinement increases the bandgap and reduces the tunneling current; here, the 2D crystals can avoid this detrimental effect, and (2) *vertical tunnel FETs* that show evolution from a side-gate geometry (left) to a vertical double-gate structures with improved performance, with 2D crystal layers. Such 2D crystals can indeed offer alternative technological embodiment and new ideas for the optimization of devices such as the EHBTFET or similar.

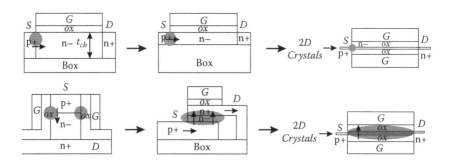

FIGURE 7.11 A schematic representation of various topologies of tunnel FETs and their evolutions toward implementations in 2D crystals for enhanced performance, after [57]. The top row shows tunnel FETs where the tunneling current flows laterally, while the bottom row shows tunnel FETs realizations in which the tunneling current flows vertically.

7.4   TUNNEL FETS FOR LOW-POWER ANALOG
AND SENSING APPLICATIONs

An aspect less explored in the literature is about the fact that tunnel FETs are open-ing interesting perspectives for analog design [58,59] of integrated circuits and appli-cations requiring high gain at ultralow voltages/currents and reduced dependence on temperature in the subthreshold region of operation.

Among the metrics that are significantly improved by tunnel FETs is the ratio between the transconductance and the drain current, g_m/I_d, which reflects the effi-ciency with which current (or power) is translated into transconductance: the greater the value of this ratio, the greater is the offered transconductance, g_m, at a given cur-rent value. It is worth mentioning that, similar to S, the ratio g_m/I_d has a fundamental limit for the MOSFET operated in the subthreshold regime, at $T = 300$ K:

$$\frac{g_m}{I_d} \cong \frac{1}{nkT/q} \le \frac{1}{(kT/q)} \cong 39 \text{ V}^{-1}. \tag{7.9}$$

It follows that tunnel FET has the ability to overpass this fundamental limit, offering significant gain improvement for analog applications operating at very low levels of drain current.

For instance, this unique property can be exploited in new design for energy-efficient analog-to digital converters (ADCs) where the power dissipation per sam-pling frequency is limited by signal-to-noise ratio (SNR), g_m/I_d, and supply voltage, V_{CC}, which cannot be scaled because of the requirements on SNR. Taking into account the relation between these parameters, power and sampling frequency, one would expect tunnel FET to open a new design space for future energy-efficient ADCs:

$$\frac{Power}{Sampling \ frequency} \approx \frac{kT \cdot SNR}{V_{CC}} \left(\frac{g_m}{I_d}\right)^{-1}. \tag{7.10}$$

In order to evaluate how tunnel FET can enhance the performance analog circuits, several building blocks such as operational transconductance amplifiers (OTAs), current mirrors, and track-and-hold circuits have been examined by Sedighi et al. [60]. They demonstrated that tunnel FETs are promising for low-power and low-voltage designs: comparing 14 nm III–V TFET-based OTAs with Si-MOSFET-based designs demonstrates up to five times reduction in the power dissipation of the ampli-fiers and more than an order of magnitude increase in their DC voltage gain.

From the temperature dependence point of view, tunnel FETs are expected to have swings with very little dependence on temperature as the only fundamental dependence is the low quasilinear bandgap dependence on the temperature. The temperature is also a very effective method to evaluate that a fabricated device is a good tunnel FET (and not a MOSFET) and also distinguish other conduction mecha-nisms like trap-assisted tunneling that have different temperature dependences.

In addition to low-power digital and analog applications, due to their structural (p/i/n) and operational similarity (in reversed bias) to photodiodes, the first intrigu-ing and nonconventional use of tunnel FETs in image sensors has been reported only recently in [61,62]. A conventional active pixel sensor (APS) uses 3–4 MOSFET transistors for cell selectivity, reset, and charge transfer functions, and there is a great incentive to increase the fill factor by reducing the number of components while improving sensitivity and dynamic range, and reducing power consumption with a one transistor tunnel FET architecture, refer Figure 7.12a and b. The first APS tunnel FET proposed by Dağtekin and Ionescu [61,62] has a partially gated p/i/n structure, as illustrated in Figure 7.12c. They were fabricated on a fully depleted (FD) SOI substrate by CEA-LETI. The extension of the intrinsic silicon (L_{IN}) on the n+ doped region that is coated with Si_3Ni_4 is principally the region where light is absorbed. The gate stack, on the other hand, due to its metal and poly-Si stack, absorbs a large fraction of the optical power, preventing carrier generation along the channel under gate 1 (L_G). Depending on the bias conditions, BTBT occurs at p/i or n/i junction, giving rise to N- and P-mode operation, respectively (ambipolar conduction).

The transfer characteristics of a 1 tunnel FET APS under various visible light intensities (E) and V_{DS} are reported in Figure 7.12d. The ambipolar characteristics indicate that BTBT takes place at either or both junctions depending on V_{DS} sign. These characteristics reveal a few interesting properties: (1) I_{off} increases linearly with E; (2) when the tunneling is relatively low, I_D is elevated due to photocurrents as in a PIN diode; (3) in P-mode operation, I_D is observed to be not only a function of V_{G1} but also of E, even at high tunneling rates. The optical excitation causes I_D to decrease, giving rise to a phototransistor with two regions: first, with a positive illumination coefficient, I_C ($=\Delta I/\Delta E$) (point B), as a conventional PIN junction, and, second, with negative I_C (point C) when operated as a TFET, with optical gain. Between these two regions, there is a unique bias point where the illuminance has no impact on I_D (called aero illumination coefficient [ZIC]), due to opposite dependences of leakage and tunneling on the charging of the potential well of this structure, with optically generated carriers [62]. Overall, compared to CMOS, the 1T tunnel FET pixel offers high sensitivity, with a detection limit <2 pW/μm^2 in visible light [62], low power operation, improved temperature stability, and high compactness.

7.5 CONCLUSIONS

Tunnel FET stands as the most promising steep slope switch candidate to reduce the supply voltage below 0.3 V and offer significant power dissipation savings together with an extension of the low-power design space compared to traditional CMOS. An optimized complementary tunnel FET technology is still under exploration and heterostructure tunnel FETs exploiting III-V/structures on silicon and/or strained (Si) Ge source devices seem to be the most promising embodiments. New concepts like the density-of-state switch (EHBTFET) exploiting the effect of dimensionality (in ultrathin films or in new architectures based on 2D materials) to tune its operation performance could offer supplementary future paths for a deep subthermal switch aiming at sub-0.1 V operation.

Additionally, tunnel FETs have unique characteristics such as gain at ultralow levels of current and voltage, which could extend the analog design of integrated circuits into a space than was never explored before and highly demanded by sensor nodes for Internet-of-Things applications. Moreover, their structure and voltage-sensitive characteristics made them also very interesting candidates for integrated sensing.

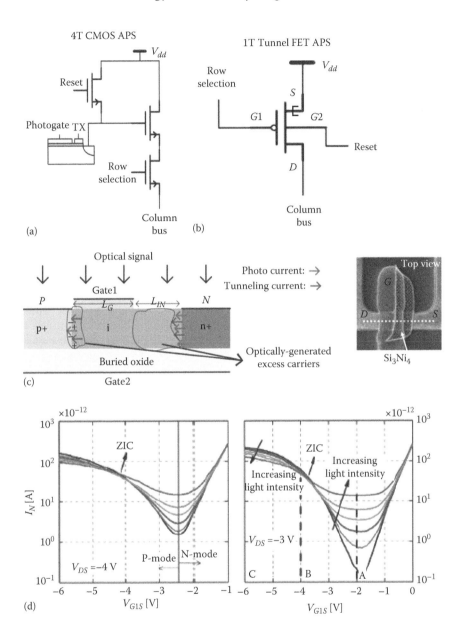

FIGURE 7.12 (a) A schematic diagram of a conventional CMOS APS with four transistors and a photogate, and (b) tunnel FET APS with 1 transistor. (c) Structure of the measured tunnel FET APS conceived by EPFL and fabricated by CEA-LETI on SOI. Main carrier injection mechanisms are indicated with arrows. SEM picture on the right shows the top view of the imager. The size of the designed and investigated 1-TFET APSs are $10 \times 0.8 \ \mu m^2 < A < 10 \times 1.24 \ \mu m^2$. (d) Measured transfer characteristics of 1T APS using tunnel FET at $V_{G2} = 0$ V, $V_{DS} = 0, -1.5, -3, -4$ V and at various illumination intensities demonstrating phototransistor behavior with a less than 2 pW/μm^2 detection limit.

ACKNOWLEDGMENT

The contributions of the partners of the STEEPER and E^2-SWITCH projects funded by the European Commission in the preparation of materials and results supporting this publications are greatly acknowledged.

REFERENCES

1. A.M. Ionescu, H. Riel, Tunnel field-effect transistors as energy-efficient electronic switches, *Nature*, 479, 2011, 329–337.
2. K. Gopalakrishnan, P.B. Griffin, J.D. Plummer, I-MOS: A novel semiconductor device with a subthreshold slope lower than kT/q, *IEEE International Electron Devices Meeting (IEDM)*, 2002, pp. 289–292.
3. S. Salahuddin, S. Datta, Use of negative capacitance to provide voltage amplification for low power nanoscale devices, *Nano Letters*, 8, 2008, 405–410.
4. A. Rusu, G.A. Salvatore, D. Jimenez, A.M. Ionescu, Metal-ferroelectric-meta-oxide-semiconductor field effect transistor with sub-60 mV/decade subthreshold swing and internal voltage amplification, *IEEE International Electron Devices Meeting (IEDM)*, 2010, pp. 16.3.1–16.3.4.
5. A.M. Ionescu, V. Pott, R. Fritschi, K. Banerjee, M.J. Declercq, P. Renaud, C. Hibert, P. Fluckiger, G.A. Racine, Modeling and design of a low-voltage SOI suspended-gate MOSFET (SG-MOSFET) with a metal-over-gate architecture, *International Symposium on Quality Electronic Design*, 2002, pp. 496–501.
6. N. Abelé, R. Fritschi, K. Boucart, F. Casset, P. Ancey, A.M. Ionescu, Suspended-gate MOSFET: Bringing new MEMS functionality into solid-state MOS transistor, *IEEE International Electron Devices Meeting (IEDM)*, 2005, pp. 479–481.
7. Y. Zhou, X. Chen, C. Ko, Z. Yang, C. Mouli, S. Ramanathan, Voltage-triggered ultra-fast phase transition in vanadium dioxide switches, *IEEE Electron Dev. Lett.*, 34, 2013, 220–222.
8. W.A. Vitale, C.F. Moldovan, A. Paone, A. Schüler, A.M. Ionescu, CMOS-compatible abrupt switches based on VO$_2$ metal-insulator transition, *2015 Joint International EUROSOI Workshop and International Conference on Ultimate Integration on Silicon (ULIS)*, 2015.
9. A.C. Seabaugh, Q. Zhang, Low-voltage tunnel transistors for beyond CMOS logic, *Proc. IEEE*, 98(12), 2010, 2095–2110.
10. K. Boucart, A.M. Ionescu, Double-gate tunnel FET with high-κ gate dielectric, *IEEE Trans. Electron Dev.*, 54, 2007, 1725–1733.
11. S. Agarwal, E. Yablonovitch, Fundamental tradeoff between conductance and sub-threshold swing voltage for barrier thickness modulation in tunnel field effect transistors, EECS Technical Report No. UCB/EECS-2014-6, University of California, Oakland, CA, 2014.
12. J. Knoch, S. Mantl, J. Appenzeller, Impact of the dimensionality on the performance of tunneling FETs: Bulk versus one-dimensional devices, *Solid-State Electron.*, 51(4), 2007, 572–578.
13. A.M. Ionescu, Tunnel FETs and emerging device concepts for subthermal switching, IEDM Short Course, 2013.
14. T. Baba, Proposal for surface tunnel transistors, *Jpn. J. Appl. Phys.*, 31(4B), 1992, L455–L457.
15. W. Reddick, G. Amaratunga, Silicon surface tunnel transistor, *Appl. Phys. Lett.*, 67(4), 1995, 494–496.
16. J. Koga, A. Toriumi, Three-terminal silicon surface junction tunneling device for room temperature operation, *IEEE Electron Dev. Lett.*, 20, 1999, 529–531.

17. K. Boucart, Simulation of double-gate silicon tunnel FETs with a high-k gate dielectric, PhD thesis no. 4729, EPFL, 2010.

18. K. Boucart, W. Riess, A.M. Ionescu, Lateral strain profile as key technology booster for all-silicon tunnel FETs, *IEEE Electron Dev. Lett.*, 30, 2009, 656–658.

19. O. Nayfeh, J. Hoyt, D. Antoniadis, Strained-Si$_{1-x}$Ge$_x$/Si band-to-band tunneling transistors: Impact of tunnel-junction germanium composition and doping concentration on switching behavior, *IEEE Trans. Electron Dev.*, 56, 2009, 2264–2269.

20. T. Krishnamohan, D. Kim, S. Raghunathan, K. Saraswat, Double-gate strained-Ge heterostructure tunneling FET (TFET) with record high drive currents and <60 mV/dec subthreshold slope, *IEDM Tech. Dig.*, December 15–17, 2008, pp. 947–949.

21. K.E. Moselund, M.T. Bjork, H. Schmid, H. Ghoneim, S. Karg, E. Lortscher, W. Riess, H. Riel, Silicon nanowire tunnel FETs: Low-temperature operation and influence of high-gate dielectric, *IEEE Trans. Electron Dev.*, 58, 2011, 2911–2916.

22. R. Gandhi, Z. Chen, N. Singh, K. Banerjee, S. Lee, CMOS-compatible vertical-silicon-nanowire gate-all-around p-type tunneling FETs with ≤50-mV/decade subthreshold swing, *IEEE Electron Dev. Lett.*, 32, 2011, 437–439.

23. P.-F. Guo, L.-T. Yang, Y. Yang, L. Fan, G.-Q. Han, G.S. Samudra, Y.-C. Yeo, Tunneling field-effect transistor: Effect of strain and temperature on tunneling current, *IEEE Electron Dev. Lett.*, 30, 2009, 981–983.

24. L. Knoll, Q.T. Zhao, S. Trellenkamp, A. Schafer, K.K. Bourdelle, S. Mantl, Si tunneling transistors with high on-currents and slopes of 50 mV/dec using segregation doped NiSi$_2$ tunnel junctions, *European Solid-State Device Research Conference (ESSDERC)*, 2012, pp. 183–156.

25. D. Kazazis, P. Jannaty, A. Zaslavsky, C. Le Royer, C. Tabone, L. Clavelier, S. Cristoloveanu, Tunneling field-effect transistor with epitaxial junction in thin germanium-on-insulator, *Appl. Phys. Lett.*, 94(26), 2009, 263508.

26. S.H. Kim, H. Kam, C. Hu, T.-J. King Liu, Germanium-source tunnel field effect transistors with record high I$_{ON}$/I$_{OFF}$, *Symposium on VLSI Technology*, 2009, pp. 178–179.

27. S. Richter, S. Blaeser, L. Knoll, S. Trellenkamp, A. Schafer, J.M. Hartmann, Q.T. Zhao, S. Mantl, SiGe on SOI nanowire array TFETs with homo- and heterostructure tunnel junctions, *14th International Conference on Ultimate Integration on Silicon (ULIS)*, 2013, pp. 25–28.

28. J. Nah, L. En-Shao, K.M. Varahramyan, D. Dillen, S. McCoy, J. Chan, E. Tutuc, Enhanced-performance germanium nanowire tunneling field-effect transistors using flash-assisted rapid thermal process, *IEEE Electron Dev. Lett.*, 31, 2010, 1359–1361.

29. K. Jeon, W.-Y. Loh, P. Patel, C.Y. Kang, J. Oh, A. Bowonder, C. Park et al., Si tunnel transistors with a novel silicided source and 46 mV/dec swing, *2010 Symposium on VLSI Technology*, 2010, pp. 121–122.

30. Q. Huang, Z. Zhan, R. Huang, X. Mao, L. Zhang, Y. Qiu, Y. Wang, Self-depleted T-gate Schottky barrier tunneling FET with low average subthreshold slope and high ION/IOFF by gate configuration and barrier modulation, *IEEE International Electron Devices Meeting (IEDM)*, 2011, pp. 16.2.1–16.2.4.

31. H.-Y. Chang, B. Adams, C. Po-Yen, J. Li, J.C.S. Woo, Improved subthreshold and output characteristics of source-pocket Si tunnel FET by the application of laser annealing, *IEEE Trans. Electron Dev.*, 60, 2013, 92–96.

32. D. Leonelli, A. Vandooren, R. Rooyackers, S. De Gendt, M.M. Heyns, G. Groeseneken, Drive current enhancement in p-tunnel FETs by optimization of the process conditions, *Solid-State Electron.*, 65–66, 2011, 28–32.

33. W.G. Vandenberghe, A.S. Verhulst, B. Sorée, W. Magnus, G. Groeseneken, Q. Smets, M. Heyns, M.V. Fischetti, Figure of merit for and identification of sub-60 mV/decade devices, *Appl. Phys. Lett.*, 102, 2013, 013510.

34. L. Knoll, S. Richter, A. Nichau, S. Trellenkamp, A. Schäfer, K.K. Bourdelle, J.M. Hartmann, Q.T. Zhao, S. Mantl, Strained Si and SiGe tunnel-FETs and complementary tunnel-FET inverters with minimum gate lengths of 50 nm, *Solid-State Electron.*, 97, 2014, 76–81.

35. L. Knoll, Q.T. Zhao, A. Nichau, S. Trellenkamp, S. Richter, A. Schäfer, Inverters with strained Si nanowire complementary tunnel field-effect transistors, *IEEE Electron Dev. Lett.*, 34, 2013, 813–815.

36. S. Richter, S. Blaeser, L. Knoll, S. Trellenkam, A. Fox, A. Schäfer, J.M. Hartmann, Q.T. Zhao, S. Mantl, Silicon–germanium nanowire tunnel-FETs with homo- and heterostructure tunnel junctions, *Solid-State Electron.*, 98, 2014, 75–80.

37. A. Villalon, C. Le Royer, P. Nguyen, S. Barraud, F. Glowacki, A. Revelant, L. Selmi et al., First demonstration of strained SiGe nanowires TFETs with I_{ON} beyond 700 μA/μm, *2014 Symposium on VLSI Technology*, 2014, pp. 1–2.

38. A.W. Dey, B.M. Borg, B. Ganjipour, M. Ek, K.A. Dick, E. Lind, C. Thelander, L. Wernersson, High-current GaSb/InAs(Sb) nanowire tunnel field-effect transistors, *IEEE Electron Dev. Lett.*, 34, 2013, 211–213.

39. G. Zhou, R. Li, T. Vasen, M. Qi, S. Chae, Y. Lu, Q. Zhang et al., Novel gate-recessed vertical InAs/GaSb TFETs with record high I_{ON} of 180 μA/μm at $V_{DS} = 0.5$ V, *IEEE International Electron Devices Meeting (IEDM)*, 2012, pp. 32.6.1–32.6.4.

40. D.K. Mohata, R. Bijesh, Y. Zhu, M.K. Hudait, R. Southwick, Z. Chbili, D. Gundlach et al., Demonstration of improved heteroepitaxy, scaled gate stack and reduced interface states enabling heterojunction tunnel FETs with high drive current and high on-off ratio, *Symposium on VLSI Technology*, 2012, pp. 53–54.

41. A. Schenk, R. Rhyner, M. Luisier, C. Bessire, Analysis of Si, InAs, and Si-InAs tunnel diodes and tunnel FETs using different transport models, *70th Device Research Conference (DRC)*, June 18–20, 2012, pp. 201–202.

42. G. Dewey, B. Chu-Kung, J. Boardman, J.M. Fastenau, J. Kavalieros, R. Kotlyar, W.K. Liu et al., Fabrication, characterization, and physics of III–V heterojunction tunneling Field Effect Transistors (H-TFET) for steep sub-threshold swing, *IEEE International Electron Devices Meeting (IEDM)*, 2011, pp. 33.6.1–33.6.4.

43. G. Dewey, B. Chu-Kung, R. Kotlyar, M- Metz, N. Mukherjee, M. Radosavljevic, III–V field effect transistors for future ultra-low power applications, *Symposium on VLSI Technology*, 2012, pp. 45–46.

44. D. Cutaia, K.E. Moselund, M. Borg, H. Schmid, L. Gignac, Ch. Breslin, S. Karg, E. Uccelli, H. Riel, Vertical InAs-Si gate-all-around tunnel FETs integrated on Si using selective epitaxy in nanotube templates, *IEEE J. Electron Dev. Soc.*, 3, 2015, 176–183.

45. E2Switch home page, Energy efficient tunnel FET switches and circuits, 2014, http://www.e2switch.org/.

46. L. Lattanzio, Innovative tunnel field-effect transistor architectures, PhD thesis no. 5691, EPFL, 2013.

47. S. Agarwal, E. Yablonovitch, Using dimensionality to achieve a sharp tunneling FET (TFET) turn-on, *69th Annual Device Research Conference (DRC)*, 2011, pp. 199–200.

48. S. Agarwal, E. Yablonovitch, Band-edge steepness obtained from esaki/backward diode current–voltage characteristics, *IEEE Trans. Electron Dev.*, 61, 2013, 1488–1493.

49. L. Lattanzio, L. De Michielis, A.M. Ionescu, Electron-hole bilayer tunnel FET for steep subthreshold swing and improved ON current, *European Solid-State Device Research Conference (ESSDERC)*, 2011, pp. 259–262.

50. L. De Michielis, L. Lattanzio, P. Palestri, L. Selmi, A.M. Ionescu, Tunnel-FET architecture with improved performance due to enhanced gate modulation of the tunneling barrier, *69th Device Research Conference*, 2011, pp. 111–112.

51. L. Lattanzio, L. De Michielis, A.M. Ionescu, The electron–hole bilayer tunnel FET, *Solid State Electron.*, 74, 2012, 85–90.

52. L. Lattanzio, L. De Michielis, A.M. Ionescu, Complementary germanium electron–hole bilayer tunnel FET for sub-0.5-V operation, *IEEE Electron Dev. Lett.*, 33, 2012, 167–169.

53. C. Alper, P. Palestri, L. Lattanzio, J.L. Padilla, A.M. Ionescu, Two dimensional quantum mechanical simulation of low dimensional tunneling devices, *44th European Solid-State Device Research Conference (ESSDERC)*, 2014, pp. 186–189.

54. A.M. Ionescu, C. Alper, J.L. Padilla, L. Lattanzio, P. Palestri, Electron-hole bilayer deep subthermal electronic switch: Physics, promise and challenges, *IEEE SOI-3D-Subthreshold Microelectronics Technology Unified Conference (S3S)*, 2014, pp. 1–3.

55. A. Revelant, A. Villalon, Y. Wu, A. Zaslavsky, C. Le Royer, H. Iwai, S. Cristoloveanu, A. We, S. Ge, Electron-hole bilayer TFET: Experiments and comments, *IEEE Trans. Electron Dev.*, 61, 2014, 2674–2681.

56. J.T. Teherani, S. Agarwal, E. Yablonovitch, J.L. Hoyt, D. Antoniadis, Impact of quantization energy and gate leakage in bilayer tunneling transistors, *IEEE Electron Dev. Lett.*, 34, 2013, 298–300.

57. D. Jena, Tunneling transistors based on graphene and 2D crystals, *Proc. IEEE*, 101, 2013, 1585–1602.

58. A.M. Ionescu, L. De Michielis, N. Dagtekin, G. Salvatore, J. Cao, A. Rusu, S. Bartsch, Ultra low power: Emerging devices and their benefits for integrated circuits, *IEEE International Electron Devices Meeting (IEDM)*, 2011, pp. 16.1.1–16.1.4.

59. C. Schulte-Braucks, S. Richter, L. Knoll, L. Selmi, Q.-T. Zhao, S. Mantl, Experimental demonstration of improved analog device performance in GAA-NW-TFETs, *44th European Solid State Device Research Conference (ESSDERC)*, 2014, pp. 178–181.

60. B. Sedighi, X.S. Hu, H. Liu, J.J. Nahas, M. Niemier, Analog circuit design using tunnel-FETs, *IEEE Trans. Circuits Syst. I*, 62, 2015, 39–48.

61. N. Dağtekin, A.M. Ionescu, Investigation of partially gated Si tunnel FETs for low power integrated optical sensing, *44th European Solid State Device Research Conference (ESSDERC)*, 2014, pp. 190–193.

62. N. Dağtekin, A.M. Ionescu, Energy efficient 1-transistor active pixel sensor (APS) with FD SOI tunnel FET, *IEEE VLSI Technology Symposium*, 2015.

8 Dopant-Atom Silicon Tunneling Nanodevices

Daniel Moraru and Michiharu Tabe

CONTENTS

8.1 Introduction 181
  8.1.1 Role of Discrete Dopants in Scaled-Down Devices 181
  8.1.2 Doping Techniques for Nanoscale Devices 183
    8.1.2.1 Uncontrolled (Random) Doping 184
    8.1.2.2 Controlled Doping 185
  8.1.3 Guidelines for Dopant-Atom Devices.. 186
8.2 Observation of Individual Dopants by LT-KPFM 187
  8.2.1 Techniques for Dopant Detection 187
  8.2.2 KPFM Setup for Dopant Detection in Devices under Normal
        Operation 189
  8.2.3 Observation of Individual Dopant Potentials and
        Single-Electron Charging 189
  8.2.4 Correlation with Electrical Characteristics 191
8.3 Single-Electron Tunneling via Dopants in Nanoscale Transistors 193
  8.3.1 Tunneling via Individual Dopants in Nanoscale Transistors........... 193
  8.3.2 Effect of Channel Pattern and Dielectric Confinement on the
        Properties of Individual Dopants... 194
  8.3.3 Interacting Dopants in Silicon Nanotransistors............................... 198
8.4 Conclusions.. 201
Acknowledgments...202
References..202

8.1 INTRODUCTION

8.1.1 ROLE OF DISCRETE DOPANTS IN SCALED-DOWN DEVICES

The invention of the transistor in 1947 represented the starting point for a fast-paced development of semiconductor technology, thus setting the grounds for the electronics industry toward continuously improved functionalities. Related technological progress supported a steady trend toward more and more compact and complex integrated circuits, which has been achieved, in particular, by the reduction of transistor's critical dimensions, as predicted by Moore's law.[1] Recently, however, it can be foreseen that fundamental limitations will seriously hinder further device miniaturization. In fact, today, commercial transistors have channel dimensions on the order of 10–20 nanometers. At such a small scale, fluctuations of device fabrication become significant and pose serious concerns in terms of device yield for industrial applications. Among such fluctuations, uncontrollability in dopant distribution starts to be dominant. The diagram shown in Figure 8.1 illustrates the role of individual dopants in different aspects of research.

One of the important problems of conventional downscaled transistors is the discreteness and randomness of dopants (impurities) introduced in the device channels. In nanometer-scale devices, the number of dopant atoms in the channel becomes low enough that small fluctuations in their exact amount and particular positions induce strong fluctuations in the electrical characteristics (measured usually as fluctuations of the threshold voltage, ΔV_{TH}),[2,3] as shown in the upper-left panel of Figure 8.1. The correlation between the randomness of dopants or even the position of individual dopants and the threshold voltage has been extensively studied during the past decades both in theory[4] and experiments.[5–7]

A different research direction was triggered by the proposals for quantum computing, based on quantum dots (QDs)[8] or based on coupled donor atoms (qubits), which suggest to encode information either in the donors' spin states[9] or charge states.[10] The schematic depictions of these approaches are shown in the upper-right panel of Figure 8.1. Advanced doping technologies and optimized nanofabrication and measurement techniques have recently allowed significant progress in this direction.[11–13]

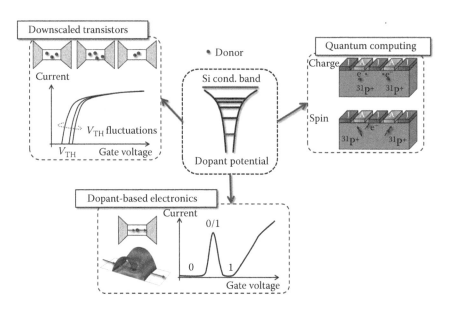

FIGURE 8.1 Research directions related to the impact of individual dopant atoms in nanodevices. Upper left: Effect of fluctuation in dopant distribution on V_{TH} fluctuations.[2–7] Upper right: Quantum computing, with information stored in the dopants' spin or charge states.[9–13] Bottom: Dopant-based electronics, with individual dopants working as natural QDs.[14–21]

From a different perspective, dopant atoms in scaled-down devices gain yet another, more "active" role in device operation, being treated as natural, ultrasmall QDs. Based on this concept, a new family of devices, called *single-dopant-atom devices* or *dopant-based devices*, has been developed. The properties of these devices have been gradually revealed through a large number of theoretical and experimental papers in recent years. As QDs, dopant atoms have attractive properties, since such dopant-induced QDs are naturally formed and have dimensions on the order of the Bohr radius, which is smaller than present capabilities of lithography techniques. Furthermore, typical dopants have limited charge occupancy, being either in an ionized state or in a neutral state. Developing dopant-atom devices using dopants in Si has an additional advantage of addressing atom-level functionalities while utilizing the well-established mainframe of Si technology. Over the past about 10 years, several research groups have deeply investigated, from various aspects, transport through individual dopants or a few coupled dopants (donors or acceptors) located in Si small-channel transistors.[14–21]

The basic principle of a single-dopant transistor operation is illustrated in the bottom panel of Figure 8.1, in an ideal design. If one dopant atom is placed in a nanoscale silicon channel, between well-defined source and drain leads, the dopant can work as a QD, mediating the charge transport between the leads. For instance, let us consider the situation of a phosphorus (P) donor atom between n-type source and drain leads. When such a system is formed in a field-effect transistor (FET) structure, a gate can control the potential of the channel and, implicitly, the potential of the donor. In particular, for negatively large gate voltages, V_G, the channel can be depleted of electrons, which means that the donor atom will also be positively ionized (p+) by removing its extra electron. Under these conditions, the ionized donor can be treated as a natural, really atomic QD because of its Coulomb well. This QD can accommodate, under regular conditions, one electron, as pointed out earlier. When V_G is gradually swept in the positive direction, the donor's ground state is aligned with the lead's Fermi level and electrons can be transported from source to drain by tunneling. In particular at low temperatures, electrical characteristics (I_D–V_G curves) exhibit a current peak, as a signature of Coulomb blockade (CB) transport mechanism. This is, basically, similar to the case of conventional single-electron transistors (SETs) with QDs artificially designed (patterned), with the key difference that, in single-dopant devices, the QD is defined by a dopant atom. Based on this concept, a variety of more complex functions can be realized, such as turnstiles or pumps, photonic devices, or even atom-level pn junction diodes, with some of these applications[22] described briefly in later sections.

For all of the earlier directions of research, it is important to develop appropriate doping techniques that could allow sufficient control of the properties of dopant-atom devices.

8.1.2 Doping Techniques for Nanoscale Devices

A number of doping techniques are already available. These techniques can be basically classified depending on the degree of control that they offer in terms of dopant position and/or number. On the one hand, conventional doping techniques,

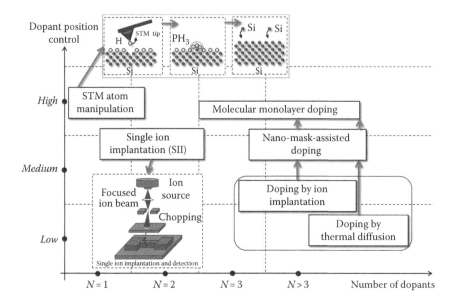

FIGURE 8.2 A schematic diagram of available doping techniques, as a function of dopant position control and dopant number. Conventional doping techniques (thermal diffusion and ion implantation) have less controllability, but they may be assisted by additional nanopatterning or molecular monolayer doping.[23] SII[5,21,29,30] has good control of dopant number and further improvement in dopant position is expected.[25–28] STM atom manipulation technique allows the precise positioning of single dopants,[20,32,35] via a relatively complex processing.

such as ion implantation and thermal diffusion, are quite practical in terms of efficiency, but they lack precise control. On the other hand, state-of-the-art techniques, such as single-ion implantation (SII) or scanning tunneling microscope (STM) atom manipulation, can allow precise control of the number and/or position of one or only a few dopant atoms. These latter techniques are, however, more complex and difficult to implement in practical applications. An overview of the available doping techniques is illustrated in Figure 8.2. Each category is briefly outlined next.

8.1.2.1 Uncontrolled (Random) Doping

Most commonly, silicon devices are doped through either (thermal) diffusion or ion implantation, with ion implantation being the main doping technique in industry for the past 40 years. Thermal-diffusion doping consists of two main steps. First, in a so-called pre-deposition step, dopant atoms are introduced into silicon by solid-phase diffusion, gas-phase diffusion, or even by ion implantation. Then, a second step, so-called drive-in, has the purpose of uniformly and deeply redistributing the dopant atoms within the volume of the semiconductor. From a macroscopic viewpoint, the distribution of dopants follows the Fick's diffusion laws. Dopant concentration gradient within the sample is a key driving force in the redistribution of the dopants.

Ion implantation is a technique which involves relatively higher energies, with dopant ions being first accelerated and then directly bombarded on the semiconductor substrate. This bombardment creates a series of lattice damages, but these can be repaired by annealing processes. As an advantage, ion implantation offers a relatively improved control of the amount and distribution of dopants. The ion beam current controls the dose of dopant ions and the acceleration energy controls their distribution in depth within the sample. It can be expected that these techniques can be improved with regard to accuracy by assisting them with precise nanolithography (e.g., electron-beam lithography) techniques or with chemical techniques, such as monolayer doping[23] or Block copolymer patterning.[24]

8.1.2.2   Controlled Doping

As devices become smaller and smaller, it is crucial to gain improved control of dopant placement in nanoscale channels, both in terms of number, as well as in terms of position. Several techniques have been proposed and developed over the past years for the aforementioned purpose. Among these, SII and atom manipulation using STM have even been used to fabricate silicon transistors with one or only a few precisely controlled dopants.[20,21]

SII is an extension of the conventional ion implantation technique, but with enhanced control of both dopant ion number and aiming accuracy. Dopant ions are extracted from a doping source one by one and directed toward a sample surface. On the way, the ion beam is focused and collimated, allowing a relatively high aiming accuracy. Additional detection circuitry permits real-time sensing of the single-dopant-ion implantation in the sample. This technique has been used to show improved control of threshold voltage (V_{TH}) fluctuations.[5] More recently, it was also used to fabricate nanoscale transistors with a few pairs of donor atoms.[21] Mainly because of the lateral and longitudinal straggles of the implanted ions, which also vary depending on the ion energy, aiming accuracy is currently still on the order of ~50 nm. Several approaches have been recently investigated for further improvement of control over dopant position using this doping technique.[25–28] Controlled deterministic implantation is also pursued by other groups, with another aim of developing this technique for quantum computer design.[29–31] With further optimization, SII may become suitable for the realization of precisely controlled dopant-atom devices, for which the implantation precision should ideally be on the order of 10 nm or less.

STM manipulation of impurity atoms in semiconductors has become an extremely precise technique for controlling the position of individual dopants in silicon devices.[32,33] When applied for the precise implantation of a phosphorus (P) atom within the Si lattice, the technique involves hydrogen-resist lithography[34] to control the dissociation of the dopant precursor, that is usually phosphine (PH_3), based on well-established chemical processes.[35] Using this technique, different types of devices with precisely controlled dopant positions have been fabricated and characterized in recent years.[20,36–39] This direction of research offers insights into the ultimate, atomic-scale functionality of silicon devices. Nevertheless, it remains relatively far from the efficiency required for most practical applications. This justifies additional studies aimed at either improving the efficiency of the state-of-the-art technology or at finding alternative solutions that could provide potential industrial applications.

8.1.3 GUIDELINES FOR DOPANT-ATOM DEVICES

So far, most reports on single-dopant transistors have focused on a low-temperature range ($T < 20$ K).[22,40] This is mainly because typical shallow dopants in Si have relatively small ionization energies (on the order of a few tens of meV), which is not sufficient to prohibit thermally activated conduction (TAC) over the dopant's tunnel barriers. This way, the component of current due to pure tunneling via the dopant, according to the CB mechanism, is quickly masked by the TAC component. Therefore, it is important to find suitable pathways to further enhance the operation temperature of dopant-atom devices using typical dopants. This approach can be treated in parallel with the challenges faced by single-electron tunneling (SET) transistors[41] for the demonstration of room-temperature operation, for which a variety of transport mechanisms must be considered as a function of temperature.[42] In the diagram shown in Figure 8.3, possible research directions and essential physics involved are schematically outlined, without assuming this to be an exhaustive overview. The cases of SETs with patterned QDs are also considered for completeness.

Single-QD transistors have been studied for decades and their operation temperature has been gradually raised by making the QDs smaller and smaller. At present, several groups demonstrated room-temperature operation of single-QD transistors. For that, fabrication of ultrasmall QDs with large charging energies has been a key technological factor.[43–46]

For single-dopant transistors, on the other hand, the operation temperature is mainly dictated by the dopant's ionization energy, which is known to be relatively small for typical dopants in Si. As a result, most reports on single-dopant transistors so far have been

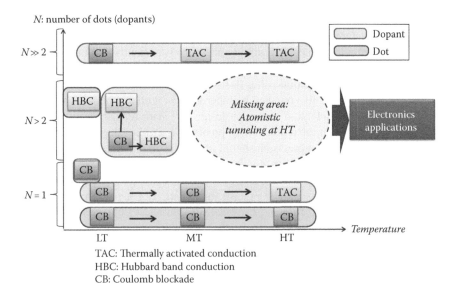

FIGURE 8.3 Research direction for conduction modes of QD-SETs and dopant-transistors as a function of number of dots (dopants) and temperature (low-, medium-, and high-temperature ranges). Representative references for each conduction mode are indicated in the diagram.

limited to a low temperature range ($T < 20$ K). As mentioned earlier, this is because, at higher temperatures, TAC becomes the dominant transport mechanism. It is, however, known that dielectric confinement and quantum confinement can strongly affect the ionization energy of dopants embedded in nanostructures, enhancing it by even one order of magnitude.[47–49] Recently, we fabricated silicon transistors with nanoscale channels specifically patterned to enhance dielectric confinement effect.[50] By doing so, we succeeded in increasing the tunnel barrier height of dopants, in the smallest devices, up to about 100 meV, which allowed us to observe SET characteristics at elevated temperatures ($T \cong 100$ K). These findings will be described in more details in a later section.

Another path toward higher-temperature operation involves not only one, but a few QDs or, respectively, dopant atoms. The properties of double QDs have been extensively treated[51–53] in correlation with single-QD transistors. In particular for semiconductor QDs, it is suggested that, depending on the interdot tunnel-coupling strength, several transport regimes can be observed.[54]

Regarding dopant-atom devices, different transport mechanisms, such as Hubbard band conduction or Anderson localization, may gain a dominant role in device behavior.[55–58] Recently, by using SII to introduce pairs of donors in nanochannels, such transport regimes have been demonstrated,[21] with transitions between them occurring either as a function of temperature or as a function of the number of dopants. For devices with higher doping concentrations, such as junctionless transistors,[59] impurity scattering would also play a critical role in the transport at elevated temperatures and should be taken into consideration.[60]

A regime which still requires more investigation is that of a few interacting dopant atoms. When a few dopants are locally doped so that they strongly interact with each other, a new QD is realized and its properties are significantly different from those of individual dopants. Based on such concept, it may be possible to access elevated operation temperatures, which may open the door for practical applications suitable for future generations of electronics. For that, it is important to deeply investigate the properties of atomistic-level tunneling in such multiple-dopant systems. We will treat some of our basic results on this topic in a subsequent section.

8.2 OBSERVATION OF INDIVIDUAL DOPANTS BY LT-KPFM

8.2.1 TECHNIQUES FOR DOPANT DETECTION

Since the role and impact of individual dopants in device operation increases in nanoscale devices, it is crucial to directly observe the spatial distribution of ionized dopant potentials and, eventually, to correlate it with the device behavior. For observation and evaluation of dopant distribution, several techniques are available, with different degrees of accuracy. Some of these techniques are suitable for rough estimations of the doping concentration in large samples, while others are optimized for precise local measurements, even of individual dopant atoms.

The basic method for evaluating the doping profile within a semiconductor film is by *secondary ion mass spectrometry* (SIMS), in which the surface of the film is sputtered by a primary ion beam and ejected secondary ions are analyzed. This technique is widely used in semiconductor film characterization. However, it has the main disadvantage of being a destructive technique, in addition to the fact that the spatial resolution is relatively low. *Atom probe tomography* (ATM) is another important technique, based on field-evaporation under high electric field, detection, and then reconstruction of the material composition.[61–63] This technique is promising for eventually resolving 3D distributions of dopants in nanostructures, but it is also destructive and, at present, its detection yield requires improvement.

Among the nondestructive techniques, *scanning tunneling microscopy* (STM) became a very accurate technique for resolving the electronic structure of the topmost monolayers in a conductive sample. Its principle is based on quantum tunneling between a metallic tip and the sample. This means, however, that special treatment of the surface is necessary and, furthermore, sensitivity to deeper charges within the structure cannot be expected.[34,64,65] Electrostatic capacitance between a tip and the sample can also be used to detect dopant charges, in a technique labeled as *scanning capacitance microscopy* (SCM).[66] Although this is, in principle, a suitable technique for detecting even small charges in a sample, it requires, however, that the surface is extremely flat and its spatial resolution must also be further improved.

One attractive technique that could overcome most fundamental limitations is *Kelvin probe force microscopy* (KPFM),[67] which allows, in principle, measurements with higher spatial resolution and higher sensitivity of charges located deeper in the device structure. We already demonstrated the possibility of observing individual dopants (both donors and acceptors) in the channel of devices under normal operation by using our specially designed Kelvin probe force microscopy (KPFM) technique.[68–70]

Our KPFM technique allows surface potential measurements in ultrahigh vacuum chamber at temperatures from 13 to 300 K, with the possibility of biasing the devices with regular FET external biasing circuit, as illustrated in Figure 8.4a.[68]

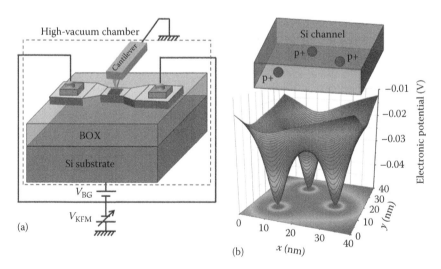

FIGURE 8.4 (a) Measurement setup for KPFM system, with variable temperature and external bias circuit. (b) For a transistor containing a few P donors in the channel, by applying negative V_{BG}, ionized P donors are expected to be observed. (After Anwar, M. et al., *Appl. Phys. Lett.*, 99, 213101-1, 2011.)

We have analyzed the effect of a gate voltage (substrate voltage for the KPFM devices) on the charge occupancy of individual donors in a nanoscale channel.[71] By comparing results at 13 and 300 K, we could observe not only electronic potential wells induced by the phosphorus (P) donors, but also distinct effects of electron injection in P donors, reflecting different electron transport mechanisms. These results will be described in more detail in the following sections.

8.2.2 KPFM Setup for Dopant Detection in Devices under Normal Operation

Our KPFM system is designed and optimized for measurements of the potential profile induced by individual dopant atoms located in the channels of devices under normal operation. The main difficulty with regular KPFM measurements consists in the presence of free carriers in the sample, which may screen the signal coming from the underlying dopant ions. At room temperature and without biases, these free carriers remain in the channel and optimal conditions for dopant detection cannot be achieved. In our system, low-temperature (~13 K) measurements are possible, which significantly limits the amount of free carriers in the channel. Furthermore, the devices can be regularly biased from the exterior and applied bias can be used to deplete the free carriers from the channel completely. Under such conditions, as schematically shown in Figure 8.4b, individual dopant ions can be observed in their ionized charge states.

For the purpose of KPFM measurements, we fabricated silicon-on-insulator (SOI) FETs without top gate, with the p-type Si substrate ($N_A \cong 1 \times 10^{15}$ cm^{-3}) working as back gate. The structure and circuit of these SOI-FETs are schematically shown in Figure 8.4a. Channel thickness is <15 nm, with length and width of about 500 and 200 nm, respectively. Top Si layer was uniformly doped with phosphorus, in a structure similar to that of a junctionless transistor, at the concentration $N_D \cong 1 \times 10^{18}$ cm^{-3}, which corresponds to an average interdonor distance of ~10 nm. The sample was inserted in the KPFM measurement chamber in ultrahigh vacuum (<5 × 10^{-7} Pa) and the electrodes were connected to external voltage sources. We measured the KPFM surface potential images in the channel region of the device. Source and drain electrodes were grounded in this experiment, which allows us to study the static charge distribution in the channel in the absence of current flow.

8.2.3 Observation of Individual Dopant Potentials and Single-Electron Charging

Using a measurement setup as described in the previous section, we measured the KPFM electrostatic potential maps for different back gate voltages, V_{BG}. At a negatively large voltage, $V_{BG} = -3$ V, the channel is expected to be fully depleted of electrons. A KPFM potential image for $V_{BG} = -3$ V taken at 13 K is shown in Figure 8.5a.

The applied V_{BG} works to ionize donors and deplete electrons from the SOI channel.[68–70] The channel potential contrast is, thus, primarily formed by the ionized P donors. In the shown area, three potential wells can be seen. Each well has a spatial extension of ~10 nm and an electronic potential depth of 10–40 mV, suggesting that each well is created by a different ionized P donor. In order to inject electrons from

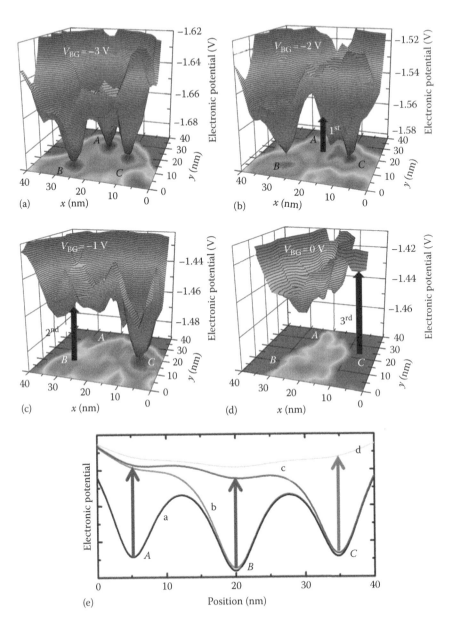

FIGURE 8.5 (See color insert.) (a)–(d) KPFM electronic potential landscapes at low temperature ($T = 13$ K) for different V_{BG}'s (−3, −2, −1, and 0 V, respectively). Three donor wells (A, B, C) are marked and vanish one by one by increasing V_{BG} in the positive direction. (e) Schematic representation of successive electron charging in individual P donors at low temperature. (After Anwar, M. et al., *Appl. Phys. Lett.*, 99, 213101, 2011.)

grounded source/drain electrodes into the channel, V_{BG} is increased from −3 to 0 V in 1 V steps. As a result, significant changes in the potential landscape can be seen. First, at V_{BG} = −2 V (Figure 8.5b), one of the potential wells (A) disappears. At V_{BG} = −1 V, a second potential well (B) successively disappears (Figure 8.5c). The last remaining potential well (C) finally disappears at V_{BG} = 0 V (Figure 8.5d). These potential changes are illustrated schematically in Figure 8.5e as successive flattening of neighboring dopant-induced potential wells.

The localized modifications in the potential landscape at low temperature are ascribed to successive single-electron filling in donors, since it is expected that each P donor potential is almost neutralized and compensated by the capture of one electron. However, this situation is significantly different at higher temperatures.[71] For a temperature of 300 K, KPFM measurements are shown in Figure 8.6. When a depleting V_{BG} is applied to the device, free carriers can be still removed from the channel and signatures of ionized P donors remain observable (Figure 8.6a). However, as V_{BG} is gradually increased in the positive direction, as seen in Figure 8.6b through d, the donor potentials become gradually smoother simultaneously. This suggests that the free carriers are not localized in individual donors anymore, but rather screen the donors' potentials uniformly. The schematic picture of this observation is shown in Figure 8.6e.

8.2.4 CORRELATION WITH ELECTRICAL CHARACTERISTICS

The results shown here indicate that it is possible to monitor not only static potentials of individual fixed charges, such as ionized dopants, but also the distribution of free carriers within the system of dopants. This can allow, in principle, a correlation between the potential landscape of a device and its electrical characteristics. In Figure 8.7, drain current (I_D) versus back gate voltage (V_{BG}) characteristics are shown for a device similar to that measured by LT-KPFM. Two different temperature regimes are analyzed, similarly to the KPFM measurements.

From the KPFM measurements, it was found that, at low temperatures, electrons tend to be localized at the location of individual dopants. This effectively shows that individual dopants work as QDs. In the I_D–V_{BG} characteristics at low temperature ($T \cong 15$ K), we can observe isolated current peaks, consistent with CB transport by SET. On the other hand, room temperature KPFM measurements suggest that thermal activation of donor electrons leads to a delocalization of the electrons in the channel, which provides more uniform screening of the dopants' potentials. The room-temperature ($T = 300$ K) I_D–V_{BG} characteristics indeed exhibit no signatures of SET anymore, but simple FET behavior.

Such correlation provides an insight into the important phenomena that must be considered in the evolution of dopant-atom devices as a function of temperature. Further detailed analysis of electrical measurements will be described in the following section.

More recently, we have further analyzed the nature of the donor-induced QDs by KPFM in different regimes of concentration. We found that, in the low-concentration regime, individual P-donors are predominantly forming the QDs in the channel, whereas at higher concentrations, QDs are formed by "clusters" of several P-donors located close to each other.[72] Furthermore, we proposed a more direct correlation between the potential features observed by KPFM and the I–V characteristics.[73]

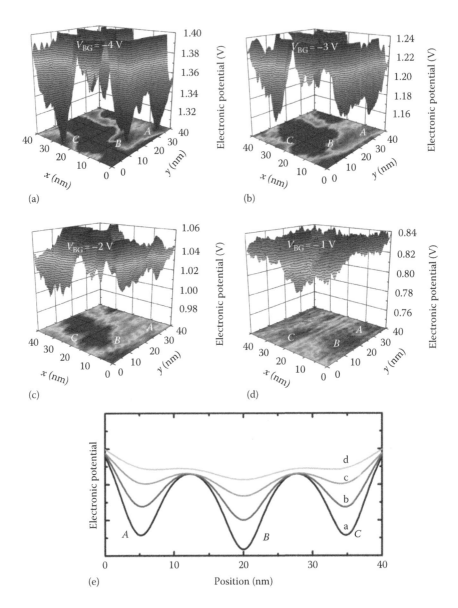

FIGURE 8.6 (a)–(d) KPFM electronic potential landscapes at high temperature (T = 300 K) for different V_{BG}'s (−4, −3, −2, and −1 V, respectively). Donor wells (A, B, C) become simultaneously smoothed by increasing V_{BG} in the positive direction. (e) Schematic representation of gradual screening of individual P donors by electrons at high temperature. (After Anwar, M. et al., *Appl. Phys. Lett.*, 99, 213101, 2011.)

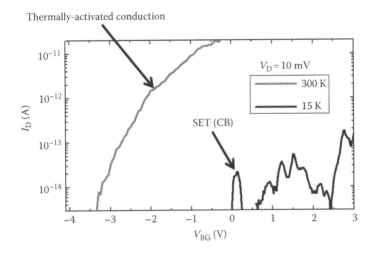

FIGURE 8.7 I_D–V_{BG} characteristics measured for a back-gated SOI-FET at low temperature ($T = 15$ K) and at high temperature ($T = 300$ K). Low temperature characteristics exhibit SET transport behavior, while TAC is the dominant transport mechanism at high temperature. (After Anwar, M. et al., *Appl. Phys. Lett.*, 99, 213101, 2011.)

8.3 SINGLE-ELECTRON TUNNELING VIA DOPANTS IN NANOSCALE TRANSISTORS

8.3.1 TUNNELING VIA INDIVIDUAL DOPANTS IN NANOSCALE TRANSISTORS

As revealed by the LT-KPFM measurements, it is possible to directly observe injection of electrons into individual dopants. This situation basically corresponds to single-electron charging of dopant-QD in tunneling transport. In an ideal principle of single-dopant transistors, illustrated in Figure 8.8, when one dopant is coupled to electrodes in a transistor structure, electrons can be transported between source and drain via the dopant-QD by successive injection into and extraction from the dopant. This is the sequential tunneling transport mechanism, similar to conventional SETs, giving rise to I_D–V_G characteristics as schematically shown in Figure 8.8c.

As stated earlier, most studies on single-dopant transistors so far have analyzed less ideal structures, in which individual dopants are either diffused from leads or randomly doped into the channel. However, even for randomly doped devices, it is possible to isolate individual dopants as QDs. This is because the lowest electronic potential is formed by the entire set of dopants through a superposition of their long-range potentials.[17] As such, there are a few (or even only one) local potential minima within the channel. When the channel minimum conduction band energy is shifted close to the source Fermi level by the gate voltage, transport occurs through the dopant-induced QD.

Transport through such individual dopant atoms is, however, mostly limited at low temperatures because of their relatively small tunnel barriers. At higher temperatures, TAC becomes a dominant transport mechanism, which is undesirable

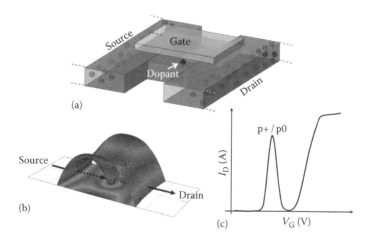

FIGURE 8.8 (a) A schematic view of a dopant atom placed in a nanochannel, controlled by a gate. (b) Dopant-induced potential well working as a QD for SET transport. (c) Ideal I_D–V_G characteristics (at low temperature) exhibiting a single current peak (due to CB transport through the dopant) before the onset of a large current.

within the typical single-dopant transistor scheme. It is, thus, critical to find pathways to enhance the tunneling operation temperature of single-dopant transistors toward more elevated temperatures. Some of the approaches that we addressed are outlined in the following sections.

8.3.2 EFFECT OF CHANNEL PATTERN AND DIELECTRIC CONFINEMENT ON THE PROPERTIES OF INDIVIDUAL DOPANTS

At elevated temperatures ($T > 100$ K), TAC, that is, transport over the tunnel barrier of thermally excited electrons, forming the tail of the Fermi–Dirac distribution, may become quickly dominant. For practical applications of single-dopant transistors, it is necessary to minimize this effect so that tunneling operation can be extended up to elevated temperatures. For that, one important condition is to enhance the tunnel barrier height of donor-QDs.

One way to modify the tunnel barriers of dopants is to tune their ionization energy. For bulk-like dopants, ionization energy is well known. However, when dopants are embedded into nanostructures, their properties significantly change. For single dopants embedded in silicon nanowires, it was recently found by simulation studies[47,48] that dielectric confinement (and, in extremely small nanostructures, even quantum confinement) plays a dominant role in redefining the properties of the dopants. In particular, it was reported that the ionization energy of dopants in nanowires is drastically increased, leading to an effective deactivation of the dopants and to a loss in conductivity, as also experimentally confirmed.[49] The significant impact of the dielectric confinement is basically due to the fact that dopants embedded in silicon nanostructures are mostly surrounded by dielectric material and the screening due to the Si matrix is thus reduced. This leads to a modification of the dopant's ground state energy, that is, a further deepening of the ground state below the conduction band edge for the case of a donor, as compared with the bulk case. This is effectively equivalent to an increase of the height of the tunnel barrier associated with the donor.

As suggested by *ab initio* simulations,[50] we focused on specific channel design that would provide optimal conditions to enhance dielectric confinement effect on P donors. We found that a promising design consists of a stub-channel, as shown by an SEM image of one of the smallest devices in Figure 8.9; a random P donor arrangement is also illustrated as a possible favorable location of individual donors. For reference, we also fabricated devices without any special pattern of the channel and the comparison is shown in Figure 8.9. It should be mentioned that, in the vertical direction, donors are embedded in an ultrathin (~5-nm-thick) Si layer (as seen from the cross-sectional transmission electron microscope [TEM] image). For both structures, donors located in the channel should experience significant dielectric confinement effect in the vertical direction. However, only when a donor is located within the edge of the stub region, dielectric confinement effect becomes strong in the lateral direction as well, because it is mostly surrounded by SiO_2. This is quite different than for the case of nonstub FETs.

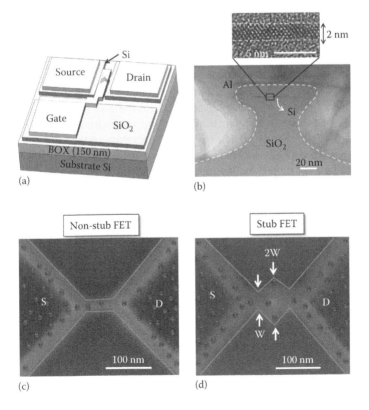

FIGURE 8.9 (a) Structure of an SOI-FET under study. (b) Cross-sectional TEM image of the channel area for a device with an ultrathin ($t_{Si} < 5$ nm) channel. (c) and (d) SEM images and possible P-donor distributions for a non-stub and, respectively, stub-channel FET. (After Hamid, E. et al., *Phys. Rev. B*, 87, 085420, 2013.)

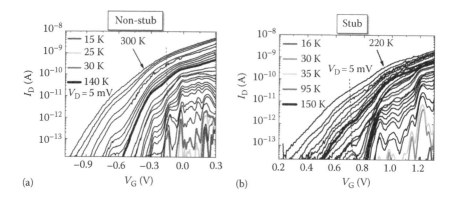

FIGURE 8.10 I_D–V_G characteristics for small source–drain bias (V_D = 5 mV) for a wide temperature range for two different types of devices: (a) with non-stub channel and (b) with stub channel. For the stub-channel FET, a current peak emerges at T = 100 K (as indicated by the dashed line), whereas for the non-stub channel FET, SET current peaks vanish at lower temperatures. (After Hamid, E. et al., *Phys. Rev. B*, 87, 085420, 2013.)

For non-stub- and stub-channel FETs, I_D–V_G characteristics were measured at a small source–drain voltage, V_D = 5 mV. Temperature was changed as a parameter, from ~15 to ~300 K. Figure 8.10a and b shows representative sets of I_D–V_G characteristics for two smallest devices, with nanowire and, respectively, stub-shaped channel. At the lowest temperatures (~15 K), the I_D–V_G characteristics exhibit a number of isolated current peaks, as seen both in Figure 8.10a and b. As argued up to this point, these peaks are due to electron tunneling transport through donor-induced QDs formed in the channel. By raising the temperature in the range of 20–100 K, several new current peaks successively emerge at more negative V_G's for both types of devices. At even higher temperatures (T > 150 K), TAC completely masks any SET features.

In Figure 8.11a and b, for clarity, only the temperature-associated lowest-V_G current peaks are extracted from the full I_D–V_G characteristics (Figure 8.10) and are plotted in the V_G–T plane. These peaks successively appear with increasing temperature and are ascribed to tunneling via P donors with deeper ground-state energies. We cannot observe the current peaks of these deep donors at low temperatures (~15 K), since the tunneling rate is too small due to the high potential barriers. (The detectable current level in the present system is around ~1 × 10⁻¹⁴ A.) With increasing temperature, however, due to broadening of the Fermi–Dirac electron distribution in the reservoir and other thermal effects, as illustrated in Figure 8.11c, the tunneling rate is enhanced and current peaks successively emerge, exceeding the detectable current level. It is found that, for the stub-channel FET, the last emerging current peak appears at T ≅ 100 K (Figure 8.11b), which is the highest temperature reported so far for single-dopant transistors operating in SET mode.[50] Moreover, the SET feature survives as a prominent hump up to ~150 K, as shown by the thick I_D–V_G curve in Figure 8.10b. At high temperatures, above ~100 K, a number of SET peaks are significantly broadened and overlap each other, marking the loss of CB as the dominant transport mechanism.

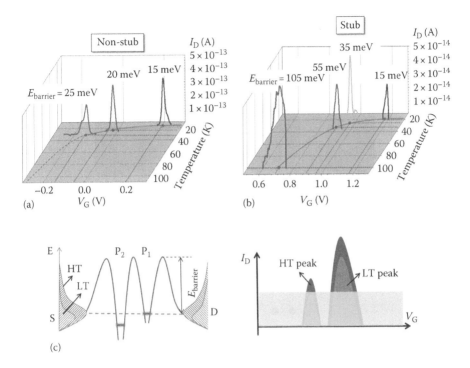

FIGURE 8.11 Current peaks observed at the lowest V_G's for different temperatures for devices with different channel patterns: (a) non-stub channel and (b) stub channel. Barrier heights, extracted from an Arrhenius plot analysis, are also indicated on top of each peak. (c) Model for the emergence of current peaks as temperature is increased. New peaks can be ascribed to tunneling transport through deeper ground-state P donors. (After Hamid, E. et al., *Phys. Rev. B*, 87, 085420, 2013.)

From the temperature dependence of the electrical characteristics, such as the data shown in Figure 8.10a and b, it is possible to extract the barrier height at different V_G's based on the analysis of Arrhenius plots. In particular, these values are most reliable as extracted from the data taken at highest temperatures, where it is certain that the thermally activated transport is the dominant transport mechanism.[14] Details of this extraction procedure and the results as a function of device dimensions can be found in our work.[50] Here, for simplicity, the values extracted for the barrier height for V_G's corresponding to the last observable current peaks are indicated in Figure 8.11a and b. It can be seen that, for the peak emerging at the highest temperature of ~100 K, the barrier height is large (~100 meV), much larger than the value known for the ionization energy of P donors in bulk Si (~44 meV).

Figure 8.12 shows the dependence on the channel width (see Figure 8.12a) for the tunneling operation temperature (Figure 8.12b) and for the extracted barrier height (Figure 8.12c). It can be seen that there is a systematic trend for both operation temperature and energy barrier to increase with reducing the channel dimensions. This tendency further supports our model, since in the narrowest devices, the impact of dielectric confinement is expected to be the strongest. These findings suggest that, in

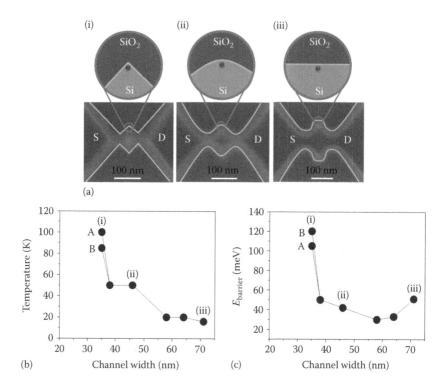

FIGURE 8.12 (a) Change of channel shape as a function of designed width, from the narrowest (i) to the widest (iii) channels. (b) and (c) Effect of channel width on tunneling operation temperature (b) and extracted barrier height (c). (After Hamid, E. et al., *Phys. Rev. B*, 87, 085420, 2013.)

stub-channel FETs, P donors most likely located in the edge of the stub region possess enhanced tunnel barriers as a result of dielectric confinement effect.

With further control of nanofabrication processes, aiming at the optimization of the channel design, it may become possible to significantly enhance the dopant's ionization energy and to challenge room-temperature tunneling operation. Nevertheless, this requires state-of-the-art technology in nanopatterning. As a consequence, this approach may present some disadvantages for practical realizations of dopant-based applications.

8.3.3 INTERACTING DOPANTS IN SILICON NANOTRANSISTORS

A different approach consists of utilizing interactions among dopants to form new dopant-induced QD with useful properties. In order to study the effect of multiple-donor interactions on the electrical characteristics of nanotransistors, we fabricated several types of devices (as illustrated in Figure 8.13) including devices with the channel selectively doped.

For the selectively doped devices, a 30 nm-wide slit was designed as a doped region separated from the lead extensions by 70 nm-wide nondoped gaps (in which

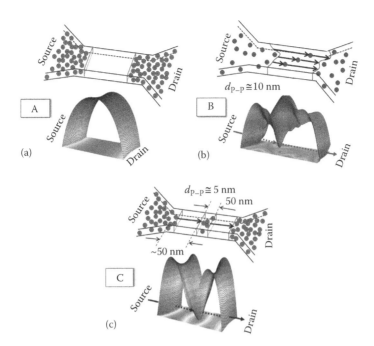

FIGURE 8.13 (**See color insert.**) Types of devices investigated, as a function of channel doping, together with a possible potential landscape: (a) nondoped channel (device type A); (b) uniformly and randomly doped channel, with relatively low concentration, $N_D \cong 1 \times 10^{18}$ cm^{-3} (device type B); and (c) selectively doped channel, with higher concentration, $N_D > 5 \times 10^{18}$ cm^{-3} (device type C).

mask oxide is preserved). Final dimensions of the slit and gap are ~50 nm each, mainly as a result of side etching of the mask oxide. Lateral diffusion of dopants was suppressed by minimizing the thermal budget during the only subsequent thermal process, gate oxide formation (a lower-temperature wet thermal oxidation was used). Higher doping concentration was chosen for these devices ($N_D > 5 \times 10^{18}$ cm^{-3}) in order to promote the formation of "clusters" containing a few strongly interacting donors. For such relatively high concentration, it is prevalent to find "clusters" of at least three donor atoms, with the donors located at distances closer than two Bohr radii from each other (i.e., strongly coupled).[74]

We measured the low-temperature ($T \leq 15$ K) electrical characteristics for several devices of different types, depending on the doping profile in the channel: (A) nondoped channel; (B) randomly discretely doped channel; and (C) selectively doped channel. Figure 8.14 shows, for comparison, representative I_D–V_G characteristics, measured for small source–drain bias ($V_D = 5$ mV), for each device type.

For the nondoped-channel FETs (device type A), there are no observable current peaks in the I_D–V_G characteristics (Figure 8.14a). For devices with randomly doped channel (device type B), isolated and irregular current peaks appear as a signature of SET transport via individual P donors (Figure 8.14b). For the FETs selectively doped with higher concentration (device type C), current peak envelopes

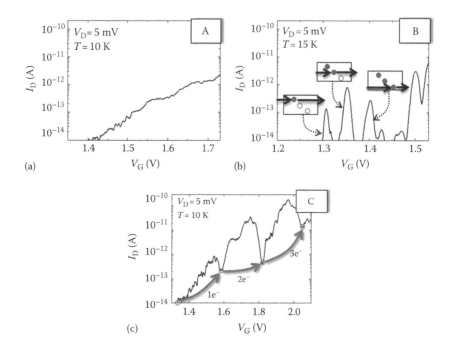

FIGURE 8.14 Typical I_D–V_G characteristics for different device types, as a function of channel doping: (a) nondoped channel (device type A); (b) uniformly and randomly doped channel (device type B); and (c) selectively doped channel (device type C).

can be observed (Figure 8.14c). These peak envelopes have a relatively periodic V_G spacing and are separated by current dips (as marked on the graph). Such features are ascribed to SET transport via a QD that can accommodate several electrons (not only one). Such a QD is formed with high probability by a few P donors located relatively close to each other, creating a "cluster"-like arrangement. In fact, multiple-donor QDs have previously been reported to be formed randomly in highly doped SOI-FETs.[75-78]

We evaluate the dimensions of the QD from the V_G period between consecutive peaks (~200 mV). Assuming a parallel-plate capacitance model, the QD radius is estimated to be approximately 8.5 ± 1.0 nm. This is significantly larger than the Bohr radius for a P donor in bulk[79,80] ($r_B \cong 2.5$ nm), suggesting that the QD covers the area of several P donors.

We now turn to a more detailed analysis of the fine features (inflections) observed embedded in each peak envelope. In Figure 8.15, I_D–V_G characteristics are shown for several low temperatures (10–20 K), which allows us to trace the current inflections. Each current inflection corresponds to a new energy level entering the source–drain bias window. For the shown device, a number of 3–5 such levels can be observed as inflections. It is known that, for interacting atoms which form a molecular-like spectrum, the number of split levels is practically equal to the number of interacting atoms.[81,82] Thus, one can evaluate that the QD is formed by 3–5 strongly interacting

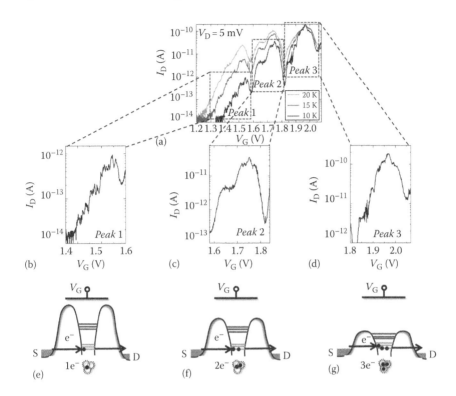

FIGURE 8.15 (a) I_D–V_G characteristics for $V_D = 5$ mV and low T for a selectively doped SOI-FET. Peak envelopes can be seen. (b)–(d) Zoom-in on consecutive peak envelopes, ascribed to SET via a multiple-dopant QD with discrete energy levels. (e)–(g) Models of the multiple-donor QD showing the ground-state multifold separated from the excited-state multifold.

P donors, which is consistent with our expectation. Furthermore, *ab initio* simulations have also confirmed the dependence of the number of split levels on the number of interacting dopants, even when embedded in extremely small nanostructures.[83]

Once the interaction between a few dopants in nanochannels can be fully clarified, by experimental and theoretical approaches and methods,[72,73,84] it may be possible to make use of such multiple-dopant QDs to challenge high-temperature operation. In principle, with optimized design of the QD and of the tunnel barriers, it is expected that the barrier height of such QD can be made much larger than for a single dopant. This can be considered as a promising approach for achieving tunneling operation at elevated temperatures.

8.4 CONCLUSIONS

In this chapter, we introduced our basic approaches for the design and operation of dopant-atom devices, in which individual dopant atoms or a few coupled dopant atoms play the dominant role in tunneling transport. For that, we first outlined the techniques available for precise doping and for the observation of dopants in nanodevices.

We then introduced our results based on Kelvin probe force microscopy measurements for the direct observation of dopant potentials and electron injection into dopants. These findings provide insights into the functionality offered by dopants at low temperatures and at room temperature in different regimes of doping concentration. We then presented our results on tunneling transport through dopants in Si nanodevices. First, we showed that specific patterning of nanochannels can allow an enhancement of the dielectric confinement effect, which, at its turn, can assist in increasing the tunnel barriers of dopants. As a result, tunneling operation can be achieved even at elevated temperatures ($T \cong 100$ K). From a different perspective, we also studied interactions within systems of a few dopant atoms by transport spectroscopy of selectively doped nanochannel transistors. With further optimization, multiple-dopant QDs may also offer an alternative pathway toward practical functionality of dopant-based devices.

ACKNOWLEDGMENTS

The authors are grateful for the collaboration and useful discussions with Prof. H. Mizuta, Prof. Y. Ono, Prof. T. Shinada, Prof. R. Jablonski, and Prof. D. Hartanto. We also acknowledge contributions to this work by many coworkers over the past few years: Mr. T. Mizuno, Dr. M. Ligowski, Dr. E. Hamid, Dr. A. Udhiarto, Dr. R. Nowak, Dr. S. Purwiyanti, and all the past and present members of our research group. This work was partially supported by Grants-in-Aid for Scientific Research from the Ministry of Education, Culture, Sports, Science and Technology of Japan (20246060, 22656082, 23226009, and 25630144) and by a Cooperative Research Project of the Research Institute of Electronics, Shizuoka University.

REFERENCES

1. Moore, G.: Cramming more components onto integrated circuits. *Electronics*, **38**, 114–117 (1965).
2. Mizuno, T., Okamura, J., and Toriumi, A.: Experimental study of threshold voltage fluctuation due to statistical variation of channel dopant number in MOSFET's. *IEEE Trans. Electron Dev.*, **41**, 2216–2221 (1994).
3. Ohtou, T., Sugii, N., and Hiramoto, T.: Impact of parameter variations and random dopant fluctuations on short-channel fully depleted SOI MOSFETs with extremely thin BOX. *Electron Dev. Lett.*, **28**, 740–742 (2007).
4. Asenov, A.: Random dopant induced threshold voltage lowering and fluctuations in sub-0.1 μm MOSFET's: A 3-D "atomistic" simulation study. *IEEE Trans. Electron Dev.*, **45**, 2505–2513 (1998).
5. Shinada, T., Okamoto, S., Kobayashi, T., and Ohdomari, I.: Enhancing semiconductor device performance using ordered dopant arrays. *Nature*, **437**, 1128–1131 (2005).
6. Pierre, M., Wacquez, R., Jehl, X., Sanquer, M., Vinet, M., and Cueto, O.: Single-donor ionization energies in a nanoscale CMOS channel. *Nat. Nanotechnol.*, **5**, 133–137 (2010).
7. Li, Y. and Yu, S.-M. Comparison of random-dopant-induced threshold fluctuation in nanoscale single-, double-, and surrounding-gate field-effect transistors. *Jpn. J. Appl. Phys.*, **45**, 6860–6865 (2006).
8. Loss, D. and DiVincenzo, D.P.: Quantum computation with quantum dots. *Phys. Rev. A*, **57**, 120–126 (1998).

9. Kane, B.: A silicon-based nuclear spin quantum computer. *Nature*, **393**, 133–137 (1998).

10. Hollenberg, L.C.L., Dzurak, A.S., Wellard, C., Hamilton, A.R., Reilly, D.J., Milburn, G.J., and Clark, R.G.: Charge-based quantum computing using single donors in semiconductors. *Phys. Rev. B*, **69**, 113301-1–113301-4 (2004).

11. Ladd, T.D., Jelezko, F., Laflamme, R., Nakamura, Y., Monroe, C., O'Brien, J.L.: Quantum computers. *Nature* (review), **464**, 45–53 (2010).

12. Morello, A. et al.: Single-shot readout of an electron spin in silicon. *Nature*, **467**, 687–691 (2010).

13. Pla, J.J., Tan, K.Y., Dehollain, J.P., Lim, W.-H., Morton, J.J.L., Jamieson, D.N., Dzurak, A.S., and Morello, A.: A single-atom electron spin qubit in silicon. *Nature*, **489**, 541 (2012).

14. Sellier, H., Lansbergen, G.P., Caro, J., Collaert, N., Ferain, I., Jurczak, M., Biesemans, S., and Rogge, S.: Transport spectroscopy of a single dopant in a gated silicon nanowire. *Phys. Rev. Lett.*, **97**, 206805 (2006).

15. Ono, Y., Nishiguchi, K., Fujiwara, A., Yamaguchi, H., Inokawa, H., and Takahashi, Y.: Conductance modulation by individual acceptors in Si nanoscale field-effect transistors. *Appl. Phys. Lett.*, **90**, 102106 (2007).

16. Lansbergen, G.P., Rahman, R., Wellard, C.J., Woo, I., Caro, J., Collaert, N., Biesemans, S., Klimeck, G., Hollenberg, L.C.L., and Rogge, S.: Gate-induced quantum-confinement transition of a single dopant atom in a silicon FinFET. *Nat. Phys.*, **4**, 656–661 (2008).

17. Tabe, M., Moraru, D., Ligowski, M., Anwar, M., Jablonski, R., Ono, Y., and Mizuno, T.: Single-electron transport through single dopants in a dopant-rich environment. *Phys. Rev. Lett.*, **105**, 016803 (2010).

18. Tan, K.Y. et al.: Transport spectroscopy of single phosphorus donors in a silicon nanoscale transistor. *Nano Lett.*, **10**, 11–15 (2010).

19. Prati, E., Belli, M., Cocco, S., Petretto, G., and Fanciulli, M.: Adiabatic charge control in a single donor atom transistor. *Appl. Phys. Lett.*, **98**, 053109 (2011).

20. Fuechsle, M., Miwa, J.A., Mahapatra, S., Ryu, H., Lee, S., Warschkow, O., Hollenberg, L.C.L., Klimeck, G., and Simmons, M.Y.: A single-atom transistor. *Nat. Nanotechnol.*, **7**, 242–246 (2012).

21. Prati, E., Hori, M., Guagliardo, F., Ferrari, G., and Shinada, T.: Anderson-Mott transition in arrays of a few dopant atoms in a silicon transistor. *Nat. Nanotechnol.*, **7**, 443–447 (2012).

22. Moraru, D., Udhiarto, A., Anwar, M., Nowak, R., Jablonski, R., Hamid, E., Tarido, J.C., Mizuno, T., and Tabe, M.: Atom devices based on single dopants in silicon nanostructures. *Nanoscale Res. Lett.*, **6**, 479-1–479-6 (2011).

23. Ho, J.C., Yerushalmi, R., Jacobson, Z.A., Fan, Z., Alley, R.L., and Javey, A.: Controlled nanoscale doping of semiconductors via molecular monolayers. *Nat. Mater.*, **7**, 62–67 (2008).

24. Farrell, R.A. et al.: Large-scale parallel arrays of silicon nanowires via block copolymer directed self-assembly. *Nanoscale*, **4**, 3228–3236 (2012).

25. Shinada, T., Kurosawa, T., Nakayama, H., Hori, M., Zhu, Y., Hori, M., and Ohdomari, I.: A reliable method for the counting and control of single ions for single-dopant controlled devices. *Nanotechnology*, **19**, 345202 (2008).

26. Hori, M., Shinada, T., Taira, K., Shimamoto, N., Tanii, T., Endo, T., and Ohdomari, I.: Performance enhancement of semiconductor devices by control of discrete dopant distribution. *Nanotechnology*, **20**, 365205 (2009).

27. Hori, M., Shinada, T., Taira, K., Komatsubara, A., Ono, Y., Tanii, T., Endoh, T., and Ohdomari, I.: Enhancing single-ion detection efficiency by applying substrate bias voltage for deterministic single-ion doping. *Appl. Phys. Exp.*, **4**, 046501 (2011).

28. Hori, M., Taira, K., Komatsubara, A., Kumagai, K., Ono, Y., Tanii, T., Endoh, T., and Shinada, T.: Reduction of threshold voltage fluctuation in field-effect transistors by controlling individual dopant position. *Appl. Phys. Lett.*, **101**, 013503 (2012).

29. Schenkel, T., Persaud, A., Park, S.J., Nilsson, J., Liddle, J.A., Keller, R., Schneider, D.H., Cheng, D.W., and Humphries, D.E.: Solid state quantum computer development in silicon with single ion implantation. *J. Appl. Phys.*, **94**, 7017 (2003).

30. Jamieson, D.N. et al.: Controlled shallow single-ion implantation in silicon using an active substrate for sub-20-keV ions. *Appl. Phys. Lett.*, **86**, 202101 (2005).

31. O'Brien, J.L., Schofield, S.R., Simmons, M.Y., Clark, R.G., Dzurak, A.S., Curson, N.J., Kabe, B.E., McAlpine, N.S., Hawley, M.E., and Brown, G.W.: Towards the fabrication of phosphorus qubits for a silicon quantum computer. *Phys. Rev. B*, **64**, 161401 (2001).

32. Schofield, S.R., Curson, N.J., Simmons, M.Y., Ruess, F.J., Hallam, T., Oberbeck, L., and Clark, R.G.: Atomically precise placement of single dopants in Si. *Phys. Rev. Lett.*, **91**, 136104 (2003).

33. Lyding, J.W., Shen, T.C., Hubacek, J.S., Tucker, J.R., and Abeln, G.C.: Nanoscale patterning and oxidation of H-passivated Si(100)-2×1 surfaces with an ultrahigh-vacuum scanning tunneling microscope. *Appl. Phys. Lett.*, **64**, 2010–2012 (1994).

34. Wilson, H.F., Warschkow, O., Marks, N.A., Curson, N.J., Schofield, S.R., Reusch, T.C.G., Radny, M.W., Smith, P.V., McKenzie, D.R., and Simmons, M.Y.: Thermal dissociation and desorption of PH_3 on Si(001): A reinterpretation of spectroscopic data. *Phys. Rev. B*, **74**, 195310 (2006).

35. Ruess, F.J., Pok, W., Reusch, T.C.G., Butcher, M.J., Goh, K.E.J., Scappucci, G., Hamilton, A.R., and Simmons, M.Y.: Realization of atomically controlled dopant devices in silicon. *Small*, **3**, 563 (2007).

36. Fuhrer, A., Fuechsle, M., Reusch, T.C.G., Weber, B., and Simmons, M.Y.: Atomic-scale all-epitaxial in-plane gated donor quantum dot in silicon. *Nano Lett.*, **9**, 707 (2009).

37. Weber, M., Mahapatra, S., Watson, T.F., and Simmons, M.Y.: Engineering independent electrostatic control of atomic-scale (~4 nm) silicon double quantum dots. *Nano Lett.*, **12**, 4001 (2012).

38. Weber, M. et al.: Ohm's law survives to the atomic scale. *Science*, **335**, 64 (2012).

39. Miwa, J.A., Mol, J.A., Salfi, J., Rogge, S., and Simmons, M.Y.: Transport through a single donor in p-type silicon. *Appl. Phys. Lett.*, **103**, 043106 (2013).

40. Koenraad, P.M. and Flatté, M.E.: Single dopants in semiconductors. *Nat. Mater.*, **10**, 91–100 (2011).

41. Kouwenhoven, L.P., Marcus, C.M., McEuen, P.L., Tarucha, S., Westervelt, R.M., and Wingreen, N.S.: Electron transport in quantum dots. In *Mesoscopic Electron Transport* (eds. Sohn, L.L., Kouwenhoven, L.P., and Schön, G.) (Kluwer, 1997).

42. Morgan, N.Y., Abusch-Magder, D., Kastner, M.A., Takahashi, Y., Tamura, H., and Murase, K.: Evidence for activated conduction in a single electron transistor. *J. Appl. Phys.*, **89**, 410–419 (2001).

43. Takahashi, Y., Nagase, M., Namatsu, H., Kurihara, K., Iwadate, K., Nakajima, Y., Horiguchi, S., Murase, K., and Tabe, M.: Fabrication technique for Si single-electron transistor operating at room temperature. *Electron. Lett.*, **31**, 136–137 (1995).

44. Saitoh, M. and Hiramoto, T.: Extension of Coulomb blockade region by quantum confinement in the ultrasmall silicon dot in a single-hole transistor at room temperature. *Appl. Phys. Lett.*, **84**, 3172–3174 (2004).

45. Shin, S.J., Jung, C.S., Park, B.J., Yoon, T.K., Lee, J.J., Kim, S.J., Choi, J.B., Takahashi, Y., and Hasko, D.G.: Si-based ultrasmall multiswitching single-electron transistor operating at room-temperature. *Appl. Phys. Lett.*, **97**, 103101 (2010).

46. Shin, S.J., Lee, J.J., Kang, H.J., Choi, J.B., Yang, S.-R.E., Takahashi, Y., and Hasko, D.G.: Room-temperature charge stability modulated by quantum effects in a nanoscale silicon island. *Nano Lett.*, **11**, 1591–1597 (2011).

47. Diarra, M., Niquet, Y.-M., Delerue, C., and Allan, G.: Ionization energy of donor and acceptor impurities in semiconductor nanowires: Importance of dielectric confinement. *Phys. Rev. B*, **75**, 045301 (2007).

48. Li, B., Slachmuylders, A.F., Partoens, B., Magnus, W., and Peeters, F.M.: Dielectric mismatch effect on shallow impurity states in a semiconductor nanowire. *Phys. Rev. B*, **77**, 115335-1–115335-10 (2008).

49. Björk, M.T., Schmid, H., Knoch, J., Riel, H., and Riess, W.: Donor deactivation in silicon nanostructures. *Nat. Nanotechnol.*, **8**, 103–107 (2009).

50. Hamid, E., Moraru, D., Kuzuya, Y., Mizuno, T., Anh, L.T., Mizuta, H., and Tabe, M.: Electron-tunneling operation of single-donor-atom transistors at elevated temperatures. *Phys. Rev. B*, **87**, 085420-1–085420-5 (2013).

51. Van der Wiel, W.G., De Franceschi, S., Elzerman, J.M., Fujisawa, T., Tarucha, T., and Kouwenhoven, L.P.: Electron transport through double quantum dots. *Rev. Mod. Phys.*, **75**, 1–22 (2002).

52. Waugh, F.R., Berry, M.J., Mar, D.J., Westervelt, R.M., Campman, K.L., and Gossard, A.C.: Single-electron charging in double and triple quantum dots with tunable coupling. *Phys. Rev. Lett.*, **75**, 705–708 (1995).

53. Yamahata, G., Tsuchiya, Y., Oda, S., Durrani, Z.A.K., and Mizuta, H.: Control of electrostatic coupling observed for silicon double quantum dot structures. *Jpn. J. Appl. Phys.*, **47**, 4820–4826 (2008).

54. Stafford, C.A. and Das Sarma, S.: Collective Coulomb blockade in an array of quantum dots: A Mott-Hubbard approach. *Phys. Rev. Lett.*, **72**, 3590–3593 (1994).

55. Anderson, P.W.: Absence of diffusion in certain random lattices. *Phys. Rev.*, **109**, 1492–1505 (1958).

56. Mott, N.F. and Twose, W.D.: The theory of impurity conduction. *Adv. Phys.*, **10**, 107–163 (1961).

57. Hubbard, J.: Electron correlations in narrow energy bands. *Proc. Roy. Soc. Lond. A*, **276**, 238–257 (1963).

58. Norton, P.: Formation of the upper Hubbard band from negative-donor-ion states in silicon. *Phys. Rev. Lett.*, **37**, 164–168 (1976).

59. Colinge, J.-P. et al.: Nanowire transistors without junctions. *Nat. Nanotechnol.*, **5**, 225–229 (2010).

60. Li, X., Han, W., Wang, H., Ma, L., Zhang, Y., Du, Y., and Yang, F.: Low-temperature electron mobility in heavily n-doped junctionless nanowire transistor. *Appl. Phys. Lett.*, **102**, 223507 (2013).

61. Thomson, K., Booske, J.H., Larson, D.J., and Kelly, T.F.: Three-dimensional atom mapping of dopants in Si nanostructures. *Appl. Phys. Lett.*, **87**, 052108-1–3 (2005).

62. Perea, D.E., Hemesath, E.R., Schwalbach, E.J., Lensch-Falk, J.L., Voorhees, P.W., and Lauhon, L.J.: Direct measurement of dopant distribution in an individual vapour-liquid-solid nanowire. *Nat. Nanotechnol.*, **4**, 315–319 (2009).

63. Inoue, K., Yano, F. Nishida, A., Takamizawa, H., Tsunomura, T., Nagai, Y., and Hasegawa, M.: Dopant distributions in n-MOSFET structure observed by atom probe tomography. *Ultramicroscopy*, **109**, 1479–1484 (2009).

64. Jäger, N.D., Urban, K., Weber, E.R., and Ebert, P.: Nanoscale dopant-induced dots and potential fluctuations in GaAs. *Appl. Phys. Lett.*, **82**, 2700–2702 (2003).

65. Nishizawa, M., Bolotov, L., and Kanayama, T.: Simultaneous measurement of potential and dopant atom distributions on wet-prepared Si(111):H surfaces by scanning tunneling microscopy. *Appl. Phys. Lett.*, **90**, 122118 (2007).

66. Goragot, W. and Takai, M. Measurement of shallow dopant profile using scanning capacitance microscopy. *Jpn. J. Appl. Phys.*, **43**, 3990–3994 (2004).

67. Nonnenmacher, M., O'Boyle, M.P., and Wickramasinghe, H.K.: Kelvin probe force microscopy. *Appl. Phys. Lett.*, **58**, 2921–2923 (1991).

68. Ligowski, M., Moraru, D., Anwar, M., Mizuno, T., Jablonski, R., and Tabe, M.: Observation of individual dopants in a thin silicon layer by low temperature Kelvin probe force microscope. *Appl. Phys. Lett.*, **93**, 142101-1–142101-3 (2008).

69. Tabe, M., Moraru, D., Ligowski, M., Anwar, M., Yokoi, K., Jablonski, R., and Mizuno, T.: Observation of discrete dopant potential and its application to Si single-electron devices. *Thin Solid Films*, **518**, S38–S43 (2010).

70. Anwar, M., Kawai, Y., Moraru, D., Nowak, R., Jablonski, R., Mizuno, T., and Tabe, M.: Single-electron charging in phosphorus donors in Si observed by low-temperature Kelvin probe force microscope. *Jpn. J. Appl. Phys.*, **50**, 08LB10-1–08LB10-5 (2011).

71. Anwar, M., Nowak, R., Moraru, D., Udhiarto, A., Mizuno, T., Jablonski, R., and Tabe, M.: Effect of electron injection into phosphorus donors in silicon-on-insulator channel observed by Kelvin probe force microscopy. *Appl. Phys. Lett.*, **99**, 213101-1–213101-3 (2011).

72. Tyszka, K., Moraru, D., Samanta, A., Mizuno, T., Jablonski, R., and Tabe, M.: Comparative study of dopant-induced quantum dots in Si nano-channels by single-electron transport characterization and Kelvin probe force microscopy, *J. Appl. Phys.*, **117**, 244307 (2015).

73. Tyszka, K., Moraru, D., Samanta, A., Mizuno, T., Jablonski, R., and Tabe, M.: Effect of selective doping on the spatial dispersion of donor-induced quantum dots in Si nanoscale transistors, *Appl. Phys. Express*, in press.

74. Thomas, G.A., Capizzi, M., De Rosa, F., Bhatt, R.N., Rice, T.M.: Optical study of interacting donors in semiconductors. *Phys. Rev. B*, **23**, 5472–5494 (1981).

75. Smith, R.A. and Ahmed, H.: Gate controlled Coulomb blockade effects in the conduction of a silicon quantum wire. *J. Appl. Phys.*, **81**, 2699–2703 (1997).

76. Augke, R., Eberhardt, W., Single, C., Prins, F.E., Wharam, D.A., and Kern, D.P.: Doped silicon single electron transistors with single island characteristics. *Appl. Phys. Lett.*, **76**, 2065–2067 (2000).

77. Tilke, A., Blick, R.H., Lorenz, H., and Kotthaus, J.P.: Single-electron tunneling in highly doped silicon nanowires in a dual-gate configuration. *J. Appl. Phys.*, **89**, 8159–8162 (2001).

78. Evans, G.J., Mizuta, H., and Ahmed, H.: Modelling of structural and threshold characteristics of randomly doped silicon nanowires in the Coulomb-blockade regime. *Jpn. J. Appl. Phys.*, **40**, 5837–5840 (2001).

79. Kohn, W. and Luttinger, J.M.: Theory of donor states in silicon. *Phys. Rev.*, **98**, 915–922 (1955).

80. Ramdas, A.K. and Rodriguez, S.: Spectroscopy of the solid-state analogs of the hydrogen atom: donors and acceptors in semiconductors. *Rep. Phys.*, **44**, 1297–1387 (1981).

81. Pimentel, G.C.: The bonding of trihalide and bifluoride ions by the molecular orbital method. *J. Chem. Phys.*, **19**, 446–448 (1951).

82. Munzarova, M.L. and Hoffmann, R.: Electron-rich three-center bonding: Role of sp interactions across the p-block. *J. Am. Chem. Soc.*, **124**, 4787–4795 (2002).

83. Moraru, D., Samanta, A., Anh, L.T., Mizuno, T., Mizuta, H., and Tabe, M.: Transport spectroscopy of interacting donors in silicon nano-transistors. *Sci. Rep.*, **4**, 6219-1–6219-6 (2014).

84. Anh, L.T., Moraru, D., Manoharan, M., Tabe, M., and Mizuta, H.: Impacts of electronic state hybridization on the binding energy of single phosphorus donor electrons in extremely downscaled silicon nanostructures, *J. Appl. Phys.*, **116**, 063705 (2014).

9 Single-Electron Transfer in Si Nanowires

Akira Fujiwara, Gento Yamahata,
and Katsuhiko Nishiguchi

CONTENTS

9.1 Introduction 207
9.2 Single-Electron Transfer...208
    9.2.1 Brief History and Some Basics..208
    9.2.2 Tunable-Barrier Single-Electron Turnstile 210
    9.2.3 Single-Electron Ratchet..213
    9.2.4 Single-Electron Transfer Using Localized States............................214
9.3 Accuracy of Single-Electron Transfer 215
    9.3.1 Evaluation of Transfer Accuracy 215
    9.3.2 Transfer Mechanism and its Impact on Transfer Accuracy.............218
9.4 Application to Information Processing...226
9.5 Conclusions..226
Acknowledgments..226
References...227

9.1 INTRODUCTION

Using state-of-the-art nanotechnology to downscale electronic devices provides us with the opportunity to control electric charges at the level of a single electron (SE). SE devices[1-4] have been studied for the past few decades in a variety of fields including low-power LSIs, high-sensitivity sensors, and solid-state quantum computing. Among the known categories of SE devices, the single-electron transistor (SET) has been the most widely investigated, because it is essentially a tiny field-effect transistor (FET), which is likely to be suitable for a wide range of applications. Accordingly, SET applications have always been considered within the framework of conventional FET-based applications.

The SE transfer device is another category that enables the ultimate one-by-one control of electric charges based on clocked transfer. Such a function cannot be performed with a conventional FET or a SET; it promises attractive device concepts toward the ultimate in electronics including current standards in metrology and an on-demand electron source for a circuit that uses an SE as a bit of information. SE transfer is also the ultimate in terms of low energy consumption, because there is no static energy dissipation and, in principle, the dynamic dissipation can be reduced to a level that is related to just the transfer of an SE. It is obvious, however, that a critical issue with SE transfer is transfer error or bit error, which must be minimized to render this technology practical. It is therefore of vital importance to evaluate the error rate quantitatively and investigate the physics that underlies the error mechanism.

This chapter reviews SE transfer technology and focuses on semiconductor-based devices. Section 9.2 begins with a brief history and some basics of SE transfer and then describes semiconductor-based SE transfer devices such as tunable-barrier SE turnstiles and pumps, and SE transfer devices that utilize localized states in semiconductors. In Section 9.3, we introduce experimental and theoretical studies of the accuracy of SE transfer. In Section 9.4, we briefly address the prospects for the application of SE transfer to information processing.

9.2 SINGLE-ELECTRON TRANSFER

9.2.1 BRIEF HISTORY AND SOME BASICS

Several years after the invention of SETs,[5,6] basic SE transfer devices such as the SE turnstile[7] and SE pump[8] were proposed and experimentally demonstrated using metal-based tunnel junctions in 1990 and 1991, respectively. Figure 9.1 shows their equivalent circuits. While a SET consists of a charge island with two tunnel junctions, the SE turnstile has three islands and four tunnel junctions. The central island has a gate, to which an AC clock signal is applied; the drain is biased in order to transfer an electron from the source to the drain. As the gate voltage is increased to some positive value, an electron enters the central island via the left island. Upon entering, it stays there because of the Coulomb blockades[1,2] created by the left and right islands; Coulomb blockade is the phenomenon that prohibits electron tunneling into a small conductive island due to the increase in charge energy that would otherwise occur. As the gate voltage is then negatively biased, the electron is ejected to the drain. Thus, an SE is transferred per cycle. The SE pump has two islands, which are gated by two phase-shifted clock signals. The signals move the SE without any drain bias by alternately activating and deactivating Coulomb blockade at the two islands.

Since the beginning of research into SE transfer, one important application target has been the metrological current standard.[9] An important milestone was recognized as the so-called quantum metrology triangle experiment,[10–12] by which one can test the consistency of three electrical quantum standards: the SE current standard, the Josephson voltage standard, and the quantum Hall resistance standard. Doing so

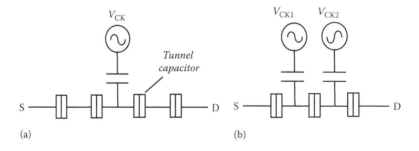

FIGURE 9.1 A schematic diagram showing equivalent circuits of (a) the SE turnstile and (b) the SE pump.

imposes two requirements[12] on the SE transfer devices used: (1) transfer current, $I = nef$ (n: integer, e: elementary charge, f: clock frequency), should be higher than sub-nanoampere level and (2) high transfer accuracy with an error rate below 10^{-8}. The first requirement means that a frequency in the GHz range is necessary if n is not very large. Since the demonstration of the first SE turnstiles and pumps, a lot of effort has been made to realize these characteristics in metal-based SE devices. In 1994, the second requirement was successfully satisfied in an experiment using seven-junction SE pumps.[13] However, the current was still in the picoampere range due to the limited frequency, on the order of MHz. Since then, raising operation frequency of the SE transfer device has been a long-standing issue.

Recently, the realization of the SE current standard has been gaining more interest with the proposal by the committee of the International Committee for Weights and Measures (CIPM) in 2011 to redefine the SI base unit ampere.[14] In the proposed new set of SI base units, kilogram, which is now defined by the prototype kilogram made of PtIr, is supposed to be defined by using Planck's constant; the constant is a measured value at present, but it will be defined as an exact value. Planck's constant is linked to the kilogram using the so-called watt-balance experiment[15] in which the weight of the mass is measured by electric current and the voltage induced in the coil in a graded magnetic field. In addition, the new ampere will be directly set based on a newly defined value of the elementary charge instead of the force between two parallel conductors. Thus, the role of electrical-based standards is increasing significantly.

We address here some of the basic physics of SE devices. As found in the textbook of SE devices,[1,2] the operation of an SE device requires that the charging energy $E_c = e^2/C$ (sometimes called the Coulomb gap energy) be larger than the thermal energy k_BT, where C is the capacitance of the charge island, k_B is Boltzmann's constant, and T is temperature. Another criterion is that the resistance of tunnel barrier R must be higher than the quantum resistance $R_Q = h/e^2 \sim 25.8$ kΩ to realize the Coulomb blockade in terms of the uncertainty principle for energy. Here, h is Planck's constant. The operating frequency, f, of SE devices is limited by the inverse of the RC time. To give a basic idea of the operating temperature and frequency of SE transfer devices, we introduce here a rough estimation of two fundamental errors in SE transfer.[2] The first one is thermal error. When n electrons are stored in the charge island coupled to an electron reservoir, the minimum error rate, ε, that the island contains $n + 1$ or $n - 1$ electrons (instead of n) when $E_c \gg k_BT$ is given by

$$\varepsilon_T \sim 2\exp\left(-\frac{E_c}{2k_BT}\right). \tag{9.1}$$

The other is the error due to a missed transition within the transfer cycle. If we assume that half of the cycle is used for one of the transitions in the SE turnstile, the error rate is given by

$$\varepsilon_F \sim \exp\left(-\frac{1/2f}{RC}\right). \tag{9.2}$$

When we assume that the error rate must be lower than 10^{-8}, Equations 9.1 and 9.2 give the conditions of $E_c/k_BT > 38$ and $f < 0.027RC^{-1}$, respectively. Thus, the capacitance of the charge island should be smaller than 0.16 aF at 300 K and 12 aF at 4 K, which correspond to the maximum frequency of 0.64 THz at 300 K and 8.6 GHz at 4 K, if we assume $R = 10R_Q$ to guarantee Coulomb blockade. Considering that nanoscale semiconductor devices can achieve the capacitance of 10 aF, these rough estimations suggest that the operating temperature of 4 K is feasible for the current standard application where such a temperature is no obstacle. On the other hand, room temperature operation is challenging, since extremely small capacitance is necessary.

9.2.2 TUNABLE-BARRIER SINGLE-ELECTRON TURNSTILE

As noted earlier, the operating frequency of the original SE transfer devices with fixed metal-oxide tunnel junctions is limited to the order of MHz due to their high R. The basic concept of the tunable-barrier SE transfer devices with gated semiconductors is to take advantage of barrier tunability to realize low resistance conditions.

Pioneering work on this type of device was done using GaAs quantum dot devices[16] in 1991, where just a single quantum dot coupled to the source and the drain via tunnel barriers is formed by split gate structures on top of a two-dimensional electron gas (2DEG) so that the tunnel barriers are tuned electrically. By modulating the height of the two barriers iteratively with a phase difference, SEs are transferred from the source to the drain. The operating frequency of this device is still limited to the order of 10 MHz, because while C is small enough, R is not, since the barrier resistance is not fully tuned.

Manipulation of SEs in Si was first demonstrated[17] in 2001 using a charge-coupled device fabricated on a silicon-on-insulator (SOI) substrate; the gate is used to form and tune the potential well rather than the barrier. The device is composed of a Si nanowire channel and a gate array, where photoexcited single holes are trapped in a potential well at the front interface under the metal-oxide-semiconductor (MOS) gates and are manipulated back and forth between the potential wells. The manipulated holes are detected by the electron current flowing on the bottom interface of the SOI channel. This demonstration proved that Si MOSFETs (metal-oxide-semiconductor field-effect transistors) have excellent ability to manipulate SEs via the tunable MOS gate potential.

Tunable-barrier-type operation in Si was then reported using Si SETs with fixed tunnel barriers formed by the bandgap modulation created by quantum confinement and oxidation-induced strain,[18,19] where MOS gates attached on the tunnel barriers additionally modulate the barrier resistances. The frequency was increased up to 10 MHz due to the small capacitance of the Si island.

In 2004, an SE transfer device with full-tunable barriers was demonstrated using a Si nanowire MOSFET.[20] The operating frequency of 100 MHz was successfully achieved. In addition to its high-frequency capability, another merit is the simplicity of its structure. The device structure and the potential diagram are shown in Figure 9.2. The device is fabricated on an SOI substrate and it features an electron-beam-lithography-defined Si nanowire channel covered by heavily doped polycrystalline Si gates. The wide upper gate is formed as a mask for ion implantation, which forms the n-type source and drain. The lower gates are used to form the potential barrier, and the upper gate is used to control the potential of the charge island formed

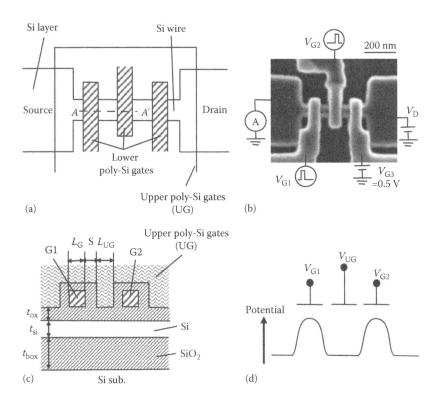

FIGURE 9.2 Device structure and potential diagram of Si tunable-barrier SE transfer device. (a) Schematic top view. (b) Scanning electron microscope image. (c) Schematic cross-section along line A–A'. (d) Schematic diagram of the potential profile and gate control. (Reprinted with permission from Fujiwara, A., Zimmerman, N. M., Ono, Y., and Takahashi, Y., Current quantization due to single-electron transfer in Si-wire charge-coupled devices, *Appl. Phys. Lett.*, 84, 1323. Copyright 2004, American Institute of Physics.)

between the lower gates. The diameter of the Si nanowire (t_{Si}) is about 20 nm and gate oxide thickness (t_{ox}) is 30 nm. The length of the lower gate (L_G) and the oxide spacer thickness (S) are 40 and 30 nm, respectively. The potential of the channel region with length $L_{UG} = 40$ nm, which acts as a charge island, is controlled by the upper gate voltage. The island capacitance is estimated to be on the order of 10 aF. As was discussed in the previous subsection, therefore, the device is promising for highly accurate SE transfer at $T = 4$ K and $f = 1$ GHz.

SE turnstile operation is described in Figure 9.3. Two alternate and repetitive pulses are applied to the two adjacent lower gates. In contrast to previous work, the potential is tuned in such a way that the barrier height changes between zero and high, which is a key for realizing higher-frequency operation. The charge island is first connected to the source. Since there is no barrier, Coulomb blockade does not exist and the number of electrons in the island can fluctuate. Then, as the barrier is raised, Coulomb blockade becomes effective and a fixed number of electrons can be captured. The island is next coupled to the drain and the similar process occurs.

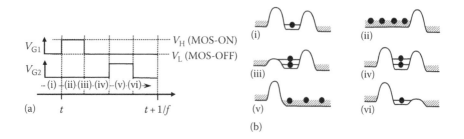

FIGURE 9.3 (a) Pulse sequence for turnstile operation. (b) Schematic potential diagram of the SE transfer procedure. (Reprinted with permission from Fujiwara, A., Zimmerman, N. M., Ono, Y., and Takahashi, Y., Current quantization due to single-electron transfer in Si-wire charge-coupled devices, *Appl. Phys. Lett.*, 84, 1323. Copyright 2004, American Institute of Physics.)

We describe here the principle of the SE turnstile based on the model of the conventional single-electron box.[21] Note that the model does not hold true in some cases, which indicates that electron capture occurs in a dynamic process as described later. The number of electrons, N_1, captured from the source, which is grounded, is given by

$$N_1 = C_{UG}V_{UG},$$

(9.3)

where
  C_{UG} is the upper gate capacitance
  V_{UG} is the upper gate voltage

The number of electrons, N_2, captured from the drain is given by

$$N_2 = C_{UG} (V_{UG} - V_D),$$

(9.4)

where V_D is the drain voltage. Thus, the net number of transferred electrons per cycle from source to drain, N, and the resultant transfer current, I, is given by

$$I = Nef = (N_1 - N_2)ef.$$

(9.5)

Thus, the theoretically predicted diagram of the transfer current as a function of V_{UG} and V_D is as shown in Figure 9.4a; it takes account of the Coulomb blockade giving rise to the quantization of N_1, N_2, and N as well as the depletion of the charge island made of semiconductor. Measured results are shown in Figure 9.4b. The structures of the current plateaus, *Nef*, show reasonable agreement between the theory and the experiment. It is seen in the diagram that current staircases form as a function of V_{UG} when V_D is large. Figure 9.4c shows an example of the current staircases measured for a device different from that used for Figure 9.4b. A similar type of tunable-barrier SE turnstile was also successfully built on a GaAs device[22] in 2007. That study was the first to demonstrate GHz operation, which proved that the tunable-barrier scheme is very feasible for high-frequency operation.

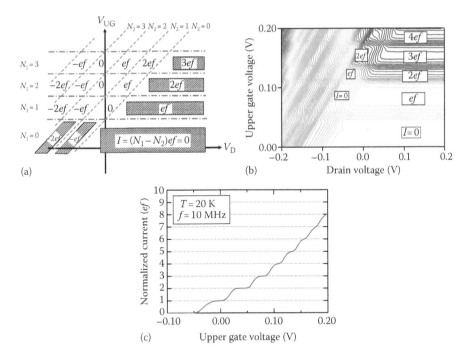

FIGURE 9.4 **(See color insert.)** (a) Stability diagram for the number of transferred electrons and the current plateaus. (b) Contour plot of the transfer current measured at $T = 20$ K and $f = 5$ MHz. (Reprinted with permission from Fujiwara, A., Zimmerman, N. M., Ono, Y., and Takahashi, Y., Current quantization due to single-electron transfer in Si-wire charge-coupled devices, *Appl. Phys. Lett.*, 84, 1323. Copyright 2004, American Institute of Physics.) (c) Current staircases due to SE transfer.

9.2.3 SINGLE-ELECTRON RATCHET

While two AC pulses are used for the turnstile operation described earlier, it is also possible to perform SE transfer just with one AC signal by utilizing the cross-capacitance between the barrier gate and the charge island. This operation was reported for Si and GaAs devices almost simultaneously in 2008.[23,24] We name this simple operation scheme the "single-electron ratchet," because the current so established is always directional and, furthermore, it can exist even at zero drain bias. A schematic diagram of its operation is shown in Figure 9.5a. A fixed negative voltage is applied to one of the lower gates, forming a potential barrier with fixed height. As an AC pulse signal is applied to the other gate, SEs are captured by the island and then ejected over the fixed barrier. Electron ejection becomes possible due to the capacitive coupling between the lower gate and the charge island, because the island's potential is further raised after it captures the electron so that it exceeds the height of the fixed barrier. The measured transfer current, shown as a function of V_{UG} and V_D in Figure 9.5b, clearly shows directional transport over a wide range of V_D. From an analysis of the shape of the current plateaus and various gate and bias voltage dependences, the total capacitance, the upper gate capacitance, the lower gate capacitances, and the drain capacitance are roughly

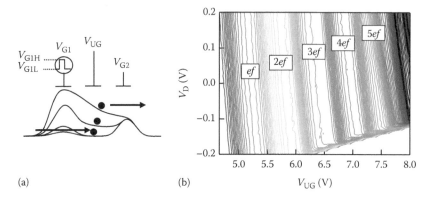

(a)   (b)

FIGURE 9.5 (a) A schematic potential diagram of the operation of the SE ratchet. (b) Contour plot of transfer current as a function of the upper gate voltage and the drain voltage at $T = 20$ K and $f = 50$ MHz. The current is directional regardless of the polarity of the drain voltage. (Reprinted with permission from Fujiwara, A., Nishiguchi, K., and Ono, Y., Nanoampere charge pump by single-electron ratchet using silicon nanowire metal-oxide-semiconductor field-effect transistor, *Appl. Phys. Lett.*, 92, 042102. Copyright 2008, American Institute of Physics.)

estimated to be 13, 0.28, 3.8, 1.4, and 0.12 aF, respectively. It is noteworthy that the V_D range is much larger than the charging energy of island, estimated to be about 10 meV. Such a wide V_D range is hard to achieve using metal-based SE devices. This is because the barrier formed by the MOS gate is highly robust to the drain bias and so, the SE ratchet works as a constant current source even for large drain bias; it can operate like a current source with high output impedance. With this SE ratchet scheme, nanoampere SE transfer was demonstrated as shown in Figure 9.6. Thus, nanowire MOSFET techniques allow us to transfer SEs at high frequencies comparable to LSI clock frequencies. In this study, we also discovered that the current staircase has significantly different shape from that predicted by the simple single-electron box model. This point will be addressed more in detail in Section 9.3.

To achieve high transfer currents, different approaches have been tried. One is to parallelize several SE transfer devices. This approach was actually tested for metal[25] and GaAs[26] devices; ten and two devices were fabricated and operated in parallel. This approach is simple, but it is logical as it is generally difficult for a single device to transfer a large number of electrons, because, for example, the charging energy of the semiconductor charge island depends on the number of electrons and is likely to be too small for larger numbers.

Another approach is to utilize the surface acoustic wave (SAW) driven at GHz frequencies for trapping and conveying SEs in GaAs[27,28]; however, the rise in device temperature remains an issue with this technique.

9.2.4 SINGLE-ELECTRON TRANSFER USING LOCALIZED STATES

An alternative approach to SE transfer is the use of localized states such as impurity states[29-34] or trap states in semiconductors.[35,36] One of the expected merits is high-temperature operation due to the high ionization energy of localized states. Another

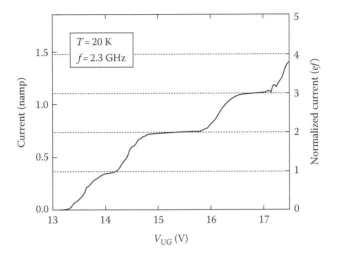

FIGURE 9.6 Nanoampere charge pumping by the SE ratchet at $f = 2.3$ GHz and $I = 3ef$. (Reprinted with permission from Fujiwara, A., Nishiguchi, K., and Ono, Y., Nanoampere charge pump by single-electron ratchet using silicon nanowire metal-oxide-semiconductor field-effect transistor, *Appl. Phys. Lett.*, 92, 042102. Copyright 2008, American Institute of Physics.)

advantage is that a large number of electrons can be transferred in each transfer cycle by using many localized states, each of which captures and emits one electron. This concept is similar to parallel pumping mentioned in the previous subsection. The first attempt at implementing Si nanodevices was made with doped nanowire channels (phosphorus dopants with a concentration of 1×10^{18} cm^{-3}); SE transfer was observed at the frequency of 1 MHz, which was attributed to the multiple random islands formed by phosphorous donors.[29–31] The experiments used uniformly doped wires covered by a large gate electrode. As an approach with better controllability, SE transfer via arsenic dopants (3×10^{17} cm^{-3}) was reported using selective area doping in the charge island of the Si tunable-barrier device with transfer gates.[32] Current plateaus of SE transfer via the dopants in the island of up to $6ef$ were observed at 2.5 MHz. The conventional SE pump scheme was adapted to the two-donor coupled system in Si, which yielded 10 MHz operation.[33] Very recently, high-frequency operation has been demonstrated using a donor in Si at 1 GHz[34] and a trap state in Si at 3.5 GHz,[36] which indicates the potential of these kinds of devices for the current standard application.

9.3 ACCURACY OF SINGLE-ELECTRON TRANSFER

9.3.1 Evaluation of Transfer Accuracy

In the usual measurement setup, the transfer current is measured with a DC current meter. Although the modern commercially available current meter provides the resolution of femtoampere, accuracy is typically on the order of 10^{-2} and 10^{-3}

in the measurement range of pico- and nanoampere, respectively. As noted in Section 9.2.1, an important milestone for evaluating transfer accuracy on the order of 10^{-8} is the metrological triangle experiment, where a cryogenic current comparator with large current amplification should be specially prepared and used.

A powerful and direct technique for evaluating accuracy is the electron-counting scheme,[13] which was proposed and experimentally demonstrated with metal-based SE pumps in 1996. In this method, electrons are transferred to/from the relatively large charge node and the change in the number of electrons in the node is counted by a charge sensor with SE sensitivity, which is capacitively coupled to the node. The measured error rate was 1.5×10^{-8} at $f = 5$ MHz and $T = 35$ mK, which is still the best error rate ever reported though the frequency is low.

In 2004, a charge sensor with SE resolution was introduced for Si device application; the charge sensor has been integrated with an SE box[37] and an SE turnstile.[38] Figure 9.7a and b shows a device structure that consists of the charge sensor and the SE tunable-barrier turnstile. With this device configuration, the number of electrons transferred to the charge (memory) node is counted by the Si nanowire charge sensor that is fabricated close to the node. The charge sensor has a sensitivity of 1.3×10^{-3} e/$\sqrt{\text{Hz}}$ at 1 Hz and room temperature.[39] Therefore, it is possible to demonstrate SE transfer and detection at room temperature[40] as shown in Figure 9.7c. Recently, it has been further demonstrated that the sensitivity of 2×10^{-4} e/$\sqrt{\text{Hz}}$ can be achieved at the bandwidth of 20 MHz by placing the charge sensor in an LC resonator circuit.[41] The method is similar to the so-called RF-SET technique.[42] The reflectometry is carried out with a carrier signal at around the resonant frequency of 100 MHz, and a small and modulated input signal to the sensor can be detected as sideband signals to the carrier signal.

Based on this Si nanowire charge sensing technique, an evaluation of the transfer errors possible in electron counting was performed for Si tunable-barrier SE transfer devices.[35] Figure 9.8a shows the schematic diagram of the device examined. In addition to the two lower gates (LG1, LG2) for the turnstile operation, the device has a third lower gate (LG3), which is used to electrically form the charge node that stores the transferred electrons. The charge sensor S-FET detects the number of electrons in the node. In the normal DC current measurement mode, positive gate bias is applied to LG3 so that FET3 is turned on and SE turnstile operation is performed using LG1 and LG2. In the electron counting mode, FET3 is turned off to form the charge node and SEs are transferred back and forth from/to the node by applying voltage pulses to LG1, LG2, and source S. Figure 9.8b shows the real-time monitoring of an SE shuttle at $T = 17$ K. It is seen in the figure that one error was detected during the SE shuttle. From longer measurements, the error rate was evaluated to be 2.1×10^{-2}. Note that real-time monitoring is conducted at a rather low frequency due to the slow speed of the charge sensor. However, the rise time of the pulse is 2 ns, which means that electron transfer takes only a short time, corresponding to the clock frequency on the order of 100 MHz. The error rate was improved to 10^{-4} by cooling the device in a dilution refrigerator[43] though the effective temperature of the device was several Kelvins due to the heating created by the AC signals. This is the best value confirmed by the electron counting scheme for semiconductor-based SE

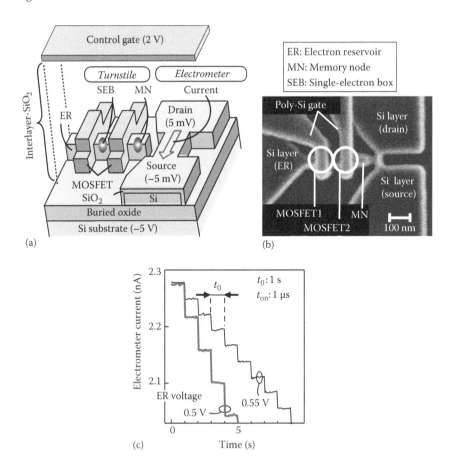

FIGURE 9.7 (a) A schematic view of the SE turnstile and charge sensor (electrometer). (b) Scanning electron microscope image of the device. (c) Change in electrometer current when transfer cycles were repeated to inject one electron and two electrons into the charge node. (Reprinted with permission from Nishiguchi, K., Inokawa, H., Ono, Y., Fujiwara, A., and Takahashi, Y., Room-temperature-operating data processing circuit based on single-electron transfer and detection with metal-oxide-semiconductor field-effect transistor technology, *Appl. Phys. Lett.*, 88, 183101. Copyright 2006, American Institute of Physics.)

transfer devices. Error counting has been also reported for GaAs devics[44]; the measured error rate was about 10^{-2}.

Another way of evaluating transfer accuracy rather precisely is to compare the SE transfer current with the one produced by electrical apparatus traceable to the primary voltage and resistance standards.[45,46] With this method, it was found that the error rate of GaAs devices is smaller than 10^{-6} at $f = 0.945$ GHz, which is the best accuracy reported for semiconductor-based SE transfer devices.[46] In this experiment, the shape of the clock pulse was optimized using an arbitrary waveform generator and a high magnetic field of 14 T was used to improve device performance. For Si, error rates as low as 5×10^{-5} at $f = 500$ MHz have been confirmed by the traceability provided by primary standards.[47]

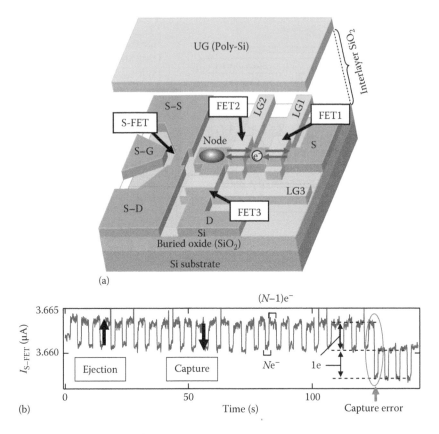

(a)

(b)

FIGURE 9.8 **(See color insert.)** (a) A schematic view of the device used to evaluate the accuracy of SE transfer. (b) Sensor current as a function of time during the SE shuttle transfer. (Reprinted with permission from Yamahata, G., Nishiguchi, K., and Fujiwara, A., Accuracy evaluation of single-electron shuttle transfer in Si nanowire metal-oxide-semiconductor field-effect transistors, *Appl. Phys. Lett.*, 98, 222104. Copyright 2011, American Institute of Physics.)

9.3.2 Transfer Mechanism and its Impact on Transfer Accuracy

The transfer mechanism of tunable-barrier SE transfer devices has been theoretically and experimentally investigated.[23,43,44,48–50] At an early stage, it was pointed out[48] that they offer substantial advantages over metal-based devices: (1) the maximum frequency at which Coulomb blockade can be maintained is high, up to the inverse of the time constant, $R_Q C$; (2) error due to cotunneling[51] can be significantly suppressed due to the high R of one of the barriers. Cotunneling is high-order electron tunneling and is unwanted in SE transfer devices; it is a major error source in metal-based SE transfer devices.[52] The other critical factor, which is now the subject of much interest, is the dynamic way in which electron capture in the charge island takes place.[23,49,50] As described in Section 9.2.2, the simplest model is an SE box where the barrier resistance is increased to capture electrons, while the box potential stays at a constant level. However, this condition is only satisfied when neither the SE box nor the charge island is capacitively coupled to the barrier-forming gate. In most cases, cross-capacitive coupling between the island and the gate is inevitable, so the island's potential is also raised during the capture process. Accordingly, electron escape to the source must be taken into account. Schematic diagrams of these thermal-equilibrium and nonequilibrium capture processes are shown in Figure 9.9. In the nonequilibrium case, as depicted in the figure, the electron-number-dependent escape rate[53] is an important factor; one electron remains in the island, while the second electron likely escapes to the source, because the two-electron level is higher, by charging energy E_c, than the one-electron level.

A simplified one-electron model has been developed to describe nonequilibrium electron capture with electron escape by thermal hopping.[23] The model was further improved and expanded to cover, in a unified manner, the tunneling regime.[43] Let us consider that, at the initial state, an SE box is formed under thermal equilibrium that contains one electron with $P_1 = 1$; P_1 is the probability that the island contains one electron. Here the electron level of the island is far below the Fermi level of the source. As the barrier is raised, the island potential also rises. When the level of the island is aligned with the Fermi level of the source, the electron begins to escape. Let us reset time t to 0 at this instant. If we assume thermal hopping over the barrier or quantum tunneling through a parabolic barrier, the escape rate, Γ_{out}, is exponentially decreased as the negative barrier gate voltage V_{G1} is increased;

$$\Gamma_{out} = \Gamma_1 \exp\left[\frac{(\alpha_B - \alpha_1)V_{G1}(t)}{k_B T^*}\right], \tag{9.6}$$

where

Γ_1 is the escape rate at $t = 0$

$(\alpha_B - \alpha_1)V_{G1}$ is the change in the barrier height seen from the island

α_B and α_1 are voltage-to-energy conversion factors for the barrier and the island, respectively, and α_1 is determined by the cross-capacitive coupling

T^* is temperature T for thermal hopping and $T^* = T_0$ for tunneling; T_0 is a parameter for the energy dependence of the tunneling probability.

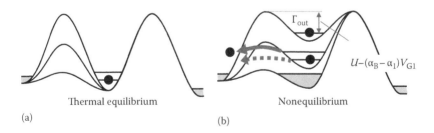

Thermal equilibrium   Nonequilibrium

(a)   (b)

FIGURE 9.9 (a) Schematic potential diagram of the thermal-equilibrium electron capture. (b) Schematic potential diagram of the nonequilibrium electron capture[23,43,49] (decay cascade model[50]). (Reprinted with permission from Yamahata, G., Nishiguchi, K., and Fujiwara, A., Accuracy evaluation and mechanism crossover of single-electron transfer in Si tunable-barrier turnstiles, *Phys. Rev. B*, 89, 165302 (ibid. 90, 039908(E), 2014). Copyright 2014, American Physical Society.)

When we assume the linear ramp of V_{G1}, Γ_{out} is given as $\Gamma_{out} = \Gamma_1 \exp(-\Gamma_{inc}t)$ where $\Gamma_{inc} = (\alpha_B - \alpha_I)|dV_{G1}/dt|/k_B T^*$. Γ_{out} should also depend on the island gate voltage, that is, the upper gate voltage V_{UG} in Figure 9.5, because it lowers the island potential and increases the barrier height with respect to the island. The rate equation for P_1 is given by

$$\frac{dP_1(t)}{dt} = -\Gamma_{out}P_1, \tag{9.7}$$

where $\Gamma_{out} = \Gamma_1 \exp(-\Gamma_{inc}t)\exp(-\alpha V_{UG}/k_B T^*)$, $\alpha = \alpha_{UG}\alpha_B/\alpha_I$, $\alpha_{UG} = eC_{UG}/C_\Sigma$. This equation is analytically solved as

$$P_1(t) = \exp\left\{A(t)\exp\left[\frac{-\alpha V_{UG}}{k_B T^*} + \ln\left(\frac{\Gamma_1}{\Gamma_{inc}}\right)\right]\right\}, \tag{9.8}$$

where $A(t) = \exp(-\Gamma_{inc}t) - 1$. We can let $A(t) \approx -1$, since Γ_{inc} is estimated to be on the order of 10^{10}–10^{11} s^{-1} in the experiment[23] and is much larger than f. Thus, the electron is captured with

$$P_1 = \exp\left\{-\exp\left[\frac{-\alpha V_{UG}}{k_B T^*} + \ln\left(\frac{\Gamma_1}{\Gamma_{inc}}\right)\right]\right\}. \tag{9.9}$$

One of the important results from this analysis is that the rise curve of the current plateau should be a double exponential function of the island gate voltage. Figure 9.10 shows the successful fit to the experimental data using the first and second derivatives of Equation 9.9, suggesting that nonequilibrium capture is a good model for this data. Thus, the asymmetric peak of the first derivative is a good indicator of nonequilibrium capture. Another point seen in Equation 9.9 is that Γ_1/Γ_{inc} is an important parameter determining the V_{UG} at which the electron is captured. This is reasonable, because, if Γ_1 is extremely high, electron capture is unlikely and only a larger V_{UG} makes the capture possible. A numerical simulation considering multiple electron number states[49] has also been conducted as shown in Figure 9.11. It is seen that probability P_m of the m-electron state evolves over time. It follows at first the thermal-equilibrium value, but then gradually deviates. Finally, it stays constant, meaning that the system is quenched in a nonequilibrium state. The impact of nonequilibrium capture on the transfer error is shown in Figure 9.12, where P_m of nonequilibrium capture is plotted as a function of V_{UG} in comparison to the initial thermal-equilibrium characteristic. Reflecting the double exponential function in Equation 9.9, P_m is also asymmetric for V_{UG}. Strikingly, the minimum error when the variance of the number of transferred electrons is minimized is much lower than the thermal-equilibrium case. This is because the error is no longer dominated by the standard thermal error in Equation 9.1, but it is dominated by the ratio of escape rates between different electron number states Γ_{m+1}/Γ_m as is described in the following.

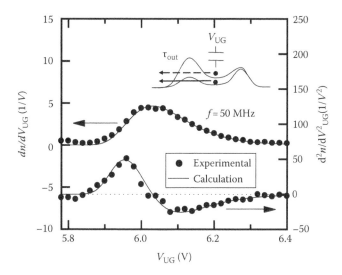

FIGURE 9.10 Calculated first and second derivatives by the nonequilibrium electron capture model fitted to the experimental results of the current plateaus of the SE ratchet. (Reprinted with permission from Fujiwara, A., Nishiguchi, K., and Ono, Y., Nanoampere charge pump by single-electron ratchet using silicon nanowire metal-oxide-semiconductor field-effect transistor, *Appl. Phys. Lett.*, 92, 042102. Copyright 2008, American Institute of Physics.)

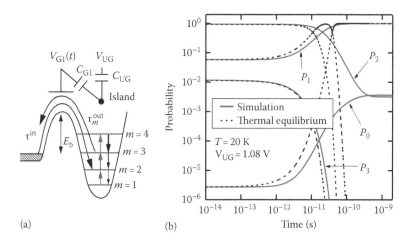

FIGURE 9.11 (a) Model of nonequilibrium electron capture with multiple electron number states. The electron number (m) states are depicted. The island is coupled to UG (C_{UG}) and G1 (C_{G1}). V_{G1} is linearly ramped to negative bias. (b) Simulated time evolution of the probability P_m of the electron number state. Broken lines represent P_m of the thermal-equilibrium state at each instance. (Reprinted with permission from Fujiwara, A., Miyamoto, S., Nishiguchi, K., Ono, Y., and Zimmerman, N. M., Dynamics of single-electron capture in Si nanowire MOSFETs, *IEEE Silicon Nanoelectronics Workshop 2008*, DOI: 10.1109/SNW.2008.5418469, © 2008, IEEE.)

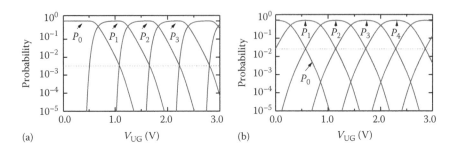

(a)   (b)

FIGURE 9.12 Simulated P_m as a function of V_UG. (a) Nonequilibrium electron capture. (b) Initial thermal-equilibrium states. (Reprinted with permission from Fujiwara, A., Miyamoto, S., Nishiguchi, K., Ono, Y., and Zimmerman, N. M., Dynamics of single-electron capture in Si nanowire MOSFETs, *IEEE Silicon Nanoelectronics Workshop 2008*, DOI: 10.1109/SNW.2008.5418469, © 2008, IEEE.)

The "decay cascade" model is a sophisticated theory[50] of nonequilibrium capture with multiple electron number states. At the ideal limit, the theoretical formula for the current plateaus turns out to be simply given by combining Equation 9.9 for the multiple states as[43,50,54]

$$\frac{I}{ef} = \sum_{m=1}^{\infty} \exp\left\{-\exp\left[\frac{-\alpha V_\mathrm{UG}}{k_\mathrm{B}T^*} + \ln\left(\frac{\Gamma_m}{\Gamma_\mathrm{inc}}\right)\right]\right\}, \tag{9.10}$$

where Γ_m is the escape rate for the m-electron state. An approximate expression[50] for the probability of the m-electron states based on the three-state model is given by

$$\{P_{m-1}, P_m, P_{m+1}\} = \left\{ \begin{array}{l} 1 - P_m - P_{m+1}, \exp\left\{-\exp\left[\frac{-\alpha V_\mathrm{UG}}{k_\mathrm{B}T^*} + \ln\left(\frac{\Gamma_m}{\Gamma_\mathrm{inc}}\right)\right]\right\} \\ \\ -P_{m+1}, \exp\left\{-\exp\left[\frac{-\alpha V_\mathrm{UG}}{k_\mathrm{B}T^*} + \ln\left(\frac{\Gamma_{m+1}}{\Gamma_\mathrm{inc}}\right)\right]\right\} \end{array} \right\}. \tag{9.11}$$

Note that Equation 9.10 is different from the one for thermal-equilibrium electron capture, which is the same with the conventional SE box model expressed as

$$\frac{I}{ef} = \sum m \left\{ \frac{\exp[-(E_\mathrm{m} - m\alpha_\mathrm{UG} V_\mathrm{UG})/k_\mathrm{B}T]}{\sum \exp\left[-(E_\mathrm{m} - m\alpha_\mathrm{UG} V_\mathrm{UG})/k_\mathrm{B}T\right]} \right\}, \tag{9.12}$$

where $E_\mathrm{m} = (me)^2/2C_\Sigma$. Note that thermal error rate in Equation 9.1 is derived from Equation 9.12.

For nonequilibrium capture, it is reasonably speculated from Equation 9.11 that lower Γ_m and higher Γ_{m+1} make P_m larger and P_{m-1} and P_{m+1} smaller, thereby leading to higher accuracy. It is actually shown that the critical error parameter[50] is

$$\delta = \ln\left(\frac{\Gamma_{m+1}}{\Gamma_m}\right). \qquad (9.13)$$

There is no analytical solution to link δ with the error rate in the decay cascade model, but an approximate expression would be useful in calculating it. That is given by

$$\varepsilon_D \approx 6.2\exp(-0.94\times\delta) \quad (\delta > 10). \qquad (9.14)$$

Let us now discuss how we can enlarge δ to reduce the error both in the thermal-hopping regime and the tunneling regime. It is also useful to clarify the conditions for the thermal-equilibrium and nonequilibrium cases. We introduce the cross-coupling constant g[43] as

$$g = \frac{\alpha_I}{(\alpha_B - \alpha_I)}. \qquad (9.15)$$

It is the ratio of the rise of the island to the rise of the barrier as seen from the island. From a theoretical consideration, we can map three different regions with regard to the transfer mechanism as a function of g and T as shown in Figure 9.13a. It should be noted that T_0 defined in Equation 9.6 corresponds to the transition temperature from thermal hopping to tunneling as regard with the parabolic barrier. When g is smaller than unity and T is higher than gT_0, thermal-equilibrium capture is dominant for both thermal hopping and tunneling. However, it changes to "tunneling cascade" capture as T falls under gT_0. This transition has been experimentally observed as shown in Figure 9.13b and c. As the device temperature falls, the peak shape that appears in the first derivative of the rise of the current plateau varies from symmetric to asymmetric. When g is larger than unity, the decay cascade is dominant for the whole temperature regime, but it changes from "hopping cascade" to "tunneling cascade" at $T = T_0$. Since $m + 1$ and m electron states are separated by E_c in energy, δ in Equation 9.13 is expressed as

$$\delta = \frac{E_c}{k_B T} \quad \text{(hopping cascade)},$$

$$\delta = \frac{E_c}{k_B T_0} \quad \text{(tunneling cascade)}. \qquad (9.16)$$

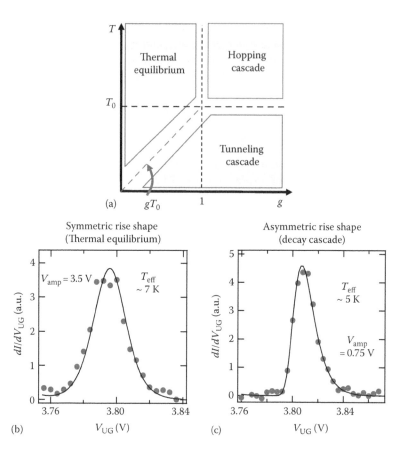

FIGURE 9.13 (a) A schematic map of the transfer mechanism as a function of the cross-coupling constant g and temperature. Mechanism cross-over observed in the first derivative of the current plateau between (b) thermal-equilibrium capture at a high temperature and (c) nonequilibrium capture at a low temperature. (Reprinted with permission from Yamahata, G., Nishiguchi, K., and Fujiwara, A., Accuracy evaluation and mechanism crossover of single-electron transfer in Si tunable-barrier turnstiles, *Phys. Rev. B*, 89, 165302 (ibid. 90, 039908(E), 2014). Copyright 2014, American Physical Society.)

It is noteworthy that δ is still dominated by E_c/kT for the hopping cascade, as is similarly seen in the thermal-equilibrium case in Equation 9.1, but the prefactor in Equation 9.14 is different and the error rate is much lower than that in the thermal-equilibrium case. In contrast, δ is constant for the tunneling cascade, because it is a parameter determined by barrier shape. Such a behavior is schematically shown in Figure 9.14 where the error rate, ε_D, is plotted as a function of temperature for both large-g and small-g cases. When $T > T_0$, the error rate scales as T decreases and large g gives a lower error rate, but it finally saturates at the tunneling-cascade-determined limit. From the theoretical fit to the experimental data in Figure 9.13c, δ is estimated to be around 20 and the corresponding error rate is on the order of 10^{-8}.

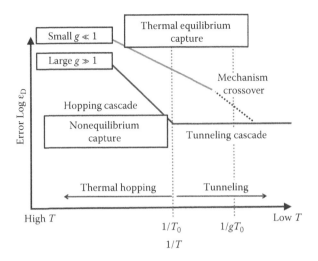

FIGURE 9.14 A schematic diagram of transfer error rate ε_D as a function of temperature T. The two cases of small cross-coupling ($g \ll 1$) and large cross-coupling ($g \gg 1$) are shown.

Here we show a guideline to improve the accuracy of tunable-barrier SE transfer. A straightforward way is to increase charging energy E_c by further shrinking the island size, which is always effective in any transfer regime. Lowering the temperature is also helpful, but it is effective only at thermal-equilibrium and the hopping-cascade regime. In the tunneling-cascade regime, T_0 should be reduced, which is demanded to increase the energy dependence of the tunneling probability. Therefore, barrier shape, for example, potential curvature, and carrier effective mass are important factors. In the case of the parabolic barrier given by $U(x) = U - ax^2/2$ (U and a are constants), T_0 is expressed as $\hbar(a/m)^{1/2}/2\pi k_B$, where m is the carrier effective mass; smaller curvature at the barrier top and heavier mass favor the realization of smaller T_0. Thus, barrier shape engineering is crucial to realize high accuracy in the tunneling-cascade regime. Works on GaAs SE pumps have shown that high magnetic fields drastically improve the current plateau characteristics.[54-56] The mechanism responsible for this is not fully clarified at present, but a theoretical simulation[56] suggests that the magnetic field increases the energy dependence of the tunneling probability by changing the potential curvature.

We should add two remarks: one is that there may be other possible error sources different from those described earlier; one of them is nonadiabatic excitation of electrons in a charge island, since the potential shape of the island may vary rapidly at a higher clock frequencies, which could excite the electron at the ground state into upper energy states, causing an increase in effective electron temperature. Signatures of such a behavior have been observed for GaAs[57] and Si[47] pumps, although further study is necessary to quantitatively clarify its impact on transfer error. The other remark is that error-correction schemes based on SE detection have begun to be discussed[58] and some have been tested in preliminary experiments.[59,60]

As described earlier, semiconductor-based SE transfer devices are now being intensively studied both theoretically and experimentally. The devices used are basically the tunable-barrier type, but there have been reports on various types of Si devices made by different fabrication processes such as the one fabricated on a Si bulk wafer[61] and those with metallic silicide islands.[62]

9.4 APPLICATION TO INFORMATION PROCESSING

A challenging research subject is to develop a way to use SE transfer devices in information processing circuits.[63] While low-temperature operation is acceptable for the metrological application, room temperature operation is strongly demanded for circuit application. If we think of employing an SE as a bit of information, detecting SEs is a crucial requirement. As noted in Section 9.3.2, SE detection is achieved at room temperature. Hence, it can be utilized to demonstrate a digital-to-analog converter based on SE transfer and detetion.[40] For practical application, however, there is a critical issue with regard to the bit error created by imperfect SE transfer, which inevitably requires fault compensation systems. One approach of interest bases error correction on SE detection as mentioned in the previous section. This would be a new and interesting research subject, but it remains a significant challenge to explore totally new circuit designs and engineering approaches. Another possible approach that differs from precise SE transfer is to employ the random motion of SEs to create physically random numbers. It has been demonstrated that such numbers can be generated from a purely Poisson process[64] and it is applicable to stochastic information processing for pattern matching and recognition.[65,66]

9.5 CONCLUSIONS

SE transfer devices are the one ultimate in electronic device application, since they allow the manipulation and transfer of individual charges. A near-term and feasible application target is the metrological current standard, for which the tunable-barrier SE transfer device based on semiconductor technology is promising due to its high frequency and highly accurate operation capabilities. It is of significant importance to understand the physics of electron capture dynamics and to optimize the operation and device structure to achieve the desired performance. The low-power nature of the devices suggests their use in integrated systems for information processing, but long-term efforts will be needed to develop suitable circuit designs and error correction protocols. Si nanowires, which offer excellent gate control, scalability, and CMOS compatibility, are promising candidates for realizing these near-term and long-term goals of the ultimate electron device.

ACKNOWLEDGMENTS

The authors thank Y. Takahashi, H. Inokawa, Y. Ono, G. P. Lansbergen, and N.M. Zimmerman for the collaboration in the work introduced in this chapter. A part of the work was supported by the Funding Program for Next Generation World-Leading Researchers of JSPS (GR103).

REFERENCES

1. Averin, D. V. and Likharev, K. K., Single electronics: A correlated transfer of single electrons and Cooper pairs in systems of small tunnel junctions, in: Altshuler, B. L., Lee, P. A., and Webb, R. A., eds., *Mesoscopic Phenomena in Solids* (Elsevier Science, New York, 1991), p. 173.
2. Grabert, H. and Devoret, M. H., eds., *Single Charge Tunneling* (Plenum, New York, 1992).
3. Likharev, K. K., Single-electron devices and their applications, *Proc. IEEE*, 87, 606, 1999.
4. Takahashi, Y., Ono, Y., Fujiwara, A., and Inokawa, H., Silicon single-electron devices, *J. Phys.: Condens. Matter*, 14, R995, 2002.
5. Kuzmin, L. S. and Likharev, K. K., Direct experimental observation of discrete correlated single-electron tunneling, *JETP Lett.*, 45, 495, 1987.
6. Likharev, K. K., Single-electron transistors: Electrostatic analogs of the DC SQUIDS, *IEEE Trans. Mag.*, 23, 1142, 1987.
7. Geerligs, L. J., Anderegg, V. J., Holweg, P. A. M., Mooji, J. E., Pothier, H., Esteve, D., Urbina, C., and Devoret, M. H., Frequency-locked turnstile device for single electrons, *Phys. Rev. Lett.*, 64, 2691, 1990.
8. Pothier, H., Lafarge, P., Urbina, C., Esteve, D., and Devoret, M. H., Single electron pump fabricated with ultrasmall normal tunnel junctions, *Physica B*, 169, 573, 1991.
9. Pekola, J. P., Saira, O.-P., Maisi, V. F., Kemppinen, A., Möttönen, M., Pashkin, Y. A., and Averin, D. V., Single-electron current sources: Toward a refined definition of the ampere, *Rev. Mod. Phys.*, 85, 1421, 2013.
10. Likharev, K. K. and Zorin, A. B., Theory of the Bloch-wave oscillations in small Josephson junctions, *J. Low Temp. Phys.*, 59, 347, 1985.
11. Keller, M. W., Current status of the quantum metrology triangle, *Metrologia*, 45, 102, 2008.
12. Feltin, N. and Piquemal, F., Determination of the elementary charge and the quantum metrological triangle experiment, *Eur. Phys. J. Spec. Top.*, 172, 267, 2009.
13. Keller, M. W., Martinis, J. M., Zimmerman, N. M., and Steinbach, A. H., Accuracy of electron counting using a 7-junction electron pump, *Appl. Phys. Lett.*, 69, 1804, 1996.
14. See http://www.bipm.org/.
15. Steiner, R., History and progress on accurate measurements of the Planck constant, *Rep. Prog. Phys.*, 76, 016101, 2013.
16. Kouwenhoven, L. P., Johnson, A. T., van der Vaart, N. C., Harmans, C. J. P. M., and Foxon, C. T., Quantized current in a quantum-dot turnstile using oscillating tunnel barriers, *Phys. Rev. Lett.*, 67, 1626, 1991.
17. Fujiwara, A. and Takahashi, Y., Manipulation of elementary charge in a silicon charge-coupled device, *Nature*, 410 560, 2001.
18. Ono, Y. and Takahashi, Y., Electron pump by a combined single-electron/field-effect transistor structure, *Appl. Phys. Lett.*, 82, 1221, 2003.
19. Ono, Y., Zimmerman, N. M., Yamazaki, K., and Takahashi, Y., Turnstile operation using a silicon dual-gate single-electron transistor, *Jpn. J. Appl. Phys.*, 42, L1109, 2003.
20. Fujiwara, A., Zimmerman, N. M., Ono, Y., and Takahashi, Y., Current quantization due to single-electron transfer in Si-wire charge-coupled devices, *Appl. Phys. Lett.*, 84, 1323, 2004.
21. Lafarge, P., Pothier, H., Williams, E. R., Esteve, D., Urbina, C., and Devoret, M. H., Direct observation of macroscopic charge quantization, *Z. Phys. B: Condens. Matter*, 85, 327, 1991.
22. Blumenthal, M. D., Kaestner, B., Li, L., Giblin, S., Janssen, T. J. B. M., Pepper, M., Anderson, D., Jones, G., and Ritchie, D. A., Gigahertz quantized charge pumping, *Nat. Phys.*, 3, 343, 2007.

23. Fujiwara, A., Nishiguchi, K., and Ono, Y., Nanoampere charge pump by single-electron ratchet using silicon nanowire metal-oxide-semiconductor field-effect transistor, *Appl. Phys. Lett.*, 92, 042102, 2008.
24. Kaestner, B., Kashcheyevs, V., Amakawa, S., Blumenthal, M. D., Li, L., Janssen, T. J. B. M., Hein, G., Pierz, K., Weimann, T., Siegner, U., and Schumacher, H. W., Single-parameter nonadiabatic quantized charge pumping, *Phys. Rev. B*, 77, 153301, 2008.
25. Maisi, V. F., Pashkin, Y. A., Kafanov, S., Tsai, J. S., and Pekola, J. P., Parallel pumping of electrons, *New J. Phys.*, 11, 113057, 2009.
26. Wright, S. J., Blumenthal, M. D., Pepper, M., Anderson, D., Jones, G. A. C., Nicoll, C. A., and Ritchie, D. A, Parallel quantized charge pumping, *Phys. Rev. B*, 80, 113303, 2009.
27. Shilton, J. M., Talyanskii, V. I., Pepper, M., Ritchie, D. A., Frost, J. E. F., Foad, C. J. B., Smith, C. G., and Jones, G. A. C., High-frequency single-electron transport in a quasi-one-dimensional GaAs channel induced by surface acoustic waves, *J. Phys.: Condens. Matter*, 8, L531,1996.
28. Ebbecke, J., Fletcher, N. E., Janssen, T. J. B. M., Ahlers, F.-J., Pepper, M., Beere, H. E., and Ritchie, D. A., Quantized charge pumping through a quantum dot by surface acoustic waves, *Appl. Phys. Lett.*, 84, 4319, 2004.
29. Moraru, D., Ono, Y., Inokawa, H., and Tabe, M., Quantized electron transfer through random multiple tunnel junctions in phosphorus-doped silicon nanowires, *Phys. Rev. B*, 76, 075332, 2007.
30. Moraru, D., Ligowski, M., Yokoi, K., Mizuno, T., and Tabe, M., Single-electron transfer by inter-dopant coupling tuning in doped nanowire silicon-on-insulator field-effect transistors, *Appl. Phys. Exp.*, 2, 071201, 2009.
31. Yokoi, K., Moraru, D., Ligowski, M., Mizuno, T, and Tabe, M., Single-gated single-electron transfer in nonuniform arrays of quantum dots, *Jpn. J. Appl. Phys.*, 48, 024503, 2009.
32. Lansbergen, G. P., Ono, Y., and Fujiwara, A., Donor-based single electron pumps with tunable donor binding energy, *Nano Lett.*, 12, 763, 2012.
33. Roche, B., Riwar, R.-P., Voisin, B., Dupont-Ferrier, E., Wacquez, R., Vinet, M., Sanquer, M., Splettstoesser, J., and Jehl, X., A two-atom electron pump, *Nat. Commun.*, 4, 1581, 2013.
34. Tettamanzi, G. C., Wacquez, R., and Rogge, S., Charge pumping through a single donor atom, *New. J. Phys.*, 16, 063036, 2014.
35. Yamahata, G., Nishiguchi, K., and Fujiwara, A., Accuracy evaluation of single-electron shuttle transfer in Si nanowire metal-oxide-semiconductor field-effect transistors, *Appl. Phys. Lett.*, 98, 222104, 2011.
36. Yamahata, G., Nishiguchi, K., and Fujiwara. A, Gigahertz single-trap electron pumps in silicon, *Nat. Commun.,* 5, 5038, 2014.
37. Nishiguchi, K., Inokawa, H., Ono, Y., Fujiwara, A., and Takahashi, Y., Multilevel memory using an electrically formed single-electron box, *Appl. Phys. Lett.*, 85, 1277, 2004.
38. Nishiguchi, K., Inokawa, H., Ono, Y., Fujiwara, A., and Takahashi, Y., Multilevel memory using single-electron turnstile, *Electron. Lett.*, 40, 229–230, 2004.
39. Nishiguchi, K., Koechlin, C., Ono, Y., Fujiwara, A., Inokawa, H., and Yamaguchi, H., Single-electron-resolution electrometer based on field-effect transistor, *Jpn. J. Appl. Phys.*, 47, 8305, 2008.
40. Nishiguchi, K., Inokawa, H., Ono, Y., Fujiwara, A., and Takahashi, Y., Room-temperature-operating data processing circuit based on single-electron transfer and detection with metal-oxide-semiconductor field-effect transistor technology, *Appl. Phys. Lett.*, 88, 183101, 2006.
41. Nishiguchi, K., Yamaguchi, H., Fujiwara, A., van der Zant, H. S. J., and Steele, G. A., Wide-bandwidth charge sensitivity with a radio-frequency field-effect transistor, *Appl. Phys. Lett.*, 103, 143102, 2013.

42. Schoelkopf, R. J., Wahlgren, P., Kozhevnikov, A. A., Delsing, P., and Prober, D. E., The radio-frequency single-electron transistor (RF-SET): A fast and ultrasensitive electrometer, *Science*, 280, 1238, 1998.

43. Yamahata, G., Nishiguchi, K., and Fujiwara, A., Accuracy evaluation and mechanism crossover of single-electron transfer in Si tunable-barrier turnstiles, *Phys. Rev. B*, 89, 165302, 2014; ibid. 90, 039908(E), 2014.

44. Fricke, L., Wulf, M., Kaestner, B., Kashcheyevs, V., Timoshenko, J., Nazarov, P., Hohls, F. et al., Counting statistics for electron capture in a dynamic quantum dot, *Phys. Rev. Lett.*, 110, 126803, 2013.

45. Giblin, S. P., Wright, S. J., Fletcher, J. D., Kataoka, M., Pepper, M., Janssen, M. T. J. B., Ritchie, D. A., Nicoll, C. A., Anderson, D., and Jones, G. A. C., An accurate high-speed single-electron quantum dot pump, *New J. Phys.*, 12, 073013, 2010.

46. Giblin, S. P., Kataoka, M., Fletcher, J. D., See, P., Janssen, T. J. B. M., Griffiths, J. P., Jones, G. A. C., Farrer, I., and Ritchie, D. A., Towards a quantum representation of the ampere using single electron pumps, *Nat. Commun.*, 3, 930, 2012.

47. Rossi, A., Tanttu T., Tan, K. U., Iisakka, I., Zhao, R., Chan, K. W., Tettamanzi, G. C., Rogge, S., Dzurak, A. S., and Möttönen, M., An accurate single-electron pump based on a highly tunable silicon quantum dot, *Nano Lett.*, 14, 3405, 2014.

48. Zimmerman, N. M., Hourdakis, E., Ono, Y., Fujiwara, A., and Takahashi, Y., Error mechanisms and rates in tunable-barrier single-electron turnstiles and charge-coupled devices, *J. Appl. Phys.*, 96, 5254, 2004.

49. Fujiwara, A., Miyamoto, S., Nishiguchi, K., Ono, Y., and Zimmerman, N. M., Dynamics of single-electron capture in Si nanowire MOSFETs, *IEEE Silicon Nanoelectronics Workshop*, Honolulu, HI, 2008, pp. 1–2. DOI: 10.1109/SNW.2008.5418469.

50. Kashcheyevs, V. and Kaestner, B., Universal decay cascade model for dynamic quantum dot initialization, *Phys. Rev. Lett.*, 104, 186805, 2010.

51. Averin, D. V. and Nazarov, Yu. V., Virtual electron diffusion during quantum tunneling of the electric charge, *Phys. Rev. Lett.*, 65, 2446, 1990.

52. Jensen, H. D. and Martinis, J. M., Accuracy of the electron pump, *Phys. Rev.*, B 46, 13407, 1992.

53. Miyamoto, S., Nishiguchi, K., Ono, Y., Itoh, K. M., and Fujiwara, A., Escape dynamics of a few electrons in a single-electron ratchet using silicon nanowire metal-oxide-semiconductor field-effect transistor, *Appl. Phys. Lett.*, 93, 222103, 2008.

54. Kaestner, B., Leicht, C., Kashcheyevs, V., Pierz, K., Siegner, U., and Schumacher H. W., Single-parameter quantized charge pumping in high magnetic fields, *Appl. Phys. Lett.*, 94, 012106, 2009.

55. Wright, S. J., Blumenthal, M. D., Gumbs, G., Thorn, A. L., Pepper, M., Janssen, T. J. B. M. et al., Enhanced current quantization in high-frequency electron pumps in a perpendicular magnetic field, *Phys. Rev. B*, 78, 233311, 2008.

56. Fletcher, J. D., Kataoka, M., Giblin, S. P., Park, S., Sim, H.-S., See, P., Ritchie, D. A. et al., Stabilization of single-electron pumps by high magnetic fields, *Phys. Rev. B*, 86, 155311, 2012.

57. Kataoka, M., Fletcher, J. D., See, P., Giblin, S. P., Janssen, T. J. B. M., Griffiths, J. P., Jones, G. A. C., Farrer, I., and Ritchie, D. A., Tunable nonadiabatic excitation in a single-electron quantum dot, *Phys. Rev. Lett.*, 106, 126801, 2011.

58. Wulf, M., Error accounting algorithm for electron counting experiments, *Phys. Rev.*, B 87, 035312, 2013.

59. Fricke, L., Hohls, F., Ubbelohde, N., Fricke, B., Wulf, M., Kaestner, B., Hohls, F. et al., Self-referenced single-electron quantized current source, *Phys. Rev. Lett.*, 112, 226803, 2014.

60. Chida, K., Nishiguchi, K., Yamahata, G., Tanaka, H., and Fujiwara, A., Thermal-noise suppression in nano-scale Si field-effect transistors by feedback control based on single-electron detection, *Appl. Phys. Lett.*, 107, 073110, 2015.

61. Chan, K. W., Möttönen, M., Kemppinen, A., Lai, N. S., Tan, K. Y., Lim, W. H., and Dzurak, A. S., Single-electron shuttle based on a silicon quantum dot, *Appl. Phys. Lett.*, 98, 212103, 2011.

62. Jehl, X., Voisin, B., Charron, T., Clapera, P., Ray, S., Roche, B., Sanquer, M. et al., Hybrid metal-semiconductor electron pump for quantum metrology, *Phys. Rev. X*, 3, 021012, 2013.

63. Ono, Y., Fujiwara, A., Nishiguchi, K., Inokawa, H., and Takahashi, Y., Manipulation and detection of single electrons for future information processing, *J. Appl. Phys.*, 97, 031101, 2005.

64. Nishiguchi, K., Ono, Y., and Fujiwara, A., Single-electron counting statistics of shot noise in nanowire Si metal-oxide-semiconductor field-effect transistors, *Appl. Phys. Lett.*, 98, 193502, 2011.

65. Nishiguchi, K., Ono, Y., Fujiwara, A., Inokawa, H., and Takahashi, Y., Stochastic data processing circuit based on single electrons using nanoscale field-effect transistors, *Appl. Phys. Lett.*, 92, 062105, 2008.

66. Nishiguchi, K., and Fujiwara, A., Single-electron counting statistics and its circuit application in nanoscale field-effect transistors at room temperature, *Nanotechnology*, 20, 175201, 2009.

10 Coupled Si Quantum Dots for Spin-Based Qubits

Tetsuo Kodera and Shunri Oda

CONTENTS

10.1 Introduction 231
10.2 Device 232
10.3 Measurement Results and Discussions 233
    10.3.1 Charge Stability Diagram of Coupled Si QDs 233
    10.3.2 Pauli-Spin Blockade in Coupled Si QDs 233
    10.3.3 Charge Sensing of Few-Electron States in Si QDs 238
    10.3.4 Si QDs Charge Sensitivity of Si Single-Electron Transistors 240
    10.3.5 CSs for Coupled Si QDs 247
    10.3.6 Adding Gating Function to CSs for Coupled Si QDs 248
10.4 Summary 251
Acknowledgments 251
References 252

10.1 INTRODUCTION

Research regarding electron spin-based quantum bits (qubits) has primarily been performed using coupled quantum dots (QDs) in a GaAs/AlGaAs heterostructure, a system for which the spin coherence physics is well understood and fundamental technologies for controlling quantum states have been developed [1–5]. This system has a small electron effective mass, m^*, (equal to $0.067m_0$, where m_0 is the mass of a free electron) and thus, the QD size required to obtain quantum effects is relatively large (~100 nm) and can be readily obtained using electron beam lithography. However, additional research into quantum operations has found defects associated with decoherence of the electron spin. The main source of this decoherence is the hyperfine interaction with the nuclear spins of the Ga and As host materials [6].

To reduce this hyperfine interaction, the use of group IV semiconductors such as C, Si, and SiGe has been investigated, since these materials have isotopes with zero nuclear spin [7–11]. It is also possible to completely remove the hyperfine interaction by controlling the isotopic composition of the host [12]. Si, in particular, is advantageous for future applications because of its potential for compatibility with conventional Si metal-oxide-semiconductor (MOS) devices. However, it is necessary to fabricate much smaller QDs in Si in order to obtain quantum effects because of the larger electron effective mass in this material ($m^* = 0.19m_0$), and so, a highly sophisticated fabrication technique is required.

To properly apply qubits based on electron spins, it is necessary to observe spin-related tunneling phenomena and to reduce the electron number in individual QDs to only a few, so as to ensure energetically well-defined spin states. To date, single-electron regimes in single QDs [13,14] and two-electron regimes and Pauli-spin blockades (PSBs) in coupled QDs [15] have been observed in Si/SiGe heterostructures. Single-electron regimes in Si QDs have also been observed in MOS structures using a double top-gate design [16], MOS fine-gate structures [17], and nanowire structures [18]. Thus far, the majority of these Si QD devices have had gate-defined QD structures in which confinement potentials are generated by electric fields with fine top gates (TGs). In contrast, the lithographically defined Si QD devices described in the following do not require gates to create confinement potentials for the QDs. These devices thus represent a technological simplification owing to the reduced number of gates and may be advantageous with regard to the integration of qubit technology [19,20], since common fabrication processes normally used to produce Si MOS technologies may be applied to make the lithographically defined Si QDs.

In a previous study on lithographically defined Si QDs, a heavily doped n-type Si on insulator (SOI) substrate was used. This approach resulted in two defects. First, the presence of unintentional localized states due to fluctuations in the dopant potential was unavoidable. Second, the electron number could not be reduced down to the few-electron regime. Recently, our group was able to obtain improved device structures by applying fabrication techniques such as electron beam lithography, oxidation, and dry-etching [21,22]. A nondoped SOI substrate was employed in place of a heavily doped n-type SOI substrate to avoid the unintentional generation of localized states due to fluctuation of dopant potentials. The resulting devices represented lithographically defined coupled Si QDs with MOS structures incorporating a TG for inducing inversion carriers and side gates for controlling potentials in the QDs. We observed spin-related tunneling phenomena [23] and reduced the electron number in the QDs to the few-electron regime [24].

10.2 DEVICE

Figure 10.1a shows a schematic of a Si-coupled QD. In this device, three constrictions between the source (S) and drain (D) and five side gates were patterned by electron beam lithography and reactive ion etching on a 60 nm thick (100) SOI layer. The thickness of the buried oxide (BOX) was 400 nm. The gate oxide was formed via a 30 min thermal oxidation at 1000°C, followed by low-pressure chemical vapor deposition (LPCVD). A large poly-Si TG formed by LPCVD was subsequently used as an ion implantation mask for the formation of the n-type S and D regions. Finally, a 300 nm thick aluminum contact pad was formed by electron beam evaporation. Figure 10.1b shows a scanning electron microscope (SEM) image of the device, in which the coupled QD is defined by tunnel barriers at the three constricted regions. A 2D electron gas (2DEG) was introduced to the Si(100) surface by applying a positive TG voltage, V_{TG}. The electrochemical potentials of

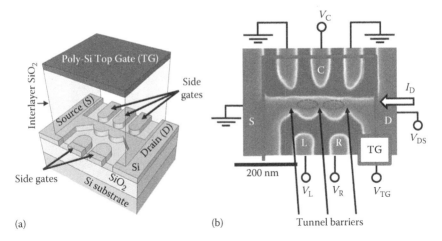

FIGURE 10.1   (a) A schematic diagram of Si-coupled QDs, showing a poly-Si TG formed on the gate oxides. (b) SEM image of the Si coupled QDs before TG formation. The two side gates located next to side gate C are grounded. Three tunnel barriers are formed at the constricted regions.

the left and right QDs were tuned by applying voltages V_L and V_R to the side gates L and R and the tunnel coupling between the two QDs was controlled by a voltage, V_C, applied to the side gate C.

10.3 MEASUREMENT RESULTS AND DISCUSSIONS

10.3.1 CHARGE STABILITY DIAGRAM OF COUPLED SI QDS [23]

The honeycomb-like charge stability observed at a temperature of 250 mK is shown in Figure 10.2, which reflects the formation of coupled QDs in series [23,25]. The charging energies of the left and right QDs were estimated to be 10.7 and 11.0 meV, respectively, from the spacing of the Coulomb peaks. Based on the distribution of the current peaks due to resonant tunneling at charge triple point A in Figure 10.2, the quantum level spacing values, ΔE, of the left and right QDs were estimated to be 310 and 260 μeV, respectively. ΔE can be approximated as $\Delta E = h^2/8\pi m^*A$, where m^* is the effective mass, h is Planck's constant, and A is the area of the QD, with spin and valley degeneracies included. From this equation, ΔE is determined to be between 260 and 380 μeV for our device geometry, which is in good agreement with the experimental estimation. We conclude that the QD was formed as intended between the two constricted regions indicated by the two ovals in Figure 10.1b.

10.3.2 PAULI-SPIN BLOCKADE IN COUPLED SI QDS [23]

Current rectification in coupled QDs due to PSB appears at a triple point with only one bias polarity [26]. We observed current rectification in the coupled Si QDs when applying a negative bias voltage at triple point B in Figure 10.2, as indicated by the trapezoid in Figure 10.3a. In contrast, no current rectification was observed with a positive bias, as can be seen in Figure 10.3b [23]. The current rectification was lifted

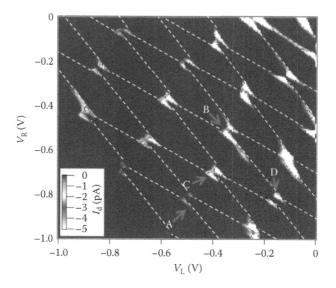

FIGURE 10.2 Charge stability diagram of the Si coupled QDs as a function of V_L and V_R at zero magnetic field, where $V_{ds} = -2$ mV, $V_{TG} = 0.90$ V, and $V_C = -1.72$ V. The white dotted lines indicate the boundaries of the stable charge states.

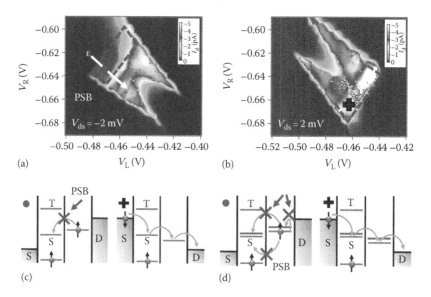

FIGURE 10.3 (See color insert.) (a) Magnified plot of the triple point B shown in Figure 10.2 with a negative bias, where $V_{ds} = -2$ mV, $V_{TG} = 0.97$ V, and $V_C = -1.76$ V. The PSB appears only at this polarity. (b) The same triple point as in (a) under a positive bias ($V_{ds} = 2$ mV). (c) Energy diagrams of a Si coupled QD at the red dot in (a) and at the blue cross in (b), at which the valley degeneracy is assumed to be lifted. (d) The same diagram as (c) without assuming minimal valley degeneracy lifting. Intradot and interdot tunneling between different valleys are assumed to be weak such that the PSB is not lifted.

along the outer edge of the PSB regime, as indicated by the circle in Figure 10.3a, because of electron exchanges between the coupled QD and the S or D regions.

Si QDs normally have doubly degenerate valleys due to confinement in the direction perpendicular to the Si surface, and this valley degeneracy could potentially lift the PSB. However, the fact that a PSB is observed indicates either a lifting of the valley degeneracy or weak tunneling between two valleys [27]. In the former case, once two spins occupy the (1,1) triplet state, as pictured in Figure 10.3c, the current is suppressed because of the PSB until relaxation from the (1,1) triplet to the (1,1) singlet occurs. In the latter case, even if degenerate valleys exist as shown in Figure 10.3d, the blockade is not lifted, because the intradot transitions and interdot tunneling between different valleys are weak. PSB-like characteristics are also observed at the triple points (B, C, and D in Figure 10.2). This is not expected for a simple spin-1/2 PSB. Since a coupled QD has many electrons, spin-3/2 ground states are possible, leading to consecutive PSBs [28]. A blockade in which valley degeneracy plays a role can also lead to consecutive PSB-like phenomena. Even when a spin doublet is formed in the coupled QDs, the current may be suppressed because of the weak tunneling between valleys noted earlier.

Figure 10.4a presents the leakage current in the PSB regime at triple point C in Figure 10.2 as a function of the magnetic field, B, applied normal to the coupled

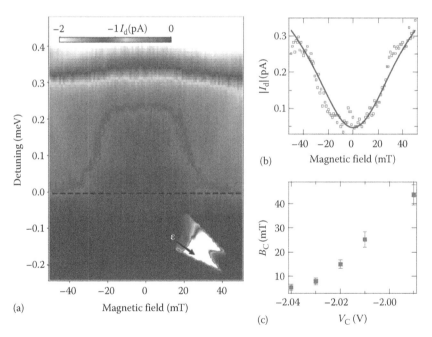

FIGURE 10.4 (a) Leakage current in the PSB regime as a function of perpendicular magnetic field and detuning, where $V_{ds} = -2$ mV, $V_{TG} = 0.97$ V, and $V_C = -1.99$ V. Inset: Magnified plot of triple point C in Figure 10.2. Here the arrow corresponds to the detuning axis ε in the main figure. (b) Current values as a function of magnetic field along the dashed line in (a), denoted by squares, and the fitting curve to the measured data, as indicated by the solid line. (c) B_C values extracted from the fit in (b) as a function of V_C. Large V_C values correspond to large interdot tunnel coupling, t.

QDs, along with a detuning, ε, indicated by the arrow [23]. A strong current dip is observed at $B = 0$, whereas the current with the opposite bias does not change as a function of B. Similar current dips have been observed for coupled QDs in InAs nanowires [29] and carbon nanotubes [7]. Leakage current dips can be attributed to spin-orbit-induced relaxation [30], which is suppressed at $B = 0$ due to Van Vleck cancellation [29,31]. In such cases, a Lorentzian line shape defined by $I_{fit} = I_{max}\{1 - 8B_C^2/9(B^2 + B_C^2)\}$, with characteristic width B_C, is predicted theoretically [30]. The data points plotted as squares in Figure 10.4b correspond to the absolute values of the leakage current in the PSB regime, as measured along the dashed line in Figure 10.4a. A fit to the Lorentzian form, corresponding to the solid line in Figure 10.4b, shows good agreement with these experimental data. As the interdot tunneling between the two QDs is enhanced by increasing V_C values, the value of B_C extracted from the fit increases, as is evident from the plot in Figure 10.4c. This result is also consistent with theoretical expectations, which predict that B_C will be proportional to the extent of interdot tunneling coupling [30]. These results suggest that the effects of spin-orbit interactions dominate spin relaxation in such devices, even though spin-orbit interactions are usually weak in Si. Another possible mechanism leading to the observed dip in leakage current around $B = 0$ is spin-valley blockade with short range disorder [32], which predicts a current dip as a function of magnetic field–induced valley splitting. There is, however, no independent evidence that the required B-dependent valley splitting exists. The valley physics in Si coupled QDs thus requires further experimental and theoretical studies.

At some triple points, a peak is observed rather than a dip in the PSB leakage current on a larger field scale. As an example, the field dependence of the leakage current at triple point A in Figure 10.2 is shown in Figure 10.5a. The arrow in the magnified plot of triple point A presented in Figure 10.5b corresponds to the detuning axis in Figure 10.5a. Among the 15 triple points that exhibit PSB-like behavior (Figure 10.2), nine show a zero-field current dip and two show a peak. In some cases, it is also possible to observe current peaks outside a current dip. In coupled GaAs QDs, zero-field peaks in the leakage current are attributed to hyperfine-induced spin relaxation [33], although the contribution of the hyperfine interaction should be small in Si systems, because the dominant isotope, ^{28}Si, has zero nuclear spin. Based on the 4.7% natural abundance of ^{29}Si and the lithographic device dimensions, the number, N, of nuclear spins in a Si QD is expected to be $(2 - 3) \times 10^4$, corresponding to a fluctuating Overhauser field magnitude $B_{nuc} = |A|/g\mu_B\sqrt{N} \sim 10\text{–}15\ \mu T$, where μ_B is the Bohr magneton, the hyperfine coupling constant $|A| \sim 0.2\ \mu eV$ [34], and $g \sim 2$ for the electrons in Si. Since the peak width in Figure 10.5c is greater than B_{nuc} by a factor of 10^4, the mechanism by which the current peaks at $B = 0$ are generated is not explained by hyperfine interaction. Similar peaks were also reported for Si QDs in Reference 35, in which case the peaks were fully explained by spin-flip cotunneling [36]. When $k_BT > t$ (where k_B is Boltzmann's constant and t is the interdot tunneling coupling), the spin-flip cotunneling current is given by $I_{cot} = 4ecg\mu_B B/3\sinh(g\mu_B B/k_BT)$, where $c = h\{[\Gamma_R/(\Delta - \varepsilon)]^2 + [\Gamma_L/(\Delta + \varepsilon - 2U' - 2eV_{ds})]2\}/\pi$, in which e represents the electron charge, $\Gamma_{L(R)}$ is the coupling between the lead and the left (right) dot, Δ is the energy depth of the

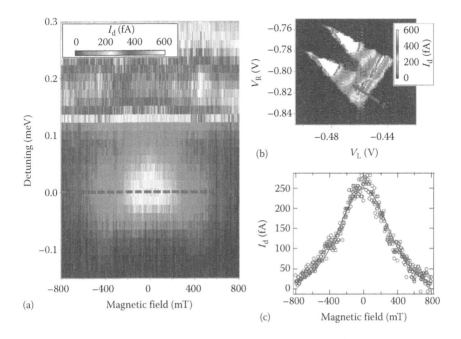

FIGURE 10.5 (a) Leakage current in the PSB regime as a function of perpendicular magnetic field and detuning, where V_{ds} = 2 mV, V_{TG} = 0.968 V, and V_C = −1.925 V. (b) Magnified plot of triple point A in Figure 10.2. Here the arrow corresponds to the detuning axis ε in (a). (c) Current values as a function of magnetic field along the dashed line in (a), denoted by circles, and the fitting curve to the measured data, as indicated by the solid line.

two-electron level, and U' is the interdot charging energy [37]. Since we observed clear resonant tunneling peaks, $\Gamma_{L(R)}$ was larger than t in the case of our experimental device. In addition, if $\Gamma_{L(R)} > t > k_BT$ = 21 μeV, the current would be much larger than the observed current shown in Figure 10.5b. Thus, $k_BT > t$, such that I_{cot} can be used to fit the current peak. The blue curve in Figure 10.5c plots the I_{cot} values and is in good agreement with the data obtained using a T value of 250 mK, yielding g = 2.3 and c = 54 kHz/μeV. Since the current exhibits only minimal variation along the base of the triangle in Figure 10.5b, we can assume that Γ_L = Γ_R ≡ Γ. Using the equation for c given earlier, together with the values Δ = 1 meV, ε = 0 meV, U' = 1 meV, and eV_{ds} = 2 meV estimated from the bias triangle shown in Figure 10.5b, we can calculate Γ = 26 μeV. In addition, using the unblocked resonant tunneling peak current (0.6 pA) in conjunction with Equation 15 in Reference 25, t can be determined to be approximately 0.3 μeV. These estimated values are similar to those reported in Reference 35, and the differences between the present values and those in the reference may reasonably be attributed to experimental variations, such that spin-flip cotunneling processes are most likely the mechanism by which the peak is generated. It should, however, be noted that spin-valley blockade with disorder could also explain the peak, but again, we have at present no evidence of the required B-field dependent valley splitting.

10.3.3 Charge Sensing of Few-Electron States in Si QDs

In order to fully understand the physics of tunneling through the QDs, it is necessary to know the number of electron in each QD. If the drain current through the
QDs can be determined as a function of the side gate voltage, or Coulomb oscillation, up to the pinch-off in the QDs, one can determine the absolute number of
electron in the QDs simply by counting the Coulomb peaks. However, it is sometimes difficult to directly observe the current from drain to source via QDs when a
large negative voltage is applied to the side gates, since this large side gate voltage
reduces the tunneling rate from the drain to source to a value that is too small
to be directly measured. In such cases, it is necessary to develop charge sensors
(CSs) composed of QDs in order to determine the absolute number of electrons
in the QDs. We thus prepared two QDs electrostatically coupled in parallel, as
presented schematically in Figure 10.6. In this device, a change in the number of
electrons in QD1 will change the Coulomb interaction between the two QDs and,
as a result, the current through QD2 will vary. QD2 thus functions as a CS for
QD1, allowing charge sensing.

To demonstrate the charge sensing in operation, the device presented in
Figure 10.7a was fabricated, consisting of lithographically defined Si QDs electro-
statically coupled in parallel with an MOS structure on an undoped SOI substrate.
This device was made using the same process described in Section 10.2, but with a
Si layer thickness of approximately 23 nm. Figure 10.7b shows an equivalent circuit
model of the two QDs electrostatically coupled in parallel, in which the two QDs are
capacitively coupled to each other via C_m. The side gate G1 (G2) primarily couples to
QD1 (QD2) via C_{G11} (C_{G22}), and finite cross-capacitance is obtained between G1 (G2)
and QD2 (QD1) via C_{G12} (C_{G21}). The drain currents I_{QD1} and I_{QD2} through QD1 and
QD2, respectively, are simultaneously measured when sweeping a side gate voltage
V_{G1}. The source voltages (V_{S1} and V_{S2}), drain voltages (V_{D1} and V_{D2}), and the other
side gate voltage V_{G2} are all fixed.

Figure 10.8a plots I_{QD1} and I_{QD2} as functions of V_{G1}, measured simultaneously at
a temperature of 4.2 K. Here, $V_{S1} = 0$ V, $V_{S2} = 0$ V, $V_{D1} = 1.5$ mV, and $V_{D2} = -2.5$ mV.
Sharp Coulomb peaks, related to charge transport through QD1, are observed in
I_{QD1}, while broad Coulomb peaks attributed to QD2 are observed as a function
of V_{G1} due to the cross-capacitance, C_{G12}. Sharp kinks are also seen in the broad

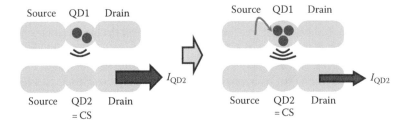

FIGURE 10.6 Schematics of charge sensing using two QDs electrostatically coupled in
parallel. When the number of electrons in QD1 changes, the Coulomb interaction between
two QDs varies, leading to a change in current through QD2.

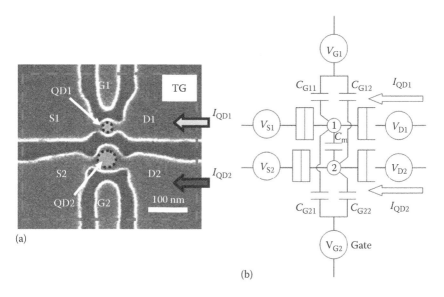

FIGURE 10.7 (a) SEM image of two QDs electrostatically coupled in parallel. (b) Equivalent circuit model. The two QDs are capacitively coupled to one another via C_m and to a side gate G_1 (G_2) via C_{G11} and C_{G12} (C_{G21} and C_{G22}).

Coulomb oscillation in I_{QD2} at the V_{G1} position of the Coulomb peaks of I_{QD1}, as indicated by dotted lines. This phenomenon occurs, because the number of electrons (N) confined in the QD changes at the V_{G1} position of the Coulomb peaks in I_{QD}, which in turn varies the electrochemical potential of QD2. Subsequently, shifts in the Coulomb oscillation in I_{QD2} on the V_{G1} axis occur, resulting in the kinks in I_{QD2}. These measurements demonstrate the successful detection of changes in the number of electrons in QD1 using QD2 as the CS. Furthermore, the observed kinks are present throughout the entire I_{QD2} region, while the I_{QD1} Coulomb peaks are not observed when a large negative gate voltage is applied, as shown in the left side of Figure 10.8a. These results indicate that the number of electrons associated with QD1 changes, but this change is not reflected in the Coulomb peaks due to the overly low tunneling rate between the source and drain. Therefore, CS represents a powerful tool for detecting changes in the number of electrons in QD1 even in cases in which the direct measurement of the Coulomb peaks of QD1 is not possible.

A few-electron regime in the QDs is also observed when applying a large negative V_{G1}. Figure 10.8b shows contour plots of I_{QD1} and dI_{QD2}/dV_{G1} as functions of V_{G1} and V_{G2}, respectively, measured simultaneously. In the upper panel, a Coulomb peak is observed at a V_{G1} value of approximately 4 V. At the same position in the lower panel, an abrupt change in dI_{QD2}/dV_{G1} occurs, representing a charge-sensing signal line. Here the broad lines running from the upper left to lower right are the Coulomb peaks of QD2 resulting from the cross-capacitance, C_{G12}, noted earlier, and can be ignored. Another charge-sensing signal line is evident at a V_{SG1} value of approximately −4.7 V and no further transition occurs in the range $V_{SG1} < −4.7$ V,

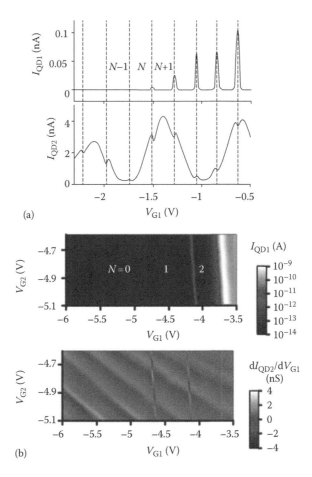

(a)

(b)

FIGURE 10.8 (a) Currents through QD1 (I_{QD1}) and QD2 (I_{QD2}) as functions of the side gate voltage V_{G1}, measured simultaneously. Here, $V_D = 0.3$ mV, $V_{DSET} = 6$ mV, $V_{TG} = 5.414$ V, and $V_{G2} = -4$ V. (b) Contour plots of I_{QD1} and dI_{QD2}/dV_{G1} as functions of V_{G1} and V_{G2}, measured simultaneously. Here, $V_D = 1$ mV, $V_{DSET} = 5$ mV, and $V_{TG} = 3.8$ V.

indicating that QD1 contains no electrons below this gate voltage. Thus, we are able to determine the few-electron regime in QD1 as well as the absolute number of electrons in QD1 by assessing the CS signal.

10.3.4 Si QDs Charge Sensitivity of Si Single-Electron Transistors [24]

Single-electron transistors (SETs) are currently one of the most widely studied nano-electronic components that can be realized using QD structures. In particular, due to their high charge sensitivities, SETs can be used as CSs for reading spin or charge qubits [38–40]. As such, many experiments have been performed using QDs and quantum point contacts (QPCs) in 2DEG systems such as GaAs/AlGaAs and Si/SiGe heterostructures [41]. Some recent studies have focused on the use of QD-SET structures as a means of improving the signal-to-noise ratio (SNR) during qubit readout [42]. The charge sensitivity of SETs has been investigated theoretically [43] and experimentally [44] to evaluate their effectiveness as CSs for dc or radio frequency (RF) measurements. However, the SNR during qubit readout depends not only on the charge sensitivity of the SET itself but also on the capacitive coupling between the QDs and the SET [45]. Since both of these factors depend in turn on the configuration of the SET, including its dimensions and distance from the QD, it is necessary to take the configuration into consideration when designing an effective SET CS. Our group therefore investigated the effects of the SET configuration on both the charge sensitivity and capacitive coupling, using the same device described in Section 10.3.3.

Stability diagrams for both QD1 and QD2 are presented in Figure 10.9a and b, respectively. The appearance of regular Coulomb diamonds indicates the presence of single QDs without unintentional localized states. Figure 10.10a and b shows contour plots of I_{QD1} and I_{QD2}, respectively, as functions of V_{TG} and V_{G1}, measured simultaneously. The Coulomb peak positions of I_{QD2} in Figure 10.10b have a zigzag structure with shifts appearing at identical voltages to the Coulomb peaks of I_{QD1}. Similarly, in

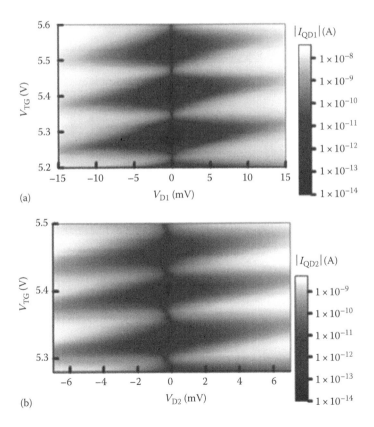

FIGURE 10.9   (a, b) Stability diagrams for QD1 and QD2, respectively. (a) $V_{G1} = 0$ V and $V_{G2} = -4$ V, (b) $V_{G1} = 0$ V and $V_{G2} = -4$ V. From the observed Coulomb diamonds, the charging energy values E_{QD1} and E_{QD2} are estimated to be 14 and 6.2 meV.

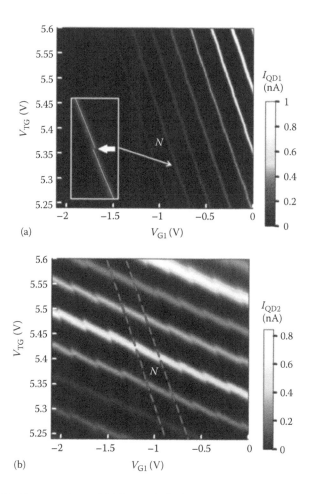

FIGURE 10.10 Contour plots of (a) I_{QD1} and (b) I_{QD2} as functions of V_{TG} and V_{G1}. Here, $V_{D1} = 1$ mV, $V_{D2} = 1$ mV, and $V_{G2} = -4$ V. The two dashed lines in (b) indicate the positions of two Coulomb peaks in (a). Inset to (a): magnified plot of a shift of a Coulomb peak.

Figure 10.10a, the Coulomb peaks of I_{QD1} exhibit shifts at voltages corresponding to the Coulomb peaks of I_{QD2}. Figure 10.11a is a schematic diagram of the twin single QD structure, with the various electrical parameters indicated. The electrochemical potential of an N-electron QD1 (μ_{QD1}) and an n-electron QD2 (μ_{QD2}) can be expressed as

$$\mu_{QD1}(N,n) = \frac{E_{QD1}}{e}\left(Ne - \frac{e}{2} - C_{L1}V_{L1} - C_{R1}V_{R1} - \sum_i C_i V_i\right)$$

$$+ \frac{E_{QD1}}{e}k_{QD1}\left(ne - C_{L2}V_{L2} - C_{R2}V_{R2} - \sum_j C_j V_j\right) + E_N, \quad (10.1)$$

and

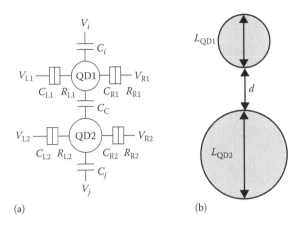

(a)   (b)

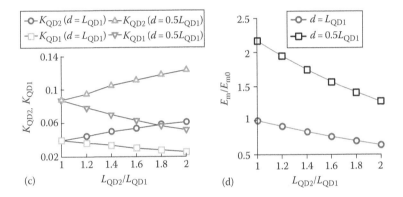

(c)   (d)

FIGURE 10.11 (a) Schematic diagrams of QD1 and QD2 and various associated electrical parameters. (b) Schematic dimensions of twin QDs used in numerical calculations. (c) Calculation results for k_{QD1} and k_{QD2} as functions of L_{QD2}/L_{QD1}. L_{QD1} and the distance, d, between the edges of QD1 and QD2 are fixed. (d) Calculated E_m/E_{m0} values as a function of L_{QD2}/L_{QD1} obtained by the same method as used for the data shown in (c). Here E_{m0} is the value of E_m when $L_{QD2} = L_{QD1} = d$. In (c) and (d), the calculation results are shown for $d = L_{QD1}$ and $d = 0.5L_{QD1}$.

$$\mu_{QD2}(N,n) = \frac{E_{QD2}}{e}\left(ne - \frac{e}{2} - C_{L2}V_{L2} - C_{R2}V_{R2} - \sum_j C_j V_j\right)$$

$$+ \frac{E_{QD2}}{e}k_{QD2}\left(Ne - C_{L1}V_{L1} - C_{R1}V_{R1} - \sum_i C_i V_j\right) + E_n, \qquad (10.2)$$

where

$C_{QD1} = C_C + C_{L1} + C_{R1} + \Sigma C_i$ and $C_{QD2} = C_C + C_{L2} + C_{R2} + \Sigma C_j$ are the total capacitances of QD1 and QD2, respectively

ΣC_i (ΣC_j) is the capacitance between QD1 (QD2) and other gates or environments for voltage V_i (V_j)

E_N and E_n are the energies of the topmost filled single-particle states for QD1 and QD2, respectively

$E_{QD1} = \alpha e^2/C_{QD1}$ and $E_{QD2} = \alpha e^2/C_{QD2}$ are the charging energies of QD1 and QD2, respectively, where $\alpha = (1 - C_C^2/C_{QD1}C_{QD2})^{-1}$

Equations 10.1 and 10.2 are equivalent to the electrochemical potentials of two QDs coupled in series, as described in Reference 25, except that there is no tunnel coupling between the parallel QDs. The term $ek_{QD2} = eC_C/C_{QD1}$ corresponds to the charge induced in QD2 when the number of electrons in QD1 decreases from N to $N - 1$. Similarly, $ek_{QD1} = eC_C/C_{QD2}$ is equivalent to the charge induced in QD1 when the number of electrons in QD2 decreases from n to $n - 1$. In the I_{QD2} Coulomb peaks seen in Figure 10.10b, the value S_{QD2} is defined as the ratio of the peak shift magnitude to the peak-to-peak spacing in terms of V_{TG}. This can be expressed as $S_{QD2} = k_{QD2}(1 + \Delta E_n/E_{QD2})^{-1}$, where $\Delta E_n = E_n - E_{n-1}$. A value for $k_{QD2} = S_{QD2}$ of 0.088 ± 0.005 may be calculated from Figure 10.10b, assuming that ΔE_n is negligibly small in the large-n regime. Similarly, a $k_{QD1} = S_{QD1}$ value of 0.038 ± 0.003 is obtained from Figure 10.10a, where $S_{QD1} = k_{QD1}(1 + \Delta E_N/E_{QD1})^{-1}$ is identical to k_{QD1} because of the large value of N. The value of k_{QD2} is larger than that of k_{QD1} in this instance, since the QD2 diameter, L_{QD2}, of approximately 70 nm is larger than the QD1 diameter, L_{QD1}, of approximately 35 nm. Both k_{QD1} and k_{QD2} may be calculated as functions of L_{QD2} by assuming that QD1 and QD2 correspond to 2D conductive disks. Here the distance, d, between their edges is constant, as is L_{QD1} (see Figure 10.11b). Figure 10.11c summarizes the results obtained using a 3D Poisson equation. The results indicate that increasing L_{QD2} causes k_{QD2} to increase and k_{QD1} to decrease, and that both k_{QD2} and k_{QD1} are larger at $d = 0.5L_{QD1}$ than at $d = L_{QD1}$. Inserting the device parameters ($L_{QD1} \approx 35$ nm, $L_{SET1} \approx 70$ nm and $d \approx 25$ nm) into the calculations gives $k_{QD2} = 0.086$ and $k_{QD1} = 0.036$. These are comparable to the measured values, confirming the validity of the calculation method. The calculation results are valid even if the materials surrounding the two QDs have different dielectric constants in vertical directions, and are thus adaptable to other 2DEG systems such as GaAs/AlGaAs and Si/SiGe heterostructures.

In order to study the charge sensitivity of the QDs when used as SET CSs, the SET current, I_{SET}, and the low-frequency spectral density of the SET shot noise, S_I, were calculated using master equations [43,46,47] for charge states corresponding to $N - 1$ ($n - 1$) and $N(n)$ excess electrons in QD1 (QD2). Assuming $R_{L(R)}$ is much greater than $R_Q = \pi\hbar/2e^2$ and the quasistationary condition in which ω is much less than I_{SET}/e [46,48], the relationships

$$I_{SET} = e\frac{\Gamma_L\Gamma_{R+} - \Gamma_{L+}\Gamma_{R-}}{\Gamma_+ + \Gamma_-}, \tag{10.3}$$

and

$$S_I = 2e^2\frac{\sum_\pm \Gamma_\mp^2\left(2\Gamma_{L\mp}\Gamma_{R\mp} + \Gamma_{L\mp}\Gamma_{R\mp} + \Gamma_{L\mp}\Gamma_{R\mp}\right)}{(\Gamma_+ + \Gamma_-)^3}, \tag{10.4}$$

are obtained. Here, the electron tunneling rates are $\Gamma_{L(R)\pm} = (1/e^2R_{L(R)})\gamma(\pm[-eV_{L(R)} - \mu_{SET}(N, n)])$, $\Gamma_+ = \Gamma_{L+} + \Gamma_{R+}$, and $\Gamma_- = \Gamma_{L-} + \Gamma_{R-}$, where $\gamma(\varepsilon) = \varepsilon/[1 - \exp(-\varepsilon/k_BT)]$

(Reference 46), and the electrochemical potential of the SET, $\mu_{\text{SET}}(N, n)$, is identical to that given in Equation 10.1 for QD1 and Equation 10.2 for QD2. In the case of dc measurements, the charge sensitivity, δq, is given by $\delta q = (S_{\text{I}})^{-1/2}/(\partial I_{\text{SET}}/\partial Q)$ [49], where Q is the gate charge in the SET and can be expressed as the term $\Sigma C_i V_i (\Sigma C_j V_j)$ in Equations 10.1 or 10.2. When QD1 and QD2 have the same resistance conditions (i.e., the same parasitic and tunnel resistances, such that $R_{\text{L1}} = R_{\text{L2}}$ and $R_{\text{R1}} = R_{\text{R2}}$), the charge sensitivities, δq_1 and δq_2, are inversely proportional to the charging energies, E_{QD1} and E_{QD2}, respectively. Therefore, E_{QD1} and E_{QD2} are the key capacitive parameters required to determine δq, and so $\delta q_2/\delta q_1 = E_{\text{QD1}}/E_{\text{QD2}}$. From the Coulomb diamonds shown in Figure 10.9a and b, the calculated value for $E_{\text{QD1}}/E_{\text{QD2}}$ is approximately 2.26, which is comparable to the value of 2.38 obtained from Figure 10.11c. The charging energies depend on the diameters of the QDs and this result indicates that a smaller QD acts as a more sensitive SET CS. It should be emphasized that increasing the SET diameter increases the total amount of induced charge but reduces the sensitivity, and so there is a trade-off when optimizing these two parameters.

If the approximation that k_{QD2} is much less than unity can be made, which is usually valid for experiments working with 2DEG systems, the SNR of the charge-sensing signal due to a change in N at the dc measurement limit is given by SNR = $20\log_{10}[ek_{\text{QD2}}(\Delta f)^{-1/2}(\delta q_2)^{-1}]$, where Δf is the measurement bandwidth. Note that hereafter QD2 is assumed to function as the SET CS. By simultaneously considering the capacitive coupling, k_{QD2}, and the charge sensitivity, δq_2, the key capacitive parameter determining the SNR can be shown to be the coupling energy, $E_{\text{m}} = e^2 \alpha k_{\text{QD2}}/C_{\text{QD2}}$, because $k_{\text{QD2}}/\delta q_2$ is proportional to E_{m}. Figure 10.11d plots calculated $E_{\text{m}}/E_{\text{m0}}$ values as a function of $L_{\text{QD2}}/L_{\text{QD1}}$. Here E_{m0} is the value of E_{m} when $L_{\text{QD2}} = L_{\text{QD1}} = d$. It can be seen that increasing L_{QD2} or d decreases E_{m}: a decrease in the SET diameter leads to an increase in the SNR. It is noteworthy that decreasing the SET diameter reduces the capacitive coupling value, k_{QD2}, but also induces a greater increase in the sensitivity, leading to an increase in the SNR.

Figure 10.12a shows contour plots of I_{QD2} as a function of V_{TG} and V_{G1} obtained from another fabricated device. In this case, QD1 is in the few-electron regime. In this low-N regime, k_{QD2} values of 1.10, 0.98, and 0.93 were obtained for the $0 \leftrightarrow 1$, $1 \leftrightarrow 2$, and $2 \leftrightarrow 3$ transitions of N, respectively, although it should be noted that n was still large under these conditions. The value of k_{QD2} increased with decreasing N, because the effective size of QD1 was reduced below its lithographically defined size due to the small number of electrons. This result indicates that operating in the few-electron regime will improve the SNR when reading charge qubits in lateral QD-SET systems.

The dependence of L_{QD2} and d on the SNR was investigated for RF single-shot measurements under the conditions shown in Figure 10.12b. Here, exact resonance at the carrier frequency, $\omega = (LC_{\text{s}})^{-1/2}$, and the quasistationary condition in which ω is much less than I_{SET}/e are assumed. Also assuming a low circuit impedance, Z, approximately equal to $L/C_{\text{s}}R_{\text{d}}$ and much less than R_0, where R_{d} is the differential resistance of the SET and R_0 is the cable wave resistance and is approximately 50 Ω, the SET bias voltage can be expressed by a simple sine wave, $V_{\text{L}}(t) = V_0 + A\sin\omega t$ [48].

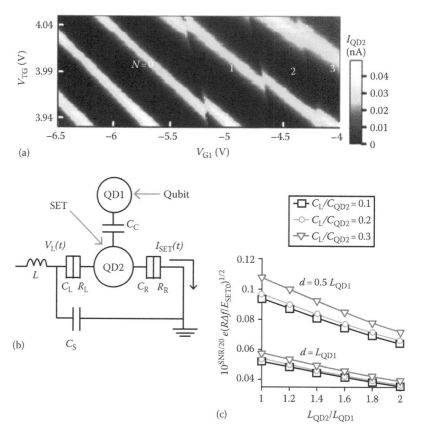

(a)

(b)

(c)

FIGURE 10.12 (a) Contour plots of I_{QD2} as a function of V_{TG} and V_{GI} for the device described in the main text. Here $V_{D1} = 1$ mV, $V_{D2} = 5$ mV, and $V_{G2} = -5.8$ V. (b) Schematic diagram of the RF measurement setup. (c) Calculation results for SNR as a function of L_{QD2} (in units of L_{QD1}) under optimized gate bias conditions. Using the y axis value $Y = 10^{SNR/20}e(R\Delta f/E_{SET0})^{1/2}$ from (c), the SNR can be expressed as SNR $= 20\log_{10}[Y/\{e(R\Delta f/E_{SET0})^{1/2}\}]$. Here, $R_L = R_R = R$, $k_BT = 0.01E_{SET0}$ (where E_{SET0} is the QD2 charging energy when $L_{QD2} = L_{QD1}$ and $d = L_{QD1}$), $A = 0.2E_{SET0}/e$, $V_0 = 0$, and L_{QD1} is fixed. Calculations were performed for $C_L/C_{QD2} = 0.1$, 0.2, and 0.3. Calculation results for $d = L_{QD1}$ and $d = 0.5L_{QD1}$ are shown.

Here V_0 is the static component and A is the amplitude of the SET bias. Restricting the analysis to the first harmonic of the reflected wave, the relationship

$$\text{SNR} = 20\log_{10}\frac{|\langle I_{SET}(t)\sin\omega t\rangle - \langle I'_{SET}(t)\sin\omega t\rangle|}{\sqrt{\Delta f\langle S_1(t)\sin^2\omega t\rangle}}, \tag{10.5}$$

is obtained. Here $\langle\cdots\rangle$ denotes the time average and I'_{SET} is the current generated when the charge of $\pm ek_{QD2}$ is induced in QD2. Calculation of the capacitive coupling between the QD and the SET demonstrates that increasing both L_{QD2} and d reduces the SNR, similar to the results obtained in the case of dc measurements.

Figure 10.12c plots the calculated SNR values as functions of L_{QD2}/L_{QD1} using the optimized gate bias conditions (included in $\Sigma C_j V_j$ in Equation 10.2). Here L_{QD1} is fixed and other calculation parameters are given in the figure caption. These plots show that increasing L_{QD2} reduces the SNR and that the SNR for $d = 0.5L_{QD1}$ is larger than that for $d = L_{QD1}$. It is also evident that increasing C_L/C_{QD2} causes a slight increase in the SNR. Decreasing L_{QD2}/L_{QD1} from a value of 2 down to 1 approximately doubles the signal power both for $d = 0.5L_{QD1}$ and $d = L_{QD1}$. Generally, experimental devices are designed such that the distance between the QD and the SET CS is as small as possible so as to enhance the capacitive coupling between the two. However, there is a lower limit to the distance that can be obtained between the QD and the SET when making such a device, and so decreasing the SET diameter can also be adopted as an effective means of improving the SNR.

This chapter focuses on the capacitive parameters of an SET CS in order to establish guidelines for improving the SNR during qubit readout. Nevertheless, it should be noted that reducing the parasitic resistance of the device is also useful with regard to improving the SNR, as is asymmetric tuning of the tunnel resistance of the SET.

10.3.5   CSs for Coupled Si QDs

Our group has also applied the charge-sensing technique to coupled Si QDs, and has fabricated coupled Si QDs (or a double QD: DQD) integrated with a CS, as shown in Figure 10.13. Currents through the DQD and the CS are denoted as I_{DQD} and I_{CS}, respectively. Figure 10.14a and b present the values of I_{DQD} and dI_{CS}/dV_{gl} as functions of V_{gl} and V_{gr}, respectively, measured simultaneously at 0.3 K. The yellow dotted lines in Figure 10.14a indicate the boundaries of the stable charge states, as described in Section 10.3.1. At the intersections of these lines, the so-called charge triple points, the current flows form a honeycomb-like pattern, which is typical of the charge stability diagram for DQDs coupled in series. No current is observed in the left side portion of Figure 10.14a, because of the low tunneling rate noted in Section 10.3.3. The image of the CS in Figure 10.14b exhibits broad blue, white, and

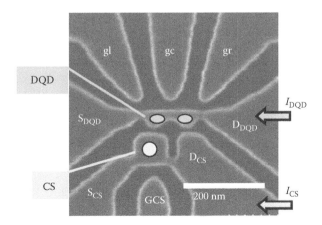

FIGURE 10.13   SEM image of Si coupled double QDs (DQD) with a CS.

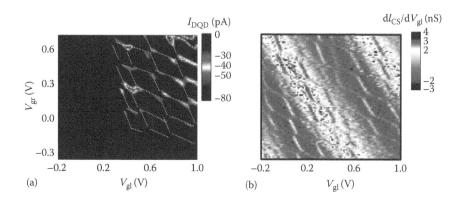

FIGURE 10.14 **(See color insert.)** (a) Charge stability diagram of the DQD with $V_{tg} = 5.55$ V and $V_{bg} = 0$ V. A honeycomb structure is observed at the right side, while no signal is seen at the left side. (b) Differential conductance of the integrated CS measured simultaneously with (a). A honeycomb pattern is observed over the entire measurement region.

red stripes over the entire surface resulting from the Coulomb peaks of the CS itself, in addition to abrupt changes in the value of dI_{CS}/dV_{gl} that form a honeycomb-like structure. The upper right region of this honeycomb pattern exactly matches the pattern described by the yellow lines in Figure 10.14a. The observed honeycomb structure thus results from detection of the single-electron transition of the DQD.

10.3.6 Adding Gating Function to CSs for Coupled Si QDs [50]

In order to achieve quantum logic operation in a device, it is necessary to integrate numerous QDs both as qubits and as CSs. The operation of all the QDs requires the integration of many gate electrodes so as to control the electrochemical potential of each QD as well as the tunnel barriers between the QDs. The requirement to physically integrate qubits, CSs, and gate electrodes in a limited device area thus becomes problematic, and so the technological improvement of QDs by multifunctionalization is desirable so as to reduce the number of gate electrodes required. As an example, it would be helpful to assign the dual functions of charge sensing and gating to a single QD.

Figure 10.15 presents an overhead view of a device structure in which Si DQDs, CSs, and gates are fabricated on an SOI substrate, together with the associated measurement apparatus [50]. In this device, a back gate (BG) is used in place of a TG to induce 2DEG at the lower Si/SiO$_2$ interface, and the thickness of the SOI (BOX) is 30 nm (145 nm). Voltages applied to the gates (V_{gl}, V_{gm}, V_{gr}, V_{gCS} and V_{bg}) control the electrochemical potential of each QD and modify the tunneling rate at each potential barrier. Normally, the source voltage of each QD is applied at a ground level as a reference voltage. In this device, however, offset voltages were applied to the DQD (V_{oDQD}) and CS (V_{oCS}) to tune the performance of each QD. Here a voltage V_{oDQD} (V_{oCS}) was applied to both the source and drain electrodes of the DQD (CS) and currents flowed through the DQD (I_{DQD}) and CS (I_{CS}) from their drain to source electrodes, as indicated by dashed arrows in the figure. There was no leakage current

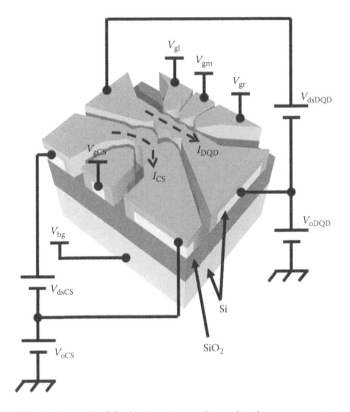

FIGURE 10.15 A schematic of device structure and associated measurement setup.

flow between the DQD and CS even when different offset voltages were applied, because these portions of the device were isolated by dry etching, thermal oxidation, and BOX. The source–drain voltages of the DQD (V_{dsDQD}) and CS (V_{dsCS}) were held constant during V_{oDQD} or V_{oCS} sweeping and a measurement temperature of 250 mK was employed.

Maintaining all the QDs on a single chip in a suitable state is difficult when these QDs are driven by a common gate, since the gate will have different capacitances with each QD, because the threshold voltages of the QDs are often different. This occurs because of variations in shape resulting from either the design of the QDs or fluctuations in the fabrication process. Applying different offset voltages to each QD can tune the energy band bending of the QDs and thus the number of electrons. For example, applying an offset voltage of V_{oCS} to the CS allows control over both the energy band bending and the number of electrons in the CS [50].

To confirm the effectiveness of this technique, modulations of the Coulomb oscillations induced by various offset voltages were measured, with the results presented in Figure 10.16a and b. Figure 10.16a shows the measured CS current as a function of the gate voltages ($V_{gl} = V_{gr}$) while varying V_{oCS} from 0 to −150 mV. Several small Coulomb peaks appeared when $V_{oCS} = 0$ mV because of the low tunneling rate through the CS. When a negative V_{oCS} was applied, the potentials of

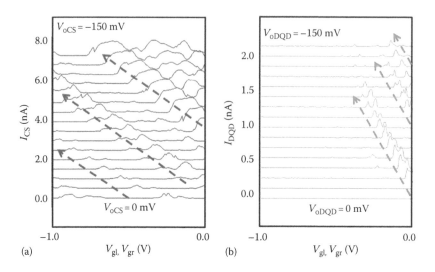

FIGURE 10.16 (a) $I_{CS} - V_{gr}$ (= V_{gl}) characteristics as V_{oCS} is varied from 0 mV (bottom trace) to –150 mV (top trace). The traces are offset for clarity. (b) $I_{DQD} - V_{gr}$ (= V_{gl}) characteristics as V_{oDQD} is varied from 0 mV (bottom trace) to –150 mV (top trace).

the CS and the barriers underwent a relative decrease, because such potentials are defined by the difference between the applied voltages of the BG (V_{bg}) and V_{oCS}, and so, many large Coulomb peaks were generated when V_{oCS} = –150 mV. Figure 10.16b plots the results of similar measurements in the case of the DQD with the same applied V_{bg} values as in Figure 10.16a. The Coulomb peaks of the DQD were evidently modulated by applying a negative V_{oDQD}. These results demonstrate the effectiveness of activating the CS and DQD independently via the application of different offset voltages.

Integrated gl, gm, gr and gCS may be fabricated in an SOI and used as gate electrodes for the DQD and CS. Electrons in these gates are induced by applying V_{bg} in the same manner as with the QDs and the number of electrons in the gates is modulated by changing the Fermi level of each gate. Thus, the QDs themselves also work as gate electrodes when an offset voltage is applied to the source and drain electrodes. Figure 10.17 summarizes the measured DQD current as a function of V_{gl} (= V_{gr}) while varying V_{oCS}. Here, moving from the bottom of the graph to top, V_{oCS} decreases from 260 to 70 mV in 8 mV steps, while V_{dsDQD} and V_{dsCS} are fixed at 5 and 10 mV. The Coulomb oscillation of the DQD is seen to shift depending on the value of V_{oCS}, demonstrating the successful operation of the gate function of the CS with regard to the DQD. The ratio of the capacitance of gl (= gr) C_{gDQD}, to that of the CS, C_{CSDQD}, associated with the DQD was calculated and a value for $C_{CSDQD}/C_{gDQD} = \Delta V_g/\Delta V_{oCS}$ of approximately 0.97 was obtained, where ΔV_g is the shift of the Coulomb peaks when V_{oCS} is changed by ΔV_{oCS}. This gating effect of the CS is sufficiently large to be useful. The dual-functional QDs serving as CSs and gates described earlier should reduce the required physical device area and enable the large-scale integration necessary for quantum information processing.

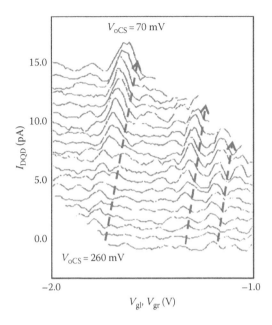

FIGURE 10.17 I_{DQD} as a function of V_{gl} ($= V_{gr}$) while varying V_{oCS}. The traces are offset for clarity. Here, from bottom to top, V_{oCS} decreases from 260 to 70 mV in 8 mV steps. V_{dsDQD} and V_{dsCS} were constant at 5 and 10 mV throughout.

10.4 SUMMARY

This chapter summarized studies our group has performed with regard to lithographically defined coupled Si QDs in MOS structures using nondoped SOI substrates. In this chapter, spin-related tunneling phenomenon were observed in Si QDs coupled in series and CSs were developed as a means of determining the absolute numbers of electrons in the QDs as well as observing the few-electron regime. Charge sensitivity was discussed herein, as was the use of CSs to perform the additional function of gating, which will be useful in terms of saving device space in order to realize the large-scale integration necessary for quantum information processing.

ACKNOWLEDGMENTS

This chapter reviews three papers published by our group: *Phys. Rev. B* **86**, 115322 (2012), *J. Appl. Phys.* **111**, 093715 (2012), and *Jpn. J. Appl. Phys.* **52**, 04CJ01 (2013). This work was financially supported by Grants-in-Aid (Kakenhi Nos. 24102703, 26709023 and 26630151) from the PRESTO program of the Japan Science and Technology Agency (JST), the Yazaki Memorial Foundation for Science and Technology, and the Project for Developing Innovation Systems of the Ministry of Education, Culture, Sports, Science and Technology (MEXT) of Japan.

REFERENCES

1. D. Loss and D. P. DiVincenzo, *Phys. Rev. A*, **57**, 120 (1998).
2. J. R. Petta, A. C. Johnson, J. M. Taylor, E. A. Laird, A. Yacoby, M. D. Lukin, C. M. Marcus, M. P. Hanson, and A. C. Gossard, *Science*, **309**, 2180 (2005).
3. F. H. L. Koppens, C. Buizert, K. J. Tielrooij, I. T. Vink, K. C. Nowack, T. Meunier, L. P. Kouwenhoven, and L. M. K. Vandersypen, *Nature*, **442**, 766 (2006).
4. M. Pioro-Ladrière, T. Obata, Y. Tokura, Y.-S. Shin, T. Kubo, K. Yoshida, T. Taniyama, and S. Tarucha, *Nat. Phys.*, **4**, 776–779 (2008).
5. T. Kodera, K. Ono, Y. Kitamura, Y. Tokura, Y. Arakawa, S. Tarucha, *Phys. Rev. Lett.*, **102**, 146802 (2009).
6. A. Khaetskii, D. Loss, and L. Glazman, *Phys. Rev. B*, **67**, 195329 (2003).
7. H. O. H. Churchill, A. J. Bestwick, J. W. Harlow, F. Kuemmeth, D. Marcos, C. H. Stwertka, S. K. Watson, and C. M. Marcus, *Nat. Phys.*, **5**, 321 (2009).
8. M. Thalakulam, C. B. Simmons, B. M. Rosemeyer, D. E. Savage, M. G. Lagally, M. Friesen, S. N. Coppersmith, and M. A. Eriksson, *Appl. Phys. Lett.*, **96**, 183104 (2010).
9. Y. Hu, H. O. H. Churchill, D. J. Reilly, J. Xiang, C. M. Lieber, and C. M. Marcus, *Nat. Nanotechnol.*, **2**, 622 (2007).
10. B. M. Maune, M. G. Borselli, B. Huang, T. D. Ladd, P. W. Deelman, K. S. Holabird, A. A. Kiselev et al., *Nature*, **481**(7381), 344 (2012).
11. W. H. Lim, F. A. Zwanenburg, H. Huebl, M. Möttönen, K. W. Chan, A. Morello, and A. S. Dzurak, *Appl. Phys. Lett.*, **95**, 242102 (2009).
12. A. M. Tyryshkin, S. Tojo, J. J. L. Morton, H. Riemann, N. V. Abrosimov, P. Becker, H.-J. Pohl, T. Schenkel, M. L. W. Thewalt, K. M. Itoh, and S. A. Lyon, *Nat. Mater.*, **11**, 143 (2012).
13. C. B. Simmons, M. Thalakulam, N. Shaji, L. J. Klein, H. Qin, R. H. Blick, D. E. Savage, M. G. Lagally, S. N. Coppersmith, and M. A. Eriksson, *Appl. Phys. Lett.*, **91**, 213103 (2007).
14. M. G. Borselli, R. S. Ross, A. A. Kiselev, E. T. Croke, K. S. Holabird, P. W. Deelman, L. D. Warren et al., *Appl. Phys. Lett.*, **98**, 123118 (2011).
15. N. Shaji, C. B. Simmons, M. Thalakulam, L. J. Klein, H. Qin, H. Luo, D. E. Savage et al., *Nat. Phys.*, **4**, 540 (2008).
16. W. H. Lim, C. H. Yang, F. A. Zwanenburg, and A. S. Dzurak, *Nanotechnology*, **22**, 335704 (2011).
17. M. Xiao, M. G. House, and H. W. Jiang, *Phys. Rev. Lett.*, **104**, 096801 (2011).
18. F. A. Zwanenburg, C. E. W. M. van Rijmenam, Y. Fang, C. M. Lieber, and L. P. Kouwenhoven, *Nano Lett.*, **9**, 1071 (2009).
19. J. Gorman, D. G. Hasko, and D. A. Williams, *Phys. Rev. Lett.*, **95**, 9, 090502 (2005).
20. T. Kodera, T. Ferrus, T. Nakaoka, G. Podd, M. Tanner, D. Williams and Y. Arakawa, *Jpn. J. Appl. Phys.*, **48**(6), 06FF15 (2009).
21. G. Yamahata, T. Kodera, H. Mizuta, K. Uchida, and S. Oda, *Appl. Phys. Express*, **2**, 095002 (2009).
22. T. Kodera, G. Yamahata, T. Kambara, K. Horibe, T. Ferrus, D. Williams, Y. Arakawa, and S. Oda, *AIP Conf. Proc.*, **1399**, 331 (2011).
23. G. Yamahata, T. Kodera, H. O. H. Churchill, K. Uchida, C. M. Marcus, and S. Oda, *Phys. Rev. B*, **86**, 115322 (2012).
24. K. Horibe, T. Kodera, T. Kambara, K. Uchida, and S. Oda, *J. Appl. Phys.*, **111**, 093715 (2012).
25. W. G. van der Wiel, S. De Franceschi, J. M. Elzerman, T. Fujisawa, S. Tarucha, and L. P. Kouwenhoven, *Rev. Mod. Phys.*, **75**, 1 (2002).
26. K. Ono, D. G. Austing, Y. Tokura, and S. Tarucha, *Science*, **297**, 1313 (2002).
27. D. Culcer, Ł. Cywiński, Q. Li, X. Hu, and S. Das Sarma, *Phys. Rev. B*, **82**, 155312 (2010).

28. A. C. Johnson, J. R. Petta, C. M. Marcus, M. P. Hanson, and A. C. Gossard, *Phys. Rev. B*, **72**, 165308 (2005).
29. A. Pfund, I. Shorubalko, K. Ensslin, and R. Leturcq, *Phys. Rev. Lett.*, **99**, 036801 (2007).
30. J. Danon and Y. V. Nazarov, *Phys. Rev. B*, **80**, 041301(R) (2009).
31. A. V. Khaetskii and Y. V. Nazarov, *Phys. Rev. B*, **64**, 125316 (2001).
32. A. Palyi and G. Burkard, *Phys. Rev. B*, **82**, 155424 (2010).
33. F. H. L. Koppens, J. A. Folk, J. M. Elzerman, R. Hanson, L. H. Willems van Beveren, I. T. Vink, H. P. Tranitz, W. Wegscheider, L. P. Kouwenhoven, and L. M. K. Vandersypen, *Science*, **309**, 1346 (2005).
34. J. Schliemann, A. Khaetskii, and D. Loss, *J. Phys.: Condens. Matter*, **15**, R1809 (2003).
35. N. S. Lai, W. H. Lim, C. H. Yang, F. A. Zwanenburg, W. A. Coish, F. Qassemi, A. Morello, and A. S. Dzurak, *Sci. Rep.*, **1**, 110 (2011).
36. W. A. Coish and F. Qassemi, *Phys. Rev. B*, **84**, 245407 (2011).
37. F. Qassemi, W. A. Coish, and F. K. Wilhelm, *Phys. Rev. Lett.*, **102**, 176806 (2009).
38. A. Shnirman and G. Schön, *Phys. Rev. B*, **57**, 15400 (1998).
39. K. W. Lehnert, K. Bladh, L. F. Spietz, D. Gunnarsson, D. I. Schuster, P. Delsing, and R. J. Schoelkopf, *Phys. Rev. Lett.*, **90**, 027002 (2003).
40. W. Lu, Z. Q. Ji, L. Pfeiffer, K. W. West, and A. J. Rimbeg, *Nature (London)*, **423**, 422 (2003).
41. C. Barthel, D. J. Peilly, C. M. Marcus, M. P. Hanson, and A. C. Gossard, *Phys. Rev. Lett.*, **103**, 160503 (2009).
42. C. Barthel, J. Medford, C. M. Marcus, M. P. Hanson, and A. C. Gossard, *Phys. Rev. Lett.*, **105**, 266808 (2010).
43. A. N. Korotkov, *Phys. Rev. B*, **49**, 10381 (1994).
44. T. Fujisawa and Y. Hirayama, *Appl. Phys. Lett.*, **77**, 543 (2000).
45. W. Lu, A. J. Rimberg, K. D. Maranowski, and A. C. Gossard, *Appl. Phys. Lett.*, **77**, 2746 (2000).
46. S. Hershfield, J. H. Davies, P. Hyldgaaed, C. J. Stanton, and J. W. Wilkins, *Phys. Rev. B*, **47**, 1967 (1993).
47. U. Hanke, Y. M. Galperin, K. A. Chao, and N. Zou, *Phys. Rev. B*, **48**, 17209 (1993).
48. A. N. Korotkov and M. A. Paalanen, *Appl. Phys. Lett.*, **74**, 4052 (1999).
49. G. Zimmerli, T. M. Eiles, R. L. Kautz, and J. M. Martinins, *Appl. Phys. Lett.*, **61**, 237 (1992).
50. T. Kambara, T. Kodera, Y. Arakawa, and S. Oda, *Jpn. J. Appl. Phys.*, **52**, 04CJ01 (2013).

11 Potential of Nonvolatile Magnetoelectric Devices for Spintronic Applications

Peter A. Dowben, Christian Binek, and Dmitri E. Nikonov

CONTENTS

11.1 Introduction 255
11.2 Background in Magnetoelectrics..257
11.3 Magnetoelectric Magnetic Tunnel Junction (ME-MTJ)..............................260
11.4 Magnetoelectric Spin Transistor..261
     11.4.1 Large Field, Very Short Channel Regime 264
     11.4.2 Low-Field Regime 264
11.5 Device Implementation..266
11.6 Magnetoelectric Memory 267
11.7 Magnetoelectric Logic..270
11.8 Conclusions...274
Acknowledgments..274
References..274

11.1 INTRODUCTION

Manipulation of magnetically ordered states by electrical means is among the most promising approaches toward developing novel spintronic devices. Incorporating a voltage-controlled nonvolatile magnetic state variable into a scalable memory device with potentially additional logical function is the Holy Grail of spintronics. The reasons are simple enough: voltage control of a nonvolatile magnetic state virtually eliminates the need for large current densities, the accompanying power consumption, and detrimental Joule heating on writing and potentially also on reading [Binek 05, Dery 07, Ney 03, Nikonov 13, Zhirnov 05]. Current densities in excess of 1 MA/m^2 are required for writing of the magnetic state in spin-transfer-torque (STT) memory elements currently under development [Jonietz 10].

Yet the inherent nonvolatility of long-range ordered magnetism adds functionality over today's random access memory (RAM) elements. The latter require refresh power during operation and the information is lost when power is lost, which is often not desirable.

Most efforts at providing magnetic memory element solid-state devices have used magnetic spin valves or magnetic tunnel junction (MTJ) type structures. A novel approach to voltage-controlled magnetic devices envisions the use of magnetoelectrics [Binek 05, Binek 13, Kleemann 13]—materials in which an applied voltage induces a magnetic moment in linear response and, above all, where sizable interface spin polarization can be electrically switched [Andreev 96, Belashchenko 10, Cao 14, He 10, Street 14, Wu 11]. While there is no implementation of a magnetoelectric (ME) device as yet, here we speculate on the possibilities and the potential of such devices. It appears reasonable to assume that, some day, the hurdles to such devices, which are largely challenges in materials science, will be overcome and such devices not only realized but also implemented.

A competitive ME device needs to fulfill a minimal set of necessary prerequisites. In the presence of an easy-to-implement stationary magnetic field, the ME coercive voltage must be small, but the internal electric field is large enough such that the product, EH, of electric field, E, and magnetic field, H, can switch between the antiferromagnetic single domain states. This switching must be achieved with sufficient speed, at least in the range of several GHz. The switching energy must be competitive with state-of the-art alternative mechanisms in multiferroics: for example, reported switching energy density of 480 $\mu J/cm^2$ [Heron 14], with minimal dissipative heating.

The electric field scales inversely proportional with the film thickness such that virtually any electric field strength below dielectric breakdown can be achieved through scaling of high-quality films with minimal leakage. In addition, the energy barriers between the two degenerate antiferromagnetic domain states are thought to scale roughly with volume, so thinner films might facilitate high switching speed in the case with decreasing activation volume. Of course, this also requires that the ME properties are retained in the thin film limit at elevated temperatures, but if this is indeed the case, the material may require a smaller critical voltage, and possibly exhibit higher switching speeds as well. There is experimental evidence that in fact, this is correct. Unfortunately, there is a limit to preserving useful ME properties, as in the thin film limit, the boundary magnetization dominates [Fallarino 14]. We have shown that for ultrathin chromia films close to the Néel temperature, a magnetic field alone can switch the entire antiferromagnetic spin structure through Zeeman coupling with the boundary magnetization [Fallarino 14]. Similarly, every exchange bias bilayer has a critical thickness for the pinning layer [Binek 03]. Below this critical thickness, pinning and thus exchange bias disappears and a potential voltage-controlled device would lose function. The thickness limit of reliable isothermal room-temperature voltage-controlled switching of an exchange coupled ferromagnetic (FM) film in a thin film geometry is of critical importance for scaling considerations. It has to be pushed to ever thinner ME film thickness below today's record achievements of about 100 nm for pure chromia pinning layers [Ashida 14, Toyoki 15].

11.2 BACKGROUND IN MAGNETOELECTRICS

ME materials, where a net magnetic moment is induced by an electric field, may be key to this approach of voltage-controlled magnetic devices. An applied electric field E results in the induction of net magnetization M [Fiebig 05]. The ME free energy can be formulated in terms of the electric polarization P and the magnetization M. The energy density should include the term of interaction with the external field E. The ME effect, and notably the linear ME effect, is derived systematically from a series expansion of the density of the Gibbs free energy,* G [Binek 13, Landau 84, Schmid 94]. From the free energy, one obtains the relationships for polarization and magnetization [Binek 13, Birol 12, Fiebig 05, Smolenskii 82, Vaz 12]. The electric polarization (here the Einstein's convention of summation over repeated indices is used) is

$$P_i(E,H) = \frac{-\partial G}{\partial E_i} = P_i^S + \varepsilon_o \varepsilon_{ij} E_j + \alpha_{ij} H_j + \frac{1}{2}\beta_{ijk} H_j H_k + \gamma_{ijk} H_j E_k + \cdots \quad (11.1)$$

and the magnetization relationship is

$$\mu_0 M_i(E,H) = \frac{-\partial G}{\partial H_i} = \mu_0 M_i^S + \mu_o \mu_{ij} H_j + \alpha_{ij} E_j + \beta_{ijk} E_j H_k + \frac{1}{2}\gamma_{ijk} E_j E_k + \cdots \quad (11.2)$$

which would imply that a very large ME constant (α_{ij}) is desirable. Its value is on the scale of 4.13 ps/m in the case of the inorganic ME Cr_2O_3 [Astrov 61]. A large ME susceptibility (α_{ij}) is desirable, in this context, regardless of whether the electric field is used to induce or switch magnetization or magnetic fields are used to induce or switch electric polarization, because (in the linear response)

$$\alpha_{ij} = \frac{\partial P_i}{\partial H_j} = \frac{\partial \mu_o M_i}{\partial E_j}. \quad (11.3)$$

Thus, it is convenient to visualize the ME coupling in terms of a coefficient α_{ij} even in the nonlinear response range.

In the device schemes and device implementations we discuss here, the voltage control of the magnetic state variable is, in fact, a nonlinear switching effect of the bulk antiferromagnetic order parameter and the rigorously coupled surface/boundary magnetization accompanying the respective antiferromagnetic single domain state [Andreev 96, Belashchenko 10, Cao 14, Dowben 11, He 10, Street 14, Wu 11]. Modeling of switching phenomena cannot be achieved by a free energy series expression in powers of the applied field.

The nonlinear switching effect between antiferromagnetic single domain states in bulk ME antiferromagnets has been experimentally investigated by Martin and

---

* Frequently in the literature, the Gibbs free energy and the Landau free energy are not carefully distinguished. An expansion of the free energy in terms of intensive field variables E and H is strictly speaking the Gibbs free energy.

Anderson [Martin 66]. In exchange bias heterostructures, switching of the pinning layer can be transferred into switching of an exchange coupled adjacent FM layer mediated via the interface magnetization or boundary magnetization that has to follow the switching of the pinning layer. A particularly pronounced boundary magnetization in single domain ME antiferromagnets has been predicted independently by Andreev [Andreev 96] and Belashchenko [Belashchenko 10]. The combination of bulk ME switching, according to [Martin 66], with the phenomenon of surface spin polarization accompanying the single domain states allows voltage-controlled switching of the boundary magnetization. In 2010, isothermal switching of boundary magnetization was utilized, for the first time, in voltage-controlled perpendicular exchange bias heterostructures. Here the voltage-controlled boundary magnetization enables isothermal and bipolar switching of the exchange bias field in a heterostructure where a FM film is coupled via quantum mechanical exchange with the boundary magnetization of the ME antiferromagnet [He 10]. In cases when the loop shift of the ferromagnet is sufficiently strong to the left and the right of the magnetic field axis, the remanent magnetization can be utilized as voltage-controlled nonvolatile state variable with strong implications for voltage-controlled spintronics discussed in this chapter.

In fact, with the onset of ME reversal of the antiferromagnetic domain state, the change in boundary magnetization and the exchange bias are seen to be sharp, and quickly saturates in some sample geometries [Cao 14, He 10]. Moreover, the detrimental effect of exchange bias training is likewise voltage-controlled in the case of chromia as a ME pinning layer and can be completely eliminated [Echtenkamp 13]. In the current experimental context, one can easily imagine an input ME device element, with a low coercive voltage, coupled to an output element with a low magnetic coercivity (as discussed in the following).

The existence of net magnetization at the surface or boundary of ME antiferromagnets has been predicted using symmetry arguments [Andreev 96, Belshchenko 10]. We have demonstrated that the interface magnetization direction [Cao 14, He 10, Street 14, Wu 11] and control of the domain size can be achieved with voltage in Cr_2O_3 [Cao 14, He 10, Street 14] and Fe_2TeO_6 [Wang 14] (another ME). The fact that this interface polarization reverses in concert with the antiferromagnetic order parameter of the ME is key to the control of the spin state and switching of an exchange coupled magnetic layer in a ME controlled device. The naïve approach, where one uses the linear ME response for isothermal electric control of exchange bias, as suggested at the occasion of ICM-2003 [Hochstrat 04], is hampered by the small value of the linear ME effect and the presence of boundary magnetization, which in the linear regime is detrimental to voltage control, because a fully polarized boundary does not experience magnetization increase on increasing the applied electric field. Isothermal switching between the single domain states is achieved when an electric field, E, and a small symmetry breaking magnetic field, H, are simultaneously applied such that the magnitude of the product EH overcomes a critical threshold. Note that the symmetry breaking H-field can be scaled down to arbitrary small values when the applied E-field is scaled up accordingly. We have demonstrated that the Earth's magnetic field is sufficient to achieve voltage control of boundary magnetization in a chromia thin film.

Early attempts in electrically controlled exchange bias tried to exploit the linear ME susceptibility of the antiferromagnetic material Cr_2O_3 as an active exchange bias pinning system [Hochstrat 04]. Later, and still in the spirit of exploiting the linear ME response, ME annealing has been utilized to achieve thermally assisted bipolar switching of exchange bias in a chromia-based perpendicular exchange bias system [Borisov 05]. Finally, the need for nonlinear switching was acknowledged to achieve the technologically relevant isothermal switching [He 10].

In a ME material, an applied electric field induces a net magnetic moment at the interface, which can be used to electrically manipulate the magnetic states of an adjacent exchange coupled FM film. The small value of the ME susceptibility of Cr_2O_3 led many researchers to the conclusion that multiferroic materials are better suited for this purpose. Such multiferroic materials have two or more ferroic order parameters, such as ferroelectric polarization and (anti)ferromagnetic order. Coupling between these order parameters has been demonstrated [Chu 08, Vaz 12, Zhao 06]; however, it is also still typically weak, and the theoretical upper limit of the ME susceptibility is rarely reached. Perhaps an exception is bismuth ferrite ($BiFeO_3$), which has been recently engineered into thin films, which allow dynamic switching of the ferroelectric order parameter accompanied by switching of a FM moment that arises from canting of the antiferromagnetic sublattices. It is important to stress that 180° switching of ferromagnetism in the absence of a magnetic field is symmetry-forbidden in equilibrium. However, dynamic switching schemes can overcome this limitation and together with the engineering of the multidomain state of $BiFeO_3$, 180° switching has been shown at room temperature [Heron 14]. Whether the weak FM moment in $BiFeO_3$ films is strong enough for applications, the domains are scalable, and the energy dissipation is sufficiently small for applications that are remain to be seen.

In our schemes, because of the nonlinear switching of the surface and interface magnetization in the single phase Cr_2O_3 material, or other ME ($LuFeO_3$ for example), we can overcome the limitation of the small linear ME susceptibility that plagues the development of voltage-controlled magnetic devices using other materials, including multiferroic materials. We achieve a nonvolatile electric control of unidirectional magnetic anisotropy in a FM layer via exchange coupling with a single-phase ME that is not a multiferroic [He 10, Wu 11]. This effect reflects the switching of the bulk antiferromagnetic domain state and the interface magnetization coupled to it. The switchable exchange bias sets in exactly at the bulk Néel temperature [He 10, Wu 11]. In fact, using a FM Pd/Co multilayer deposited on the (0001) surface of a Cr_2O_3 single crystal, we achieved reversible, room temperature isothermal switching of the exchange bias field between positive and negative values by reversing the electric field while maintaining a permanent magnetic field [He 10, Echtenkamp 13].

Not all the magneto electrics are useful for the simple devices we describe here. For example, magnetoelectrics like $GaFeO_3$ with high critical temperatures (well above room temperature) are neither particularly valuable replacements to chromia nor $LuFeO_3$ in the devices, in spite of the high critical temperatures. In the example of $GaFeO_3$, this is pyroelectric (thus not switchable) and ferrimagnetic. The spontaneous polarization is along the b axis, and magnetization is along c (the structure is $Pc2_1n$). Thus, this is a ME with two transverse components of the ME tensor allowed by symmetry: coupling H_b to E_c and only the ME coefficient coupling H_c to E_b appears to be large [Arima 04]. This coefficient corresponds to the longitudinal change of the ferrimagnetic magnetization as the electric field is applied along the pyroelectric axis. But because of site disorder [Trassin 09], ME switching is not possible, because the induced ME moment would need to exceed the saturation magnetization.

11.3 MAGNETOELECTRIC MAGNETIC TUNNEL JUNCTION (ME-MTJ)

Magnetoelectrics are a possible route for voltage control of magnetization in an MTJ structure [Bibes 08, Binek 05, Chen 06, Chen 10, Hochstrat 06, Kleeman 13], based on the familiar MTJ memory element, which consists of two FM layers separated by a nonmagnetic insulator where the device resistance is determined by the relative orientation of the magnetization of the two FM layers. Typically, the magnetization of one FM layer, in a MTJ memory element, is pinned, while the second is free to switch between two states. In conventional MTJs, the switching is typically accomplished by fringe fields, or by spin-transfer torque that require large current densities. This is a technological impediment to implementation of spin-transfer torque, with the necessary accompanying dissipation of currents and Joule heating. ME antiferromagnets, with their voltage-controllable boundary magnetization [Andreev 96, Belshchenko 10, He 10], provide much required alternative material that could lead to the development of electrically writable nonvolatile MTJ memory with high speed and high density. In such a device, a simple bias voltage rather than a large current pulse is used to write information to the element [Binek 05, Dery 07, Kleemann 13]. The magnetization of the free layer is exchange coupled to the Cr_2O_3 (or other ME) interface magnetization (Figure 11.1). A bias voltage applied across the Cr_2O_3 layer reverses the interface magnetization, which

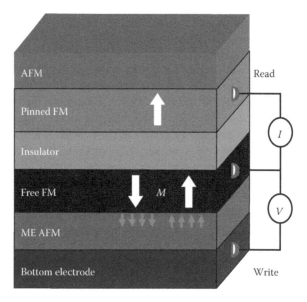

FIGURE 11.1 **(See color insert.)** Voltage-controlled MTJ where free magnetic layer magnetization is controlled by an ME interface, separating the read and write aspects of the device.

in turn switches the magnetization of the free layer [He 10, Wu 11]. The state of the element is read by detecting the resistance of the FM/insulator/FM stack and the read and write circuits are decoupled. Obviously, this is a three-terminal device, and thus, a little more complex than the standard MTJ. Our goal remains more or less the same nonetheless: magnetic memory elements controlled entirely by an applied voltage. The stack can be placed together with an MOS transistor to create an MRAM cell, but the circuit implications will be much different than the normal MRAM, which requires currents for switching the magnetic tunneling stack.

For the standard, magnetic field switched MRAM cell, where the magnetic field is generated by the write lines 1, 2 [Durlam 03], currents through two different write lines are required to be coherent in order to produce the net magnetic field needed to switch the free layer of the MTJ stack. A simpler stack is required to use current spin-torque switching of magnetization [Brosse 04]. This circuit is essentially no more complicated than a standard DRAM cell. Yet, the magnetoelectric MTJ requires, in principle, no high currents and no external fields. As outlined earlier, voltage-controlled switching of boundary magnetism in ME oxides [He 10] has the potential to enable new classes of nonvolatile/reprogrammable magnetoelectronic devices, because now high currents are not required nor are write lines to create a local magnetic field external to the MTJ. The scheme that uses write lines to generate an external magnetic field to write the free magnetic layer is a complication at several levels, including a limit to the density of devices, because of the stray magnetic fields.

We emphasize that there is at least the potential of much additional functionality, which utilizes on-demand reprogrammability. By fabricating a nanoscale magnetic gate on top of a submicron semiconductor Hall junction, researchers at NRL demonstrated how fringing magnetic fields emanating into the junction could provide the basis for a bistable [Johnson 00] nonvolatile output with the capacity to be reprogrammed within a single clock cycle. These devices have never been broadly adopted due to the absence of efficient, low-energy, schemes for switching magnetization, but may be applicable to voltage-controlled magnetoelectrics [He 10], therefore with the potential to provide a whole new class of low-power magnetoelectronics. If such devices are implemented, the magnetoresistance (MR) ratios should be at least consistent with current MTJ devices used in memory applications and, certainly not worse, as the basic MTJ remains the same.

11.4   MAGNETOELECTRIC SPIN TRANSISTOR

We start here from the basic concept of a spin modulator [Datta 90] or a spinFET [Schmidt 00, van Dijken 02], see Figure 11.2. A spinFET operation is based on FM source and drain electrodes. Electrons injected to the channel are spin polarized by one of the electrodes. The magnitude of current is modified by the orientation of magnetization (parallel or anti-parallel) of the other electrode. The magnetization of the source or the drain can be switched by an external magnetic field or by spin-transfer torque. The direction of magnetization set the nonvolatile state of a spinFET.

The spin polarization of electrons in the channel does not have to remain constant. It can be made to precess by the external magnetic field (Figure 11.3) or by the action of the gate (Figure 11.2). In the latter case, the voltage applied to the

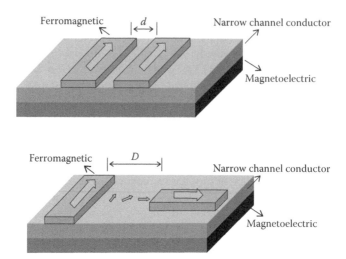

FIGURE 11.2 Slightly different conceptual schemes of a spinFET with an FM source and drain. The source and drain are ferromagnets. The gate voltage induces precession of spins of electrons conducted in the channel as a result of the interface magnetization from the manipulation of the ME gate. With a short source to drain distance (d), the electron spin would not have an opportunity to undergo precession (top) that could occur with a much greater source to drain distance (D) as seen in the bottom panel.

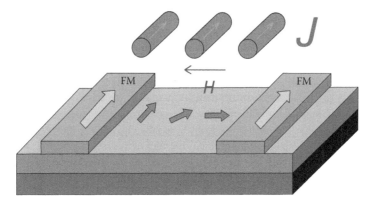

FIGURE 11.3 A different schematic of a spinFET where the spin precession is produced by the magnetic field of a current in a wire or coil around the spinFET device.

gate can vary the strength of spin–orbit coupling in the channel (due to the Rashba effect or the Dresselhaus effect). Thus, the motion of electrons in the channel is coupled to spin precession. Electrical spin injection and controlled spin precessional motion in silicon lateral transport devices using *externally* applied magnetic fields at room temperature and above has been demonstrated [Jonker 07, van't Erve 07]. Spin field effect transistors [Dery 07, Semenov 07, Semenov 10, Zaliznyak 13] have been described using graphene and other 2D channel conductors.

We can modify the spinFET idea into a magnetoelectric spinFET (MEspinFET). Its functionality is different from the original spinFET in two aspects. Firstly, the magnetization of the source or the drain is switched by applying voltage to the ME material stacks. Secondly, the ME material adjacent to the channel has surface magnetization and can cause precession of electron spins regardless of the spin–orbit coupling in the channel material (bottom panel to Figure 11.2). The MEspinFET concept [Binek 14] outlined here can use Cr_2O_3 or any other suitable ME material (e.g., $LuFeO_3$, etc.) as a gate dielectric for an MEspinFET and exploit the voltage-controlled interface magnetization to polarize the channel conductor. The device concept benefits from the key fact that ME antiferromagnets are in fact great dielectrics and create low-leakage dielectric gate barriers. The success of the whole idea behind the MEspinFET depends on whether one can polarize the channel of an FET by the gate dielectric. Can this interface polarization be voltage controlled? If yes, this concept provides a nonvolatile transistor that combines both memory and logic at low power.

Quantum mechanical exchange coupling between the boundary magnetization and the spins of the carriers in the channel of the spinFET gives rise to damped precession of the spins injected from the source into the channel. When utilizing channel materials with virtually zero spin–orbit coupling, such as graphene or Si, the effective exchange field of the voltage-controlled boundary magnetization is the sole source for spin precession. If the channel conductor is sufficiently thin, the transport channel will be spin polarized by the proximity effect, providing coherent spin transport or modulated spin precession. For a given remanent boundary magnetization, the exchange field will determine the spin state of the carriers at the FM drain in concert with the length of the channel.

The device of Figure 11.4 (and below) works because of the huge interface spin polarization of dielectric chromia or other ME (as the gate dielectric). This boundary

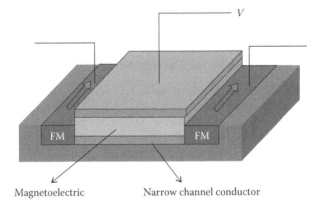

Magnetoelectric   Narrow channel conductor

FIGURE 11.4   The basic top gated MEspinFET with an FM source and drain. The thin channel conductor/semiconductor (blue) could be anything really, suitable to the task (graphene, InP, GaSb, PbS, MoS_2, WS_2, $MoSe_2$, WSe_2, etc.). If the narrow channel conductor is high Z and accompanied by a loss of inversion symmetry spin orbit coupling would be "turned on" adding functionality. Possible materials with high spin orbit coupling are InP, GaAs, PbS, WS_2, $HfSe_2$, WSe_2, etc. A narrow channel conductor such as graphene or silicon would still be spin polarized by proximity to the high interface polarization of the ME. Here the green is the dielectric chromia or other ME, with a gate electrode on top. (From Binek, Ch. et al., Magneto-electric voltage controlled spin transistors, U.S. Patent application serial no. 61766025, filed February 18, 2013, efilingAck18224783, non-provisional filing number 14182521, February 18, 2014. With permission.)

polarization is voltage controlled and described earlier [Belashchenko 10, Cao 14, He 10, Street 14, Wang 14]. The thin (2D) channel conductors (graphene, InP, GaSb, PbS, MoS_2, WS_2, $MoSe_2$, WSe_2, etc.) of the FET schemes of Figures 11.2 and 11.4 with a ME gate dielectric are polarized by this very high interface polarization, a proximity effect. The proximity effect was originally described as a mean field effect in magnetism, not directly attributed to any specific material and is embodied by the Landau–Ginzburg equation, for which solutions are known for some boundary conditions [Dowben 91, Miller 93]. In general, the conduction channel should be no thicker than 1–2 nm, thinner being better. This is why the simplest implementations to visualize involved monolayers of graphene, MoS_2, WS_2, WSe_2, In_4Se_3, etc. In all these cases, the conduction channel is 0.3–0.5 nm thickness and may be as much as 1.2 nm with one of the 2D materials placed in proximity to the ME gate as a bilayer. There are obviously ways to increase the width of the conduction channel, as in the case of the correct doping of GaAs. With the correct doping, the channel conductor could be as thick as 100 nm with GaAs, but this is a special case with doping close to the nonmetal to metal transition. In general, it is difficult to implement very thick semiconductor conduction channel, because the proximity polarization from the ME (say chromia, Cr_2O_3) interface into the semiconductor falls off with exponential decay [Dowben 91, Miller 93]. The penetration depends on the paramagnetic correlation length, and this is typically 0.3–2 nm. The paramagnetic correction length is a property of the semiconductor conduction channel, and this will be dependent on materials. Of course there is a design consideration if you place both top and bottom gate to the channel conductor to the spinFET, so the conduction channel has a top and bottom interface with a ME gate, you roughly double the allowed width of the conduction channel because of increases in the induced magnetization and/or because of increased electrostatic control of spin orbit coupling in the conduction channel. Yet we need to divide the concept into the following two types of devices.

11.4.1 LARGE FIELD, VERY SHORT CHANNEL REGIME

Conduction bands are strongly exchange-split. Spin precession length is microscopic, and spins completely decohere over small lengths, or do not precess (Figure 11.2 at top). Here we have a magnetized semiconductor sandwiched between leads in the lateral geometry. In this regime, it is sufficient to have one magnetic lead to measure magnetoresistance while the source contact can be a nonmagnetic metal. We estimate an MR ratio larger than 200% at room temperature is possible with little effort, and this can improve with better source and drain spin polarizations. In this regime, the magnetization of the electrode and of the ME's boundary magnetization should not be orthogonal.

We have recently demonstrated that we can grow such structures. The switching of only the source or drain electrode magnetization may be necessary and might be achieved at the setup of the transistor, after fabrication or later by spin-torque transfer or other voltage-controlled switching, as in a voltage-controlled MTJ or spin valve.

11.4.2 LOW-FIELD REGIME

The Datta–Das transistor is a concept that has captured our imagination for decades [Datta 90]. Spin precession is envisioned as the spins travel from the FM source to a FM drain contact, which serves as a spin analyzer (Figure 11.2 at bottom).

The applied gate voltage controls the spin precession through the electric field that is in turn controlled through the spin-orbit-dependent Rashba effect and thus controls the source–drain current. A major obstacle for the realization of an effective room temperature Datta–Das spinFET is the weakness of the Rashba effect. One approach to overcome this obstacle is the MEspinFET. This fundamentally overcomes the problem of weak Rashba spin–orbit coupling by either fully replacing or significantly supporting the Rashba effect through a magnetic field derived from a voltage-controlled ME material. If the spin-splitting is tuned in such a way that the precession length is almost equal to the channel length, we can have magnetoresistance controlled by the Hanle effect.

Here we need two magnetic electrodes magnetized orthogonally to the boundary magnetization of the ME gate (Figure 11.2 at bottom). Here, the switching of only the source or drain electrode magnetization is not necessary and might be at the setup stage of the transistor, after fabrication, but before use.

The precession-based MEspinFET works better when the source and drain are neither parallel nor antiparallel (since that way you can "tap" into the precession, or spin rotation better), because then the spin must rotate. In order to realize voltage-controlled resistance change, the angle between the electrodes should be appropriately selected (and it cannot be 180°). In this low-field scenario, it is important that spin coherence is maintained over the entire length of the channel. The MEspinFET fundamentally overcomes the problem of weak Rashba spin–orbit coupling by either fully replacing or exceeding the Rashba effect through an effective magnetic field derived from a voltage-controlled ME material. The electrically switchable and nonvolatile boundary magnetization of ME antiferromagnets, such as chromia, generates a voltage-controlled exchange field, which determines the spin precession of the carriers in the conducting channel. It is important to realize the local magnetization is present through a sort of proximity effect, and that the chromia or other ME is a dielectric, thus is the gate dielectric, in this scheme. The device scheme in its most simple form is shown in Figure 11.2, in the bottom panel, but more complex versions exist as shown in Figure 11.5.

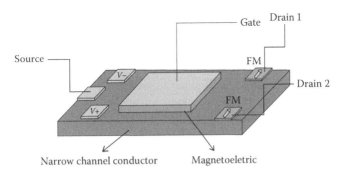

FIGURE 11.5   A multistate six-terminal MEspinFET. The chromia or other ME (green) is the gate dielectric and polarized the 2D channel conductor, ideally with spin orbit coupling "turned on" adding functionality, possibly exploiting such materials as InP, GaAs, PbS, WS_2, $HfSe_2$, WSe_2, etc. The narrow channel conductor would be polarized by proximity.

11.5 DEVICE IMPLEMENTATION

Graphene is a gapless semiconductor, so silicon or InP, GaSb, PbS, MoS_2, WS_2, $HfSe_2$, WSe_2, etc., may be more desirable. The choice of semiconductor for the channel depends on whether the MEspinFET is of type A (no or little spin–orbit coupling) or type B (where spin–orbit coupling might be desirable). Increasing the thickness of the conduction channel by increasing the proximity polarization penetration through doping of the semiconductor will not work in this scheme. While this does increase the paramagnetic correction length, quantum coherence is suppressed: only classical spin coherence lengths are increased. Other problems require more investigation. Is the coercive voltage small enough to be useful, is the switching speed high enough to be useful: all of this is unknown at the time of this writing. While decreasing the thickness of the ME gate and the transistor size would seem to be a solution, by decreasing the latent volume, ME stability at room temperature with the smaller spatial dimensions is unknown at the time of this writing.

Magnetoresistance ratios of 2–3 have been demonstrated at room temperature. Simulations of nominally half metallic materials predict that MR ratios of 20–30 are possible. Achieving the high magnetoresistance ratios requires highe-fidelity spin-polarized current injection and spin precession, as well as wave vector filtering mechanisms, but these are known effects for increasing MR ratios. This latter approach includes using magnetic tunnel barriers as the source and drain, but as discussed earlier, there are possible ME controlled MTJ schemes and combining this with the MEspinFET will lead to much increased MR ratios, although with the disadvantage of much reduced currents.

The earlier device concepts also have a major advantage in that memory and logic operations are combined: the device state is nonvolatile, but switchable. Multistate logic may be possible. Multilogic states could be envisioned through transistor architecture as in Figure 11.5. As noted earlier, the narrow channel conductor need not be graphene. If the narrow channel conductor is high Z and accompanied by a loss of inversion symmetry (in the presence of inversion symmetry, spin–orbit splitting is symmetry-forbidden), spin–orbit coupling would be "turned on" adding functionality, possibly materials such as InP, GaAs, PbS, WS_2, $MoSe_2$, WSe_2, etc. A narrow channel conductor such as graphene or silicon would still be spin polarized by proximity to the high interface polarization of the ME—acting also as the gate dielectric.

It should be understood that in all likelihood, the ME needs to be grown on top of the channel conductor with either the source or drain as a source of spin-polarized current (to either inject or sense), as shown in Figures 11.4 and 11.5. The external magnetic field of the earth, or the stray field from the FM electrodes, may be sufficient for the device operation; so, as in the aforementioned diagrams, no refresh current is needed (low power), and multilogic state memory or computing may be possible.

As stated at the outset, the implementation of any ME device could mean low operational power, but also low loss. If the current or voltage gains are too small, the logic fidelity will be suspected and the utility is diminished, so readout to any write element of a device has to overcome this intrinsic problem. The device of Figure 11.1 would result in little current and correspondingly little voltage. As noted earlier, achieving the high MR ratios in the MEspinFET will require higher-fidelity spin polarized current injection and spin precession, as well as wave vector filtering mechanisms, but hybrid devices combining possible ME controlled MTJs schemes with MEspinFET are possible. Still, another option is to combine a ME write element with an inverse ME read element [Nikonov 13]. Less limited by the no gain situation that accompanies a ME controlled MTJ (the device of Figure 11.1) is memory.

11.6   MAGNETOELECTRIC MEMORY

Most of the ME memory devices target the improvement of switching energy compared to spin-transfer torque memory (STTRAM). The scheme of magnetoelectric memory (MERAM) comprising switching of magnetization using the adjacent ME layer was envisioned in [Binek 05], see Figure 11.6. In this structure, the direction of magnetization is sensed by the MR effect, just like in STTRAM.

Voltage control of exchange bias has been experimentally demonstrated in a similar structure [Martin 07]. Implementation though might require a nonvolatile latch using the ME controlled MTJ. A nonvolatile latch based on MTJs has been described in the literature [Sakimura 08]; this makes use of current-driven devices that suffer from prohibitive power consumption, but could be modified to include

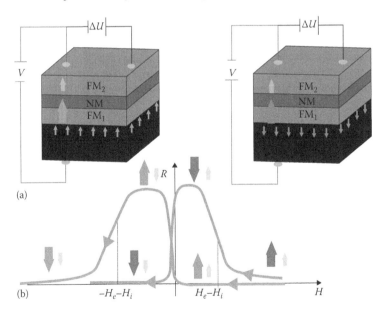

FIGURE 11.6 (a) Scheme of the two states of switched magnetization in MERAM and (b) expected resistance versus external magnetic field for ME MRAM, from [Binek 05]. (Reprinted with permission from Binek, Ch. and Doudin, B., Magnetoelectronics with magnetoelectrics, *J. Phys. Condens. Matter*, 17, L39–L44. Copyright 2005, Institute of Physics Publishing.)

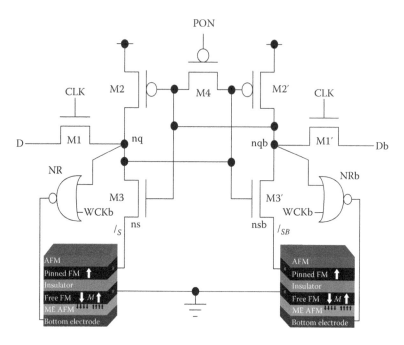

FIGURE 11.7 A nonvolatile latch using a magnetoelectric controlled MTJ [Marshall 14], such as in Figure 11.1, and based on MTJ memory latch described in the literature [Sakimura 08], where free FM represents the soft magnetic layer (middle red layer) of the MTJ controlled by voltage through the ME layer (ME AFM—in green). The top antiferromagnetic layer (brown) pins the next to top FM layer (red). The dielectric tunnel junction barrier layer is in yellow. (From Marshall, A. et al., Non-volatile latch using magneto-electric and ferro-electric tunnel junctions, U.S. Patent Pending: Application no.: 14/690,923, Filing Date: April 20, 2015. With permission.)

a ME device, requiring the conversion include a three-terminal device instead of a two-terminal MTJ. This might include a circuit as illustrated in Figure 11.7.

The SRAM cell programming state is achieved as in a conventional SRAM. Yet the voltage can be taken off the cell, either to save power, or simply to close down the chip while not in operation. Thus, the SRAM can then be turned off. When power is reapplied, the SRAM cell would be driven into the same state it was in before power-down: the lower resistance ME device will cause the source of the NMOS to hold lower, forcing turn-on of M3, nq to be held at the lower voltage, and flip SRAM cell into the prior state. Similar in spirit, but with a much easier implementation, with less complexity to the circuit, and with a much smaller spatial footprint would be a memory element like that of Figure 11.8 [Marshall 14]. We caution that careful adjustment of the input and output states, of the scheme outlined by Figure 11.8, might be necessary.

Thus far, we overviewed magnetic memories that have just two logic states.

A cell with four logic states has been experimentally demonstrated [Gajek 07]. This functionality is enabled by forming a MTJ with a FM layer, possibly $La_{0.65}Sr_{0.35}MnO_3$ (LSMO), a multiferroic or ferroelectric layer (LBMO), and a nonmagnetic layer (NM), here suggested as a good electrode material like gold (Au), as illustrated in Figure 11.9. Since the magnetization and the polarization of the multiferroic multilayer

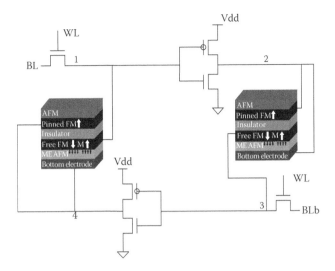

FIGURE 11.8 A nonvolatile SRAM based on the ME controlled MTJ (Figure 11.1), where the "free FM" represents the soft magnetic layer (middle red layer) of the MTJ controlled by voltage through the ME layer (ME AFM—in green). The top antiferromagnetic layer (brown) pins the next to top FM layer (red). The dielectric tunnel junction barrier layer is in yellow. (Marshall, A., Bird, J., Singisetti, U., and Nikonov, D., Non-Volatile latch using Magneto-Electric and Ferro-Electric Tunnel Junctions, U.S. Patent Pending: Application no.: 14/690,923, Filing Date: April 20, 2015. With permission.)

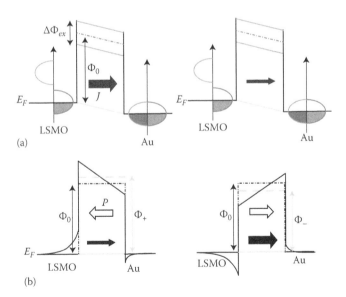

FIGURE 11.9 Energy band diagram of a multiferroic tunnel junction with four logic states. (Reproduced by permission from Macmillan Publishers Ltd. *Nat. Mater.*, Gajek, M., Bibes, M., Fusil, S., Bouzehouane, K., Fontcuberta, J., Barthelemy, A., and Fert, A., Tunnel junctions with multiferroic barriers, 6, 296–302, Copyright 2007.)

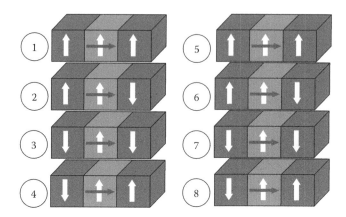

FIGURE 11.10 Scheme of a multiferroic tunnel junction with eight logic states. (Reproduced with permission from Yang, F., Tang, M.H., Ye, Z., Zhou, Y.C., Zheng, X.J., Tang, J.X., Zhang, J.J., and He, J., Eight logic states of tunneling magnetoelectroresistance in multiferroic tunnel junctions, *J. Appl. Phys.* 102, 044504, 2007. Copyright 2007, American Institute of Physics.)

(i.e., a stack containing both magnetic and ferroelectric layers in this case) can be switched independently between two states, the cell exhibits four values of resistance due to both tunneling magnetoresistance (TMR) and electroresistance (ER) effects.

This approach has been extended to eight logic states [Yang 07], as seen in Figure 11.10. The envisioned cell has an FM layer, a multiferroic layer, and another FM layer. The two magnetizations and the polarization of multiferroic can be switched independently.

An important attribute of ME switching by exchange bias is that it exerts an effective magnetic field with definite direction, which can be reversed by voltage. It is in contrast to, for example, magnetostrictive switching, which utilizes voltage-controlled anisotropy. The anisotropy is not a vector quantity, that is, it has a preferred axis, but not a direction that can be reversed. This difference makes switching magnetization by 180° with exchange bias straightforward, while switching magnetization by 180° through anisotropy switching difficult. Due to time inversion invariance of the electric field, switching is only allowed when utilizing dynamics, while it is rigorously symmetry-forbidden in equilibrium. Nevertheless, such pure electric reversal of magnetization has been experimentally demonstrated [Heron 11, Heron 14], as seen in Figure 11.11.

11.7 MAGNETOELECTRIC LOGIC

Logic devices represent more challenges. With a slight variation on the ME layer stack, one can achieve some of the functionality of logic gates. Spintronic logic devices providing complex functions are more difficult to design, and thus, fewer examples of them are found.

A majority of gates are made from a combination of memory ME device elements as in Figure 11.12, but this would provide only small output voltage steps. This is not entirely desirable, as while the overall ensemble might well be an excellent majority

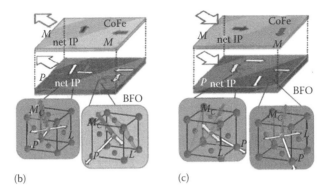

(b)   (c)

FIGURE 11.11   Scheme of switching of magnetization in CoFe by 180°, due to electric switching of polarization in $BiFeO_3$. (Reproduced with permission from Heron, J.T., Trassin, M., Ashraf, K., Gajek, M., He, Q., Yang, S.Y., Nikonov, D.E., Chu, Y.-H., Salahuddin, S., and Ramesh, R., Electric-field-induced magnetization reversal in a ferromagnet-multiferroic heterostructure, *Phys. Rev. Lett.*, 107, 217202, 2011. Copyright 2011 by the American Physical Society.)

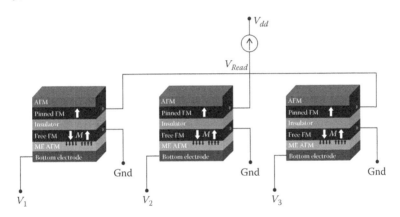

FIGURE 11.12   A simple majority gate design [Marshall 14b] where free FM represents common ground in circuit. Voltages V_{1-3} are applied to bottom electrode to switch the magnetoelectric layer (ME AFM—in green) and the output voltage V_{read} swings between Gnd and V_{dd}. The soft magnetic layer (middle red layer) of the magnetic tunnel is separated by a dielectric tunnel junction barrier layer (in yellow) from the top antiferromagnetic layer (brown)—FM layer (red) combination. (With permission of Marshall, A., Dowben, P.A., and Bird, J., Majority- and minority-gate logic schemes based on magneto-electric devices, U.S. Patent Pending; Majority-gate logic schemes based on magneto-electric devices, U.S. Patent Pending; Application no.: 62/069,138, Filing Date: October 27, 2014.)

gate, the "read" operation might be problematic while trying to cascade from one majority gate to the next. This type of logic scheme has some other significant disadvantages, as the delay time will be significantly increased by the use of three ME device elements, and the power will not be aided by the sum of three tunnel junctions—as those in an off state will still have loss.

It is better to make use of the fact that the onset of magnetic electric switching has a sharp onset voltage, and the ME is an excellent dielectric so that fringing fields can be summed to switch the boundary polarization in a single device as in Figure 11.13. This multilayer would be an inverted (with respect to Figures 11.1, 11.6 through 11.8, 11.12) ME controlled MTJ stack, but with three gates on ME layer. Here the ME switching requires biasing of two or more gates on a single multilayer stack. The integration of the majority-gate function into performance of a single tunnel junction leads to significant area and energy savings.

Spin waves can be generated and detected by a ME cell [Khitun 08]. An example of an ME cell in Figure 11.14 actually relies on another ME effect—combination of piezoelectric and magnetostrictive effects. However, the theoretical treatment can be easily extended to exchange bias interaction. Spin wave circuits provide rich functionality, such as a full adder [Khitun 11], with very few elements required, see Figure 11.15.

Magnetoelectric majority gate (MEMG) based on domain wall motion was proposed and benchmarked in [Nikonov 13], see Figure 11.16. Such a device will, however, be limited to the speed of the domain wall propagation.

ME interaction (including one involving multiferroics) was proposed to be used for clocking of nanomagnetic logic (NML) gates in [Niemier 11].

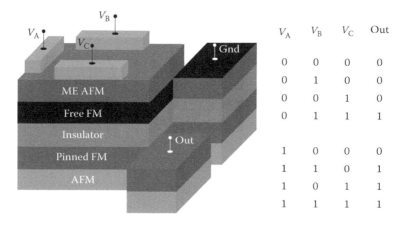

FIGURE 11.13   A magnetoelectric majority gate controlled MTJ stack [Marshall 14b], with three input voltage connections attached to the ME layer (green). A combination of two or more would be required to turn the majority gate on. Voltages V_{A-C} are applied to top to switch the magnetoelectric layer (ME AFM) and the output voltage V_{read} swings between Gnd and V_{dd}. The soft magnetic layer (darkest gray layer) of the magnetic tunnel is separated by a dielectric tunnel junction barrier layer (in lighter gray) from the bottom antiferromagnetic layer—FM layer combination. (With permission of Marshall, A., Dowben, P.A., and Bird, J., Majority- and minority-gate logic schemes based on magneto-electric devices, U.S. Patent Pending; Majority-gate logic schemes based on magneto-electric devices, U.S. Patent Pending; Application no.: 62/069,138, Filing Date: October 27, 2014.)

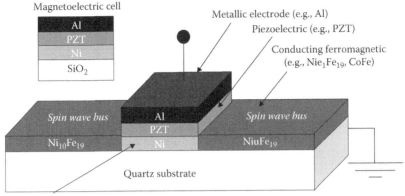

FIGURE 11.14 Scheme of an ME cell for generating and detecting spin waves. (Reproduced with permission from Khitun, A. and Wang, K.L., Non-volatile magnonic logic circuits engineering, *J. Appl. Phys.*, 110, 034306, 2011. Copyright 2011, American Institute of Physics.)

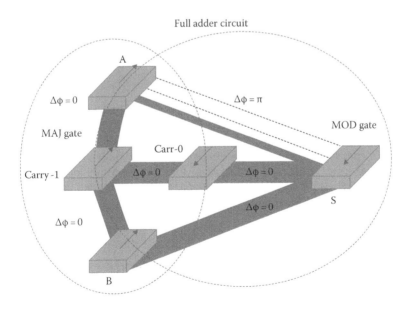

FIGURE 11.15 Scheme of a spin wave full adder based on ME cells. (Reproduced with permission from Khitun, A. and Wang, K.L., Non-volatile magnonic logic circuits engineering, *J. Appl. Phys.*, 110, 034306, 2011. Copyright 2011, American Institute of Physics.)

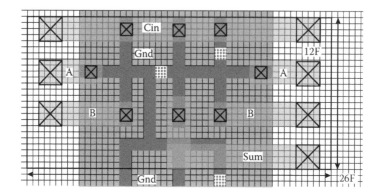

FIGURE 11.16 Scheme of a 1-bit full adder based on MEMG. (Reproduced with permission from Nikonov, D.E. and Young, I.A., Overview of beyond-CMOS devices and a uniform methodology for their benchmarking, *Proc. IEEE*, 101, 2515. © 2013 IEEE.)

11.8 CONCLUSIONS

As yet, all the aforementioned device concepts are just that: concepts. Whether or not any of them are realizable depends upon a number of advances, the outcome and potential of which are hard to gauge, but advances are being made. Higher critical temperature magnetoelectrics have been developed and continue to be developed. Better understanding of thin film growth of these magnetoelectrics is a subject of much present research, and preservation of the ME characteristics has been demonstrated in increasingly thinner films.

ACKNOWLEDGMENTS

This work was supported by the Semiconductor Research Corporation through the Center for NanoFerroic Devices (CNFD), an SRC-NRI Center under Task ID 2398.001, and by the NSF through Nebraska MRSEC DMR-1420645 and DMR 0747704. The authors are grateful to Evgeny Tsymbal, Ian Young, Sasikanth Manipatruni, Xia Hong, and Jeff Kelber for helpful discussions and Andrew Marshall, Kirill Belashchenko, and Jonathan Bird for their seminal contributions and Jon Bird and Shi Cao for help with the figures.

REFERENCES

[Andreev 96] A.F. Andreev, Macroscopic magnetic fields of antiferromagnets, *JETP Lett.* 63, 758 (1996).

[Arima 04] T. Arima, D. Higashiyama, Y. Kaneko, J.P. He, T. Goto, S. Miyasaka, T. Kimura, K. Oikawa, T. Kamiyama, R. Kumai, and Y. Tokura, Structural and magnetoelectric properties of $Ga_{2-x}Fe_xO_3$ single crystals grown by a floating-zone method, *Phys. Rev. B* 70, 064426 (2004).

[Ashida 14] T. Ashida, M. Oida, N. Shimomura, T. Nozaki, T. Shibata, and M. Sahashi, Observation of magnetoelectric effect in Cr_2O_3/Pt/Co thin film system, *Appl. Phys. Lett.* 104, 152409 (2014).

[Astrov 61] D.N. Astrov, Magneto-electric effect in chromium oxide, *Sov. Phys. JETP* 13, 729 (1961); *J. Exptl Theoret. Phys.* (USSR) 40, 1035–1041 (1961).

[Belashchenko 10] K.D. Belashchenko, Equilibrium magnetization at the boundary of a magnetoelectric antiferromagnet, *Phys. Rev. Lett.* 105, 147204 (2010).

[Bibes 08] M. Bibes, A. Barthélémy, Towards a magnetoelectric memory, *Nat. Mater.* 7, 425–426 (2008).

[Binek 03] C. Binek, *Ising-Type Antiferromagnets: Model Systems in Statistical Physics and the Magnetism of Exchange Bias*, Springer Tracts in Modern Physics, Vol. 196, Springer, Berlin, Germany, 2003.

[Binek 05] Ch. Binek, B. Doudin, Magnetoelectronics with magnetoelectrics, *J. Phys. Condens. Matter* 17, L39–L44 (2005).

[Binek 13] W. Kleemann and C. Binek, in: H. Zabel and M. Farle, eds., *Magnetic Nanostructures*, Springer Tracts in Modern Physics, Vol. 246, Springer-Verlag, Berlin, Germany, 2013, pp. 163–187.

[Binek 14] Ch. Binek, P.A. Dowben, K. Belashchenko and J. Kelber, Magneto-electric voltage controlled spin transistors, U.S. patent application serial no. 61766025, filed February 18, 2013, efilingAck18224783, non-provisional filing number 14182521, February 18, 2014.

[Birol 12] T. Birol, N.A. Benedek, H. Das, A.L. Wysocki, A.T. Mulder, B.M. Abbett, E.H. Smith, S. Ghosh, C.J. Fennie, *Curr. Opin. Solid State Mater. Sci.* 16, 227–242 (2012).

[Borisov 05] P. Borisov, A. Hochstrat, X. Chen, W. Kleemann, Ch. Binek, Magnetoelectric switching of exchange Bias, *Phys. Rev. Lett.* 94, 117203 (2005).

[Brosse 04] J. de Brosse, *Nanotechnology Symposium*, Cornell's CNS, slides available online at: http://www.cns.cornell.edu/documents/JohnDeBrosseCNSSymp504.pdf, May 14, 2004.

[Cao 14] S. Cao, X. Zhang, N. Wu, A.T. N'Diaye, G. Chen, A.K. Schmid, X. Chen, W. Echtenkamp, A. Enders, Ch. Binek, and P.A. Dowben, Spin polarization asymmetry at the surface of chromia, *New J. Phys.* 16, 073021 (2014).

[Chen 06] X. Chen, A. Hochstrat, P. Borisov, and W. Kleemann, Magnetoelectric exchange bias systems in spintronics, *Appl. Phys. Lett.* 89, 202508 (2006).

[Chen 10] X. Chen, A. Hochstrat, P. Borisov, and W. Kleemann, Magnetoresistive element, particularly memory element or logic element, and method for writing information to such an element, U.S. Patent 7,719,883 B2, May 2010.

[Chu 08] Y.-H. Chu, L.W. Martin, M.B. Holcomb, M. Gajek, S.-J. Han, Q. He, N. Balke, C.-H. Yang, D. Lee, W. Hu, Q. Zhan, P.-L. Yang, A. Fraile-Rodríguez, A. Scholl, S.X. Wang, R. Ramesh, Electric-field control of local ferromagnetism using a magnetoelectric multiferroic, *Nat. Mater.* 7, 478–482 (2008).

[Datta 90] S. Datta and B. Das, Electronic analog of the electro-optic modulator, *Appl. Phys. Lett.* 56, 665 (1990).

[Dery 07] H. Dery, P. Dalal, Ł. Cywiński, L.J. Sham, Spin-based logic in semiconductors for reconfigurable large-scale circuits, *Nature* 447, 573 (2007).

[Dowben 91] P.A. Dowben, D. LaGraffe, D. Li, A. Miller, L. Zhang, L. Dottl, and M. Onellion, Substrate-induced magnetic ordering of rare-earth overlayers, *Phys. Rev. B* 43, 3171–3179 (1991).

[Dowben 11] P.A. Dowben, N. Wu, and Ch. Binek, When measured spin polarization is not spin polarization, *J. Phys. Condens. Matter* 23, 171001 (2011).

[Durlam 03] M. Durlam, D. Addie, J. Akerman, B. Butcher, P. Brown, J. Chan, M. DeHerrera, B.N. Engel, B. Feil, G. Grynkewich, J. Janesky, M. Johnson, K. Kyler, J. Molla, J. Martin, K. Nagel, J. Ren, N.D. Rizzo, T. Rodriguez, L. Savtchenko, J. Slater, J.M. Slaughter, K. Smith, J.J. Sun, M. Lien, K. Papworth, P. Shah, W. Qin, R. Williams, L. Wise, and S. Tehrani, A 0.18 um 4Mb toggling MRAM, *IEDM Technical Digest*, IEEE Press, New York, 2003, p. 34.6.1.

[Echtenkamp 13] W. Echtenkamp and Ch. Binek, Electric control of exchange bias training, *Phys. Rev. Lett.* 111, 187204 (2013).

[Fallarino 14] L. Fallarino, A. Berger, and Ch. Binek, Giant temperature dependence of the spin reversal field in magnetoelectric chromia, *Appl. Phys. Lett.* 104, 022403 (2014).

[Fiebig 05] M. Fiebig, Revival of the Magnetoelectric effect, *J. Phys. D: Appl. Phys.* 38, R123–R152 (2005).

[Gajek 07] M. Gajek, M. Bibes, S. Fusil, K. Bouzehouane, J. Fontcuberta, A. Barthelemy, and A. Fert, Tunnel junctions with multiferroic barriers, *Nat. Mater.* 6, 296–302 (2007).

[He 10] X. He, Y. Wang, N. Wu, A.N. Caruso, E. Vescovo, K.D. Belashchenko, P.A. Dowben, Ch. Binek, Robust isothermal electric control of exchange bias at room temperature, *Nat. Mater.* 9, 579 (2010).

[Heron 11] J.T. Heron, M. Trassin, K. Ashraf, M. Gajek, Q. He, S.Y. Yang, D.E. Nikonov, Y.-H. Chu, S. Salahuddin, and R. Ramesh, Electric-field-induced magnetization reversal in a ferromagnet-multiferroic heterostructure, *Phys. Rev. Lett.* 107, 217202 (2011).

[Heron 14] J.T. Heron, J.L. Bosse, Q. He, Y. Gao, M. Trassin, L. Ye, J.D. Clarkson, C. Wang, Jian Liu, S. Salahuddin, D.C. Ralph, D.G. Schlom, J. Iniguez, B.D. Huey, and R. Ramesh, Deterministic switching of ferromagnetism at room temperature using an electric field, *Nature* 516, 370 (2014).

[Hochstrat 04] A. Hochstrat, Ch. Binek, X. Chen, and W. Kleemann, Extrinsic control of the exchange bias, *J. Magn. Magn. Mater.* 272–276, 325 (2004).

[Hochstrat 06] A. Hochstrat, X. Chen, P. Borisov, and W. Kleemann, Magnetoresistive element in particular memory element or logic element and method for writing information to such an element, International Patent Application Pct/Ep2006/002892_Wo/2006/103065 (Wo2006103065), published October 5, 2006.

[Johnson 00] M. Johnson, B.R. Bennett, P.R. Hammar, and M.M. Miller, Magnetoelectronic latching Boolean gate, *Sol. State Electron.* 44, 1099 (2000).

[Jonietz 10] F. Jonietz, S. Mühlbauer, C. Pfleiderer, A. Neubauer, W. Münzer, A. Bauer, T. Adams, R. Georgii, P. Böni, R.A. Duine, K. Everschor, M. Garst, and A. Rosch, Spin transfer torques in MnSi at ultralow current densities, *Science* 330, 1648 (2010).

[Jonker 07] B.T. Jonker, G. Kioseoglou, A.T. Hanbicki, C.H. Li, and P.E. Thompson, Electrical spin-injection into silicon from a ferromagnetic metal/tunnel barrier contact, *Nat. Phys.* 3, 542 (2007).

[Khitun 08] A. Khitun, B. Mingqiang, and K.L. Wang, Spin wave magnetic nanofabric: A new approach to spin-based logic circuitry, *IEEE Trans. Magn.* 44, 2141–2152 (2008).

[Khitun 11] A. Khitun and K.L. Wang, Non-volatile magnonic logic circuits engineering, *J. Appl. Phys.* 110, 034306 (2011).

[Kleemann 13] W. Kleemann, Magnetoelectric spintronics, *J. Appl. Phys.* 114, 027013 (2013).

[Landau 84] L. Landau, L. Pitaevskii, and E. Lifshitz, Electrodynamics of continuous media, Landau, E. Lifshitz, *Course of Theoretical Physics*, vol. 8, Pergamon Press Ltd. Oxford; 1984.

[Marshall 14] A. Marshall, J. Bird, U. Singisetti, and D. Nikonov, Non-volatile latch using magneto-electric and ferro-electric tunnel junctions, U.S. Patent Pending, Application no.: 14/690,923, Filing Date: April 20, 2015.

[Marshall 14b] A. Marshall, P.A. Dowben, and J. Bird, Majority- and minority-gate logic schemes based on magneto-electric devices, U.S. Patent Pending; Majority-gate logic schemes based on magneto-electric devices, U.S. Patent Pending; Application no.: 62/069,138, Filing Date: October 27, 2014.

[Martin 66] T.J. Martin and J.C. Anderson, Antiferromagnetic domain switching in Cr_2O_3, *IEEE Trans. Mag.* 2, 446 (1966).

[Martin 07] L.W. Martin, Y.H. Chu, Q. Zhan, R. Ramesh, S.J. Han, S.X. Wang, M. Warusawithana, and D.G. Schlom, Room temperature exchange bias and spin valves based on $BiFeO_3/SrRuO_3/SrTiO_3/Si(001)$ heterostructures, *Appl. Phys. Lett.* 91, 172513 (2007).

[Miller 93] A. Miller and P.A. Dowben, Substrate-induced magnetic ordering of rare-earth overlayers: II, *J. Phys. Condens. Matter* 5, 5459–5470 (1993).

[Niemier 11] M.T. Niemier, G.H. Bernstein, G. Csaba, A. Dingler, X.S. Hu1, S. Kurtz, S. Liu, J. Nahas, W. Porod, M. Siddiq, and E. Varga, Nanomagnet logic: Progress toward system-level integration, *J. Phys. Condens. Matter* 23, 493202 (2011).

[Ney 03] A. Ney, C. Pampuch, R. Koch, and K.H. Ploog, Programmable computing with a single magnetoresistive element, *Nature* 425, 485–487 (2003).

[Nikonov 13] D.E. Nikonov and I.A. Young, Overview of beyond-CMOS devices and a uniform methodology for their benchmarking, *Proc. IEEE* 101, 2515 (2013).

[Sakimura 08] N. Sakimura, T. Sugibayashi, R. Nebashi, and N. Kasai, Nonvolatile magnetic flip-flop for standby-power-free SoCs, *Proceedings of the IEEE, Custom Integrated Circuits Conference (CICC)*, 2008, pp. 355–358; *IEEE J. Solid-State Circuits*, 44, 2244 (2009).

[Schmid 94] H. Schmid, Multi-ferroic magnetoelectrics, *Ferroelectrics* 162, 317 (1994).

[Schmidt 00] G. Schmidt, D. Ferrand, L.W. Molenkamp, A.T. Filip, and B.J. van Wees, Fundamental obstacle for electrical spin injection from a ferromagnetic metal into a diffusive semiconductor, *Phys. Rev. B* 62, R46267 (2000).

[Semenov 07] Y.G. Semenov, K.W. Kim, and J.M. Zavada, Spin field effect transistor with a graphene channel, *Appl. Phys. Lett.* 91, 153105 (2007).

[Semenov 10] Y.G. Semenov, J.M. Zavada, and K.W. Kim, Electrical injection and detection of spin-polarized carriers in silicon in a lateral transport geometry, *J. Appl. Phys.* 107, 064507 (2010).

[Smolenskii 82] G.A. Smolenskii and I.E. Chupis, Ferroelectromagnets, *Sov. Phys. Usp.* 25, 475 (1982).

[Street 14] M. Street, W. Echtenkamp, T. Komesu, S. Cao, P.A. Dowben, and Ch. Binek, Increasing the Néel temperature of magnetoelectric chromia for voltage-controlled spintronics, *Appl. Phys. Lett.* 104, 222402 (2014).

[Toyoki 15] K. Toyoki, Yu. Shiratsuchi, A. Kobane, S. Harimoto, S. Onoue, H. Nomura, and R. Nakatani, Switching of perpendicular exchange bias in $Pt/Co/Pt/\alpha-Cr_2O_3/Pt$ layered structure using magneto-electric effect, *J. Appl. Phys.* 117, 17D902 (2015).

[Trassin 09] M. Trassin, N. Viart, G. Versini, S. Barre, G. Pourroy, J. Lee, W. Jo, K. Dumesnil, C. Dufour, and S. Robert, Room temperature ferrimagnetic thin films of the magnetoelectric $Ga_{2-x}Fe_xO_3$, *J. Mater. Chem.* 19, 8876–8880 (2009).

[van Dijken 02] S. van Dijken, X. Jiang, S.S.P. Parkin, Room temperature operation of a high output current magnetic tunnel transistor, *Appl. Phys. Lett.* 80, 3364 (2002).

[van't Erve 07] O.M.J. van't Erve, A.T. Hanbicki, M. Holub, C.H. Li, C. Awo-Affouda, P.E. Thompson, and B.T. Jonker, Electrical injection and detection of spin-polarized carriers in silicon in a lateral transport geometry, *Appl. Phys. Lett.* 91, 212109 (2007).

[Vaz 12] C.A.F. Vaz, Electric field control of magnetism in multiferroic heterostructures, *J. Phys. Condens. Matter* 24, 333201 (2012).

[Wang 14] J. Wang, J.A. Colón Santana, N. Wu, C. Karunakaran, J. Wang, P.A. Dowben, and Ch. Binek, Magnetoelectric Fe_2TeO_6 thin films, *J. Phys. Condens. Matter* 26, 055012 (2014).

[Wu 11] N. Wu, X. He, A. Wysocki, U. Lanke, T. Komesu, K.D. Belashchenko, Ch. Binek, and P.A. Dowben, Imaging and control of surface magnetization domains in a magnetoelectric antiferromagnet, *Phys. Rev. Lett.* 106, 087202 (2011).

[Yang 07] F. Yang, M.H. Tang, Z. Ye, Y.C. Zhou, X.J. Zheng, J.X. Tang, J.J. Zhang, and J. He, Eight logic states of tunneling magnetoelectroresistance in multiferroic tunnel junctions, *J. Appl. Phys.* 102, 044504 (2007).

[Zaliznyak 13] I. Zaliznyak, A. Tsvelik, D. Kharzeev, Nanodevices for spintronics and methods of using same, U.S. Patent 8,378,329 (2013).

[Zhao 06] T. Zhao, A. Scholl, F. Zavaliche, K. Lee, M. Barry, A. Doran, M.P. Cruz, Y.H. Chu, C. Ederer, N.A. Spaldin, R.R. Das, D.M. Kim, S.H. Baek, C.B. Eom and R. Ramesh, Electrical control of antiferromagnetic domains in multiferroic $BiFeO_3$ films at room temperature, *Nat. Mater.* 5, 823–829 (2006).

[Zhirnov 05] V.V. Zhirnov, J.A. Hutchby, G.I. Bourianoff, J.E. Brewer, Emerging research logic devices, *IEEE Circuits Dev.* 21, 37 (2005).

Index

A

Active pixel sensor (APS), 174–175
Air-gap parallel plate capacitor model, 125
Analog-to digital converters (ADCs), 173
Annealing processes, 185
Arrhenius plot analysis, 197
Atom probe tomography (ATM), 188

B

Back-gated SOI-FET, 193
Ballistic transport, MOSFETs
    carrier–carrier scattering, 8
    Landauer formula, 6, 8
    Poisson's equation, 6
    potential barrier, 5
    space-charge neutral region, 5
    thermal velocity, 7
Band-pass filtering behavior, 160
Band to band tunneling (BTBT) mechanisms
    ambipolar characteristics, 174
    broken band alignment, 165
    drain current dependence, 168
    tunnel field-effect transistors, 158
    2D–2D tunneling, 168
Barrier shape, 224–225
Beam displacement, transient, 135
Bias heterostructures, 258
Bias voltage, 213, 260
Biomolecular sensors, 144
Bismuth ferrite (BiFeO$_3$), 259
1-Bit full adder, 274
Bohr magneton, 236
Bohr radius, 183, 200
Boltzmann's constant, 209, 236
BOX, *see* Buried oxide (BOX)
Brillouin space, 42
Broglie wavelength
    carrier wave packet, 12
    statistical considerations, 14
BTBT, *see* Band to band tunneling (BTBT)
    mechanisms
Buried oxide (BOX)
    conventional CMOS APS, 175
    device structure, 18, 72
    SE turnstile, 217
    Si-coupled QD, 232

C

Carbon nanotubes (CNTs)
    electrical and thermal conductivities, 98
    NEM resonator sensors, 144
Carrier–carrier scattering
    ballistic transport in MOSFETs, 8
    many-body interactions, 30
    site representation, scattering of, 23
Carrier wave packet
    Broglie wavelength, 12
    classical potential, 13
    Fourier components, 11–12
Casimir force, 126, 131
Channel doping functions, 200
Channel pattern
    ab initio simulations, 195
    cross-sectional TEM image, 195
    dielectric confinement, 194
    effects, 194–198
    ionization energy, 194
    small source drain bias, 196
    stub-shaped channel, 196
    width, 197–198
Charge sensing
    Coulomb oscillation, 238
    electrostatically coupled QDs,
        238–239
    equivalent circuit model,
        238–239
    sensitivity
        SETs capacitive parameters, 247
        SNR, 241
Charge stability
    Coulomb peaks, 233
    DQD diagram, 248
    zero magnetic field, 234
Charge states, 182
Clusters, 191, 199
CNTs, *see* Carbon nanotubes (CNTs)
Coherent-state representation, 12
Complementary metal oxide
        semiconductor (CMOS)
    fabrication technology, 123
    long channel transport, 42
    Moore's law, 37
    vs. NEMS, 130
    technology boosters, 162

Controlled doping techniques, 185
Cotunneling, 218
Coulomb diamonds, 241
Coulomb oscillation, 238
Coulomb peaks, 241–242
Coulomb scattering, many-body interactions, 28
Coulomb well, 183
Counter-doped pockets, 163
Coupled dopants, 183
Coupled Si quantum dots, spin-based qubits
    defects, 232
    device, 232–233
    double top-gate design, 232
    electron beam lithography, 231
    electron effective mass, 231
    fabrication processes, 232
    hyperfine interaction, 231
    measurement results
        charge-sensing technique, 247–248
        charge stability, 233–234
        few-electron states, charge sensing of,
            238–240
        gating function to CSs, 248–251
        Pauli-spin blockade, 233–237
        SET, charge sensitivity of, 240–247
    QDs, 231
    quantum states, 231
    SOI substrate, 232
Cross-capacitance, 213
Cross-capacitive coupling, 219
Cryogenic current comparator, 216
Current density, 160
Current onset voltage (COV)
    drain current variability, 60
    percolation path, 62
    potential dividing line, 62–63

D

Datta–Das transistor, 262, 264
Debye temperature, 14
Decay cascade model, 222
DELTA structure, SOI FinFETs, 84
Density of states (DOS), 168, 170
Device-matrix-array (DMA)-test-element group
    (TEG), 54, 63–64
Device under test (DUT), 63
DIBL, see Drain induced barrier
    lowering (DIBL)
Dielectric chromia, 263
Discrete dopants
    electrical characteristics, 182
    individual dopant atoms, 182
    lithography techniques, 183
    Moore's law, 181
    quantum computing, 182
    scaled-down devices, 183

Dopant-atom silicon tunneling nano devices
    controlled doping, 185
    discrete dopants, scaled-down devices,
        181–183
    LT-KPFM, individual dopants
        detection techniques, 187–189
        electrical characteristics correlation,
            191–193
        KPFM setup under normal operation, 189
        single electron charging observations,
            189–192
    single-electron tunneling via nanoscale
        transistors
        channel pattern effect and dielectric
            confinement, 194–198
        interacting dopants, 198–201
        tunneling via individual dopants,
            193–194
    uncontrolled doping, 184–185
Dopant-induced potential well, 193–194
Dopant-QD, 193
Dopant segregation, 163–164
Doping techniques
    controlled doping, 185
    degree of control, 183
    thermal diffusion and ion implantation, 184
    uncontrolled (random) doping, 184–185
Double-gate (DG) thin-film configuration,
    161–162
Double QD (DQD), 247–250
Drain electrode magnetization, 264
Drain induced barrier lowering (DIBL)
    electrostatics of tri-gate transistors, 39
    fin doping concentrations, 40
    nFETs and pFETs, 64–65
Dresselhaus effect, 262
Dry-etching, 232
Duke tunneling formula, 6
DUT, see Device under test (DUT)
Dynamic random access memory (DRAM)
    computer hierarchy, 106
    NEM memories, 130–131
    volatile logic circuits, 105

E

Earth's magnetic field, 258
Effective potential, quantum behavior of device
    carrier wave packet
        Broglie wavelength, 12
        classical potential, 13
        Fourier components, 11–12
        Schrödinger equation, 12
    device simulation, 15
    statistical considerations
        Broglie wavelength, 14
        high-temperature, 15

Electromechanical relay devices
    CMOS device *vs.* NEM, 130
    DC transfer characteristics, 129
    micromachining technology, 128
    scaling factors, 129
    three-terminal NEM relay, 128
Electron beam evaporation, 232
Electron beam lithography, 231
Electron-counting scheme, 216
Electron-hole bilayer tunnel FET
        (EHBTFET)
    device structure, 170–171
    leakage mechanisms, 172
Electron spin, 262
Electrostatics
    force, plates, 126
    of tri-gate transistors, 39–40
11G transistors, variability of
    DMA-TEG, 63–64
    large-scale integrated circuits, 63
    measurement of
        DUT, 63
        I–V characteristics, 64
        V_{th} monitoring circuit, 63
    SS, 67–68
    V_{thc}, 64–66
    V_{thex} and Ion, 66–67
EMC, *see* Ensemble Monte Carlo (EMC)
EMC-MD approach, 28; *see also* Many-body
        interactions
Energy efficiency, 156–157
Energy-reversible (ER) NEMS switches,
        130–131
Energy-reversible nonvolatile NEM memories
    cantilever configurations, 140, 142
    double-clamped beam memory, 142
    SOI-compatible technology, 141
Ensemble Monte Carlo (EMC), 28
Equivalent circuit model, 238–239
Equivalent oxide thickness (EOT), 37–38,
        166, 168

F

Fabricated devices cross-section, 167
Fabrication processes, 232
FD SOI transistors, intrinsic channel; *see also*
        Fully depleted-silicon-on-insulator
        (FD-SOI)
    CMOS platform, 149
    FD SOTB SRAM stability, 74–75
    RDF, 72
    V_{th} and drain current variability
        I_d-V_{gs} characteristics, 72–73
        vs. potential fluctuations of transistor
            channels, 73–74
        SOTB, 72–73

Fermi–Dirac electron distribution, 196
Fermi–Dirac integrals, half-integer order, 8
Fermi level
    coupled Si QDs, 250
    donor's ground state, 183, 250
    one-electron model, 219
    single-electron tunneling, 193
Fermi window, 170
Ferroelectric polarization, 259
Ferroelectric RAM (FeRAM), 132
Ferromagnetic (FM) film, 256
Few-electron regime, 239
Fick's diffusion laws, 184
Field-effect transistor (FET), 183
Fintype field-effect transistors (FinFETs)
    analog operation, 90–91
    birds-eye view of, 85
    DC operation
        drain current *vs.* drain voltage, 89
        p+ doped silicon slab region, 89
        thermal resistance, 87
    STI, 84
    thermal analysis
        Mason's unilateral power gain, 90
        self-heating effects, 91
    3D structure MOSFET, 83
4T2MTJ cell, 111, 113
Fully depleted-silicon-on-insulator (FD-SOI),
        37–38
Fully depleted tri-gate transistors
    body/fin thickness, 38–39
    single-gate FD-SOI, 38

G

GaSb–InAsSb heterostructure, 166
Gas phase diffusion, 184
Gate voltages, 249–250
Gaussian wave packet, *see* Coherent-state
        representation
Geon insulator (GOI) structures, 163
Germanium (Ge), 163
Gibbs free energy, 257
Granularity, silicon nanodevices
    density fluctuation, 9–10
    SOI, 8–9
Graphene
    gapless semiconductor, 266
    spin field effect transistors, 262
    thin (2D) channel conductors, 264
    zero spin–orbit coupling, 263

H

Hanle effect, 265
Hard disk drive (HDD), 105
Hartree–Fock corrections, 23
High-*k* gate dielectric, 163

Homojunction silicon tunnel FET, 162
Honeycomb structure, 247–248
Hopping cascade, 223–224
Hubbard band conduction, 187
Hyperfine interaction, 231

I

Individual dopants
    channel pattern and dielectric confinement, 194
    controlled doping, 185
    nanoscale transistors, 193–194
    observation, LT-KPFM
        ATM, 188
        KPFM, 188–189
        landscape correlation, electrical
            characteristics, 191, 193
        potential changes, 189–191
        potential landscapes, high temperature,
            191–192
        SIMS, 187–188
        STM and SCM, 188
    scaled-down devices, 182
Information and communication
        technology (ICT), 131
Information processing, 208, 226
In situ doping techniques, 163
Integrated NEM resonator sensors
    beam structure, 149
    fabrication process, 148
    FD-SOI-CMOS platform, 149
Interacting dopants
    multiple donor interactions, 198
    nondoped channel, 199
    uniformly and randomly doped channel, 199
Interface magnetization, 260
Interface polarization
    device implementation, 266
    magnetoelectrics, 258
    proximity effect, 264
International Committee for Weights and
        Measures (CIPM), 209
Ion implantation techniques, 184
Ionization energy, 194
I–V characteristics, transistor variability
    drain current, 60–61
    measurement of, 64
    V_{the} variability, 64–65

J

Josephson voltage standard, 208
Joule heat
    bulk MOSFETs, 84
    ME-MTJ, 260
    nonvolatile magnetic state, 255
    thermal modeling of circuits, 97

K

Kelvin probe force microscopy (KPFM)
    electronic potential landscapes
        high temperature, 191–192
        low temperature, 189–190
    setup under normal operation, 189

L

Landauer formula
    ballistic transport, MOSFETs, 8
    simple one-dimensional theory, 4
    site representation, scattering, 22
    small MOSFETS, 2–3
Large field regime, magnetoelectric spin
        transistor, 264
Large-scale integration circuits (LSIs)
    microprocessors, 107
    MTJ-based memory, 113
    nanoscale 3D MOSFETs, 83
Lateral tunnel FETs, 172
Lattice temperature, 8
Leakage current, 235–236
Leakage mechanism, 172
Lienard–Wiechert potential, 13
Lithography
    patterning, 136
    techniques, 183
LOP SRAM, *see* Low operating-power
        (LOP) SRAM
Lorentzian line shape, 236
Low field regime, magnetoelectric
        spin transistor
    Datta–Das transistor, 262, 264
    ME gate, 262, 265
    Rashba effect, 265
    spin analyzer, 264
Low operating-power (LOP) SRAM, 120
Low-pressure chemical vapor deposition
        (LPCVD), 232
LSIs, *see* Large-scale integration circuits (LSIs)

M

Magnetic random access memories (MRAM),
        132, 261, 267
Magnetic tunnel junction (MTJ)
    CMOS latch, 115
    computer hierarchy, 106
    large field, very short channel regime, 264
    magnetoelectric majority gate controlled
        stack, 272
    magnetoelectric memory, 267–268
    solid-state devices, 256
    SRAM based, 269
    STT-MRAM cell, 113

Magnetoelectric majority gate (MEMG), 274
Magnetoelectrics (ME), 260
    annealing, 259
    antiferromagnets, 257
    electric polarization, 257
    Gibbs free energy, 257
    isothermal switching, 258
    magnetic moment, 257
    memory
        MTJ memory latch, 268
        multiferroic tunnel junction, 269–270
        nonvolatile latch, 268
        STTRAM, 267
        voltage-controlled anisotropy, 270
    multiferroic materials, 259
    naïve approach, 258
    pinning layer, 258
    spinFET
        device implementation, 266
        spin transistor, 261–263, 265
Magnetoresistance (MR) ratios
    higher-fidelity spin-polarized current
        injection, 266
    MTJ devices, 261
Many-body interactions
    carrier–carrier scattering, 30
    Coulomb interaction, 28–29
    EMC, 28
    inelastic plasmon scattering, 29
    MD, 28
Mason's unilateral power gain, 90
Mass sensitivity simulation, 145–148
MD, *see* Molecular dynamics (MD)
Mean-free path, of carriers, 3
ME cell, 272–273
MEMS, *see* Micro-electro-mechanical
        systems (MEMS)
Metal-insulator-transition (MIT), 158–159
Metallic silicide islands, 226
Metal-oxide-semiconductor (MOS)
    charge sensing operation, 238
    electron spin-based quantum bits, 231–232
    ME-MTJ, 260–261
    tunable-barrier single-electron turnstile, 210
Metal-oxide-semiconductor field-effect
        transistors (MOSFETs)
    ballistic transport, 5–8
    drain current *vs.* gate voltage characteristics, 156
    scaling, 155, 157, 162
    simple one-dimensional theory
        channel length, 3–4
        Landauer formula, 4
    small
        Landauer formula, 3
        oxide–semiconductor interface, 3
        source and drain electrodes, 2
    subthreshold swing, 156, 158, 163, 168

Micro-electro-mechanical systems (MEMS),
        123, 132
Microprocessors (MPUs)
    latch circuit, 107, 109
    power-gating method, 106–107
    state machine, 107
    32 bit NV MPU, 107–110
Microwave transmission, 128
Molecular dynamics (MD)
    carrier–carrier scattering, 30
    many-body interactions, 28
Monte Carlo simulations
    ballistic transport, 8
    device simulation, 16
    SWR-induced, 48
Moore's law, 181
    silicon integrated circuit miniaturization, 2
    tri-gate transistors, 37
MOSFETs, *see* Metal-oxide-semiconductor field-
        effect transistors (MOSFETs)
MPUs, *see* Microprocessors (MPUs), NV
        computing system
MR ratios, *see* Magnetoresistance (MR) ratios
MTJ, *see* Magnetic tunnel junction (MTJ)
Multiferroic tunnel junction, 269–270
Multiple electron number state, 220–221
Multistate six-terminal MEspinFET, 265

N

Naïve approach, 258
Nano-electro-mechanical systems
        (NEMS) devices
    memories
        energy-reversible nonvolatile, 140–142
        MEMS, 132
        NEMory cell, 140
        RAMs and NVMs, 131–132
        self-buckling floating gate, 133–137
        suspended-gate, 137–140
    resonator sensors
        integrated systems, 148–150
        mass sensitivity simulation, 145–148
    switches
        ER, 130–131
        MOSFET, 125
        pull-in and pull-out, 125–126
        relays, 127–130
        (SG-) FET, 126–127
        VLSI, 124
Nanoscale 3D MOSFETs
    FinFET structures, 84–85
    self-heating effects, 83–84
    thermal analysis
        FinFETs, 87–91
        interface thermal resistance, 86
        thermal conductivity, 85–86

thermal modeling
  circuits, 97–98
  CNTs, 98–101
  devices, 92–93
  interconnects and vias, 93–97
Nanowires (NWs), 163
n-channel tri-gate transistors, 41–42
Néel temperature, 256, 259
Negative bias temperature instability (NBTI), 76
NEM nonvolatile memory (NEMory) cell, 140
NEMS devices, *see* Nano-electro-mechanical
      systems (NEMS) devices
NMOS and PMOS transistors, 42–43
NMOSFETs, 41–42
Nonequilibrium electron capture model, 221
Nonequilibrium Green's functions, 8
Nonvolatile (NV) computing system
  computer hierarchy
    memory and logic functions, 106
    semiconductor device
      miniaturization, 105
  logic LSIs, 106
  MPU
    latch circuit, 107, 109
    power-gating method, 106–107
    state machine, 107
    32 bit NV MPU, 107–110
  recognition processor
    vs. conventional SRAM, 111, 113
    fine-grained power-gating technique, 112
    4T2MTJ cell, 113
    hardware intelligent systems, 111
    input vector, 111–112
    time-critical applications, 110
Nonvolatile magnetoelectric devices, spintronic
    applications
  device implementation, 260–267
  magnetoelectric logic
    gate design, 270–271
    magnetic electric switching, 272
    spintronic logic devices, 270
    spin waves, 272–273
  magnetoelectric memory, 267–270
  ME (*see* Magnetoelectrics (ME))
  ME-MTJ
    FM layers and stack, 260–261
    spin-transfer torque, 260
    write lines, 261
  RAM elements, 256
  scalable memory device, 255
  spin transistor
    dielectric chromia, 263
    drain electrodes, 261
    FET, 261–262
    large field, very short channel regime,
      262, 264
    low field regime, 264–265

    proximity effect, 263
    zero spin orbit coupling, 263
  STT, 255
  switching energy, 256
  voltage control, 255
Nonvolatile memories (NVMs), 131
Nonvolatile recognition processor
  vs. conventional SRAM, 111, 113
  fine-grained power-gating technique, 112
  4T2MTJ cell, 113
  hardware intelligent systems, 111
  input vector, 111–112
  time-critical applications, 110
NV computing system, *see* Nonvolatile (NV)
    computing system

O

Offset voltages, 249
1D power-gating method, 118–119
1D semiclassical treatment, of MOSFETs, 3
One-electron model, 219
1T1MTJ cell, 113; *see also* Spin-transfer-torque
    magnetoresistive random access
    memory (STT-MRAM)
Operational transconductance amplifiers
    (OTAs), 173
Oxide/nitride/oxide (ONO), 140

P

Pauli-spin blockade (PSB)
  current rectification, 233
  doubly degenerate valleys, 235
  intradot and interdot transitions,
    235–236
  leakage current, 235–236
  strong current dip, 236
  triple point, 234
P-channel field-effect transistor (PFET), 107
p+ doped silicon slab region, 89
Pelgrom and Takeuchi plot, 59–60
Percolation path, 62
Phase change RAM (PCRAM), 132
Phosphorus (P) donor atom, 183, 185, 189
Planck's constant, 233
PMOSFETs, 43–44
Poisson distribution
  quantum behavior of device, 11
  random variability, 58
  suspended-gate NEM memories, 140
Poisson process, 226
Power-gating method
  microprocessor, NV computing system,
    106–107
  STT-MRAMs, 118
Proximity effect, 263–264

Proximity polarization, 264, 266
Pull-in and pull-out switches, 125–126; *see also*
    Nano-electro-mechanical systems
    (NEMs) devices

Q

QDs, *see* Quantum dots (QDs)
Quantum behavior
    effective potential
        carrier wave packet, 11–13
        device simulation, 15
        statistical considerations, 13–15
    Poisson's equation, 11
    potential
        density gradient approach *vs.* effective
            potential approach, 17
        Schrödinger's equation, 16
    simulations
        device structure, 18–19
        site representation, scattering of, 22–24
        transmission and reflection *vs.* Fermi
            energy, 24–27
        wave function and technique, 19–22
Quantum coherence, 266
Quantum computing, 182
Quantum dots (QDs)
    scaled-down devices, 182, 231–233
    single-electron devices, 27
    spin-based quantum bits, 231
Quantum Hall resistance standard, 208
Quantum metrology triangle experiment, 208
Quantum simulations
    device structure
        BOX, 18
        Hamiltonian matrix, 18–19
    site representation, scattering of
        carrier–carrier scattering, 23
        Fermi golden rule, 23
        Fourier transform, 24
        Landauer formula, 22
    transmission and reflection *vs.* Fermi energy, 21
    wave function and technique
        recursion algorithm, 22
        Schrödinger equation, 19–21
Quantum–well FETs (QWFETs), 49
Quartz crystal microbalance (QCM), 144, 148
Quasiballistic transport regime, 50
Quasi-1D system, 4
Qubit readout, 241
QWFETs, *see* Quantum–well FETs (QWFETs)

R

Radio frequency (RF)
    measurements, 241, 245–246
    MEMS, 128

Random access memory (RAM), 131, 256
Random dopant fluctuation (RDF), 58
Rashba effect, 262, 265
Resistive RAM (ReRAM), 132
Resonant frequency, Si MEMS, 132
Resonator sensors, mass sensitivity simulation
    coating configurations, 144, 147
    QCM, 144, 148
    suspended-gate MOSFETs, 146

S

Scalability, 226
Scaled-down devices
    Coulomb blockade (CB), 183
    coupled donor atoms, 182
    dopant atoms, 183
    individual dopants, 182
Scaled MOSFETs
    11G transistors, variability of
        measurement of, 63–64
        SS, 67–68
        V_{thc}, 64–66
        V_{thex} and I_{on}, 66–67
    FD SOI transistors, intrinsic channel
        FD SOTB SRAM stability, 74–75
        V_{th} and drain current variability,
            72–74
    SNM, of SRAM cells, 68–71
    transistor variability
        classification of, 54–56
        COV, origin of, 62–63
        drain current, 60–62
        number dependency, 56–57
        random variability, origin of, 58
        self-suppression, 75–79
        65 nm transistors, 54
        size dependence, 59–60
Scanning capacitance microscopy (SCM), 188
Scanning tunneling microscope (STM),
    184–185, 188
Schottky source–drain device, 2
Schrödinger equation
    effective carrier wave packet, 12
    quantum potential, 16
    suspended-gate NEM memories, 140
    wave function and technique, 19
Secondary ion mass spectrometry (SIMS), 187
Selectively doped channel, 199
Self-buckling floating gate NEM memory
    beam displacement, 135
    device structure, 133
    electrical characteristics, 134
    electron micrographs, 136
Semiconductor-based electromechanical
    resonators, 123–124
SETs, *see* Single electron transistors (SETs)

Shallow trench isolation (STI)
   FinFETs, 84
   thermal resistance, 88
   tri-gate transistors
      stress direction and stress magnitude, 43
      3D simulation structure, 47
Sidewall roughness (SWR), 47
Signal-to-noise ratio (SNR), 173, 241, 245
Silicon bulk wafer, 226
Silicon coupled double QDs, 247
Silicon-nanodot-embedded self-buckling beam, 136
Silicon nanowires, 163
Silicon-on-insulator (SOI)
   compatible technology, 141
   FinFETs, 87
   granularity, 8
   heat dissipation paths, 84
   KPFM system, 189
   silicon nanotransistors, 200
   substrate, 210, 232, 238, 248
   thermal resistance, 88
Silicon-on-thin-BOX (SOTB), 72
Silicon tunnel FET
   additive boosters, 162
   parameters, 163
Simple one-dimensional theory, MOSFETs
   channel length, 3–4
   Landauer formula, 4
Simulated band diagram, 161–162
Single-dopant transistor operation, 182–183
Single electron (SE); *see also* Single-electron
      transfer (SET)
   current standard, 208
   ratchet scheme
      cross-capacitance, 213
      electron ejection, 213
      nanoampere charge pump, 215
      SAW approach, 214
      single-electron box model, 212, 214
      upper gate and drain voltage, 213–214
      zero drain bias, 213
Single-electron transfer (SET)
   accuracy
      transfer accuracy evaluation, 215–218
      transfer mechanism, 218–226
   applications, 207–208
   Boltzmann's constant, 209
   Coulomb blockades, 208, 210–212, 218
   equivalent circuits
      pump, 208
      SE turnstile, 208
   fundamental errors, 209
   high transfer accuracy, 209
   information processing circuits, 226
   localized states usage
      doped nanowire channels, 215
      trap states, 214

Planck's constant, 209
quantum metrology triangle experiment, 208
SE ratchet scheme, 213–215
transfer current, 209
tunable-barrier SE turnstile, 210–213
watt-balance experiment, 209
Single electron transistors (SETs), 183, 186, 193
Single-ion implantation (SII), 185, 187
Single-QD transistors, 186
16 kbit SRAM DMA-TEG, 68–69
65 nm transistors variability, 54
Small MOSFETs
   Landauer formula, 3
   oxide–semiconductor interface, 3
   source and drain electrodes, 2
SNM, *see* Static noise margin (SNM)
SNR, *see* Signal-to-noise ratio (SNR)
SOI, *see* Silicon-on-insulator (SOI)
Solid phase diffusion, 184
Solid-state drive (SSD), 105
SOTB, *see* Silicon-on-thin-BOX (SOTB)
Spatial footprint, 268
Spin field effect transistors (SpinFET),
      261–262, 264
Spin-flip cotunneling processes, 236
Spin orbit coupling, 263
Spin-orbit interactions, 236
Spin polarization, 261
Spin precession, 262
Spin transfer torque (STT), 255
Spin-transfer-torque magnetoresistive random
      access memory (STT-MRAM), 106
   fast write cycle
      chip microphotograph, 117
      CMOS latch, 115–116
      differential pair-type cell, 114–116
      MTJ-based memory LSIs, 113
      PL/SL driver, 116
   power consumption
      LOP SRAM, 120
      power-gating technique, 118
      vs. SRAM cell, 117
Spin-transfer torque memory (STTRAM), 267
Spintronics-based computing systems
   NV computing system
      computer hierarchy, 105–106
      logic LSIs, 106
      MPUs, 106–110
      recognition processor, 110–113
   STT-MRAM
      fast write cycle, 113–117
      MTJ-based memory LSIs, 113
      power consumption, benchmark of,
         117–120
SRAM, *see* Static random access
      memory (SRAM)
SS, *see* Subthreshold swing (SS)

Static noise margin (SNM)
  DIBL, 71
  measured SNM, 69
  variability of
    butterfly curves, 69–70
    16 kbit SRAM DMA-TEG, 68–69
    V_{th} variability, 71
Static random access memory (SRAM)
  computer hierarchy, 106
  vs. NV recognition processor, 111, 113
  subthreshold leakage, 105
Steeper transition, 156, 158
Steep slope switch, 125, 156
STI, see Shallow trench isolation (STI)
STM, see Scanning tunneling microscope (STM)
STM manipulation technique, 185
Strained Si NW array complementary tunnel
    FETs, 164
STT-MRAM, see Spin-transfer-torque
    magnetoresistive random access
    memory (STT-MRAM)
Subthreshold slope saturated drain bias (SS
    SAT), 39–40
Subthreshold swing (SS)
  subthreshold current, 65, 67
  3D device simulation, 67–68
Suspended-gate field-effect-transistor (SG-FET)
  electron micrograph, of device, 126
  subthreshold swing, 127
Suspended-gate silicon nanodot
    memory (SGSNM)
  data retention, 138
  FEM simulation, 139–140
  MOSFET memory, 137
  programming and erasing, 138
  simulation, 139
Switched magnetization, 267
SWR, see Sidewall roughness (SWR)

T

TEM, see Transmission electron microscope (TEM)
Templated selective epitaxy (TSE), 168
TG, see Top gate (TG)
Thermal analysis
  FinFETs
    analog operation, 90–91
    DC operation, 87–89
    interface thermal resistance, 86
    thermal conductivity, 85–86
Thermal diffusion, 184
Thermal hopping, 219, 223
Thermally activated conduction (TAC), 186–187
Thermal modeling, 3D MOSFETs
  circuits
    vs. input power of transistor, 98
    lattice temperature, 97–98

CNTs
  finite length interconnect wire vs.
    interconnect length, 100
  isolated system, 101
  thermal conductivity, 98
  devices
    equivalent circuit, FinFETs, 92
    thermal resistances, 93
  interconnects and vias
    Laplace's equation, 95
    1D temperature distribution, 96
Thermal resistances, 87–88, 93
Thermal velocity, MOSFETs
  ballistic transport, 7
  of carriers, 3
32 bit NV MPU
  internal power supply voltage, 107, 110
  local MTJ-based, 107–108
  microphotograph, 107, 109
  pipeline registers, 107, 109
3D configuration, tri-gate transistor, 41–42
3D device simulation, 67–68
3D MOSFETs
  FinFETs
    birds-eye view of, 85
    STI, 84
  self-heating effects
    heat dissipation paths, 84
    SOI, 83–84
3D Poisson equation, 244
III–V single gate tunnel FET schematic diagram,
    166–167
Three-terminal NEM relay, 128
Threshold voltage (V_{thc}) variability, 11G
  transistors
    cumulative distributions, 64
    I–V characteristics, 64–65
Top gate (TG), 232–234, 248
Total switching energy vs. supply voltage,
    156–157
Transfer accuracy
  charge sensor and SE turnstile, 216–217
  DC current meter, 215
  electron-counting scheme, 216
  metrological triangle experiment, 216
  RF-SET technique, 216–217
  scanning electron microscope image, 216–217
  sensor current, 216, 218
  SE shuttle transfer, 217
Transfer characteristics log vs. gate voltage, 159
Transfer error rate, 225
Transfer mechanism, 223–224
Transistors
  long channel
    drift mobility, 41
    electron vs. hole mobilities, 42
    NMOSFETs, 41–42

short channel
  hole mobility *vs.* stress, 46
  NMOS and PMOS transistors, 42–43
  pMOSFETs, 43–46
Transistor variability, 65 nm transistors
  classification
    DMA-TEG, 54
    schematic diagram, 54–55
    V_{th}, nFETs, 55–56
  drain current
    COV, 60
    I–V characteristics, 60–61
  number dependency, 56–57
  random variability, 58
  self suppression
    bias conditions, of pFETs, 77
    butterfly curves, 78
    NBTI, 76
  size dependence
    Pelgrom and Takeuchi plot, 59–60
    Poisson distribution, 59
Transmission electron microscope (TEM),
    58, 195
Tri-gate transistors
  CMOS, 37–38
  electrostatics, 39–40
  FD, 38–39
  silicon
    quasiballistic transport regime, 50
    QWFETs, 49
  transport
    long channel, 41–42
    short channel, 42–46
  variation of
    Monte Carlo simulation, 48
    SWR, 47
Tunable-barrier SE turnstile
  contour plot, 212–213
  conventional single-electron box, 212
  current staircases, 212–213
  device structure and potential diagram, 211
  GaAs quantum dot devices, 210
  island capacitance, 211
  low resistance conditions, 210
  Si nanowire, 211
  SOI substrate, 210
  transferred electrons and current plateaus,
      212–213
  turnstile operation, 211–212
  EG, 210
Tunnel field-effect transistors (tunnel FETs), 158
  DOS electronic switch, 2D–2D tunneling
    EHBTFET, 170–172
    electronic-grade 2D crystals, 172
    energy filtering, 170

Fermi window, 168, 170
    on-state conductivity, 170
  energy efficiency, 155
  low power analog and sensing applications,
      173–175
  MOSFETs dimensions scaling, 155
  principle and state of the art
    DG thin-film configuration, 161
    III–V materials, 165–169
    silicon and germanium, 162–165
    solid-state semiconductor, 161
  steep slope switch, 156
  voltage scaling, 156
Tunneling barrier transparency, 162, 170
Tunneling cascade capture, 223
Tunneling current, 159
Tunneling magnetoresistance (TMR), 270
Tunneling operation
  channel width, 197
  multiple-dopant QDs, 201
  nanofabrication processes, 198
  nanoscale transistors, 194
Two-dimensional electron gas (2DEG), 210, 232,
    245, 248
2D power-gating method, 118

U

Ultranarrow channels, 172
Unidirectional magnetic anisotropy, 259

V

Vanadium dioxide (VO_2), 158
van der Waals force, 126
Vertical tunnel FETs, 172
Very large scale integration (VLSI), 124
Visible light intensities, 174–175
Voltage-controlled MTJ, 260
Voltage transfer characteristics (VTC), 164
V_{thc}, *see* Threshold voltage (V_{thc}) variability, 11G
    transistors
V_{thex} and I_{on} variability, of 11G transistors
  cumulative distributions, 66–67
  nFETs and pFETs, 66

W

Watt-balance experiment, 209
Wave function and technique, 19
Wigner–Kirkwood expansion, 14

Z

Zeeman coupling, 256